Pharmacology in Caenorhabditis elegans

Dayong Wang

Pharmacology in *Caenorhabditis elegans*

Springer

Dayong Wang
Southeast University
Medical School
Nanjing, Jiangsu, China

ISBN 978-981-92-5165-0 ISBN 978-981-92-5166-7 (eBook)
https://doi.org/10.1007/978-981-92-5166-7

This Springer imprint is published by the registered company Springer Nature Singapore Pte Ltd.
The registered company address is: 152 Beach Road, #21-01/04 Gateway East, Singapore 189721, Singapore

Preface

What contributions can *Caenorhabditis elegans* make to the study of pharmacology?

The history of pharmacological study in *C. elegans* is almost as long as that of its biological research. For example, the levamisole resistance and its genetic basis were described in 1980. As a classic model animal, the well described molecular, genetic, cellular, and developmental backgrounds make *C. elegans* as a powerful model to carry out the research on drug targeting at different levels and to elucidate underlying mechanisms of observed pharmacological effects of compounds and drugs. The available physiological and pathological models in *C. elegans* build a bridge for the pharmacological study between this small model animal and the mammals. Especially, the tiny body size and the very small dosage for drug testing make it possible to perform the large scale and the high throughput drug screening in *C. elegans*.

In this book, we mainly focused on four important concerns:

1. What a pharmacological evaluation system can be established in *C. elegans*?
2. How to conduct the research on drug targeting at different levels in *C. elegans*?
3. How to determine the underlying mechanisms for pharmacological effects of compounds and drugs?
4. How to perform the high throughput drug screening in *C. elegans*?

Based on these concerns, in Chaps. 1, 2, 3, 4, and 5, we introduced the pharmacological evaluation system in *C. elegans*. In Chaps. 6, 7, 8, 9, and 10, we introduced the drug targeting at different levels in *C. elegans*. In Chaps. 11, 12, and 13, we introduced the pharmacological mechanisms of compounds and drugs in *C. elegans*. Finally, in Chap. 14, we discussed the high throughput drug screening in *C. elegans*.

Nanjing, Jiangsu, China Dayong Wang

Contents

Values of *Caenorhabditis elegans* for Pharmacological Study

1

Abstract

Caenorhabditis elegans is a classic model animal. Besides in the field of life science, *C. elegans* has been widely used in pharmacological study. We introduced and discussed several aspects of value of *C. elegans* in pharmacological study of drugs or compounds. Meanwhile, the possible limitations of *C. elegans* for pharmacological study were also discussed.

Keywords

Pharmacology · Value · Limitation · *Caenorhabditis elegans*

1.1 Introduction

Caenorhabditis elegans is the classic model animal for the study of life science. Besides this, *C. elegans* has shown the values for the study in several other research areas, including the pharmacological study [1]. Initially, due to the well-described molecular and genetic backgrounds, *C. elegans* has been used for identification and determination of certain pharmacological effects of candidate compounds. Again, with the raise of pathological models by transformation of wild-type and mutant human genes associated with occurrence of certain diseases, *C. elegans* can be used as an important alternative assay system to a certain degree for elucidating the underlying mechanisms and identifying molecular targets of drugs or compounds. Moreover, due to the small size, perfect reproductive capacity, and high sensitivity to environmental exposure, *C. elegans* has been frequently used in high-throughput screen of candidate drugs in small molecule library. During the pharmacological study in *C. elegans*, what the examined includes both the drugs and the carriers. The examined drugs can be compounds, plant extracts, or traditional Chinese medicines

D. Wang, *Pharmacology in Caenorhabditis elegans*,
https://doi.org/10.1007/978-981-92-5166-7_1

(TCMs). In this chapter, we discussed the several aspects of value of *C. elegans* in the pharmacological study of drugs or compounds.

The mainly introduced values of *C. elegans* for pharmacological study are:

1. High sensitivity to environmental exposure.
2. Design of certain exposure duration, stage, and route for drugs or compounds.
3. Detection of translocation and accumulation of compounds and their effects on target organs.
4. Detection of chemical transformation of compounds in the body.
5. Availability of developmental, physiological, and metabolic endpoints for pharmacological assessment.
6. Availability of certain pathological models for pharmacological study.
7. Elucidation of underlying mechanisms of certain pharmacological effects.
8. Identification of potential targets of drugs or compounds.
9. High-throughput screen and identification of compounds having certain pharmacological effect.
10. Examination of contributors to amplify the pharmacological effect of drugs or compounds.
11. Safety assessment of drugs or compounds.

Meanwhile, we also explained and discussed the limitations of *C. elegans* in the pharmacological study.

1.2 High Sensitivity to Environmental Exposure

Using some sublethal endpoints, *C. elegans* can be used to assess toxicity of different chemicals, pollutants, or stresses [2–11]. Especially, *C. elegans* has been proven to have the potential to be applied in evaluating toxicity of different pollutants at environmentally relevant concentration [12–18]. These backgrounds have demonstrated the high sensitivity of *C. elegans* to environmental exposures. Largely due to this high sensitivity, firstly, *C. elegans* is helpful to examine the association between dose of drugs or compounds and their pharmacological effects. Considering that the pharmacological effects of some compounds are associated with their adaptive response at low doses, secondly, *C. elegans* is useful to detect this aspect of pharmacological effects of compounds at relatively low doses. Thirdly, the high sensitivity to environmental exposure makes *C. elegans* more suitable to evaluate the effects of surface chemical modification or alteration in physicochemical properties on pharmacological effects of certain compounds. Fourthly, it can be easily to design and to perform pharmacological treatment at multiple doses covering wider scope, which will be helpful for us to judge and to predict the possible pharmacological doses of drugs or compounds in mammals.

1.3 Design of Certain Exposure Duration, Stage, and Route for Drugs or Compounds

The *C. elegans* has very short lifespan and life cycle. This allows us use relatively short time to finish the exposure duration for the examined drugs or compounds. *C. elegans* has four larval stages (L1, L2, L3, and L4) and two adult stages (young adult and adult). Based on this, we can make a choice to perform the pharmacological treatment at certain larval or certain adult stage. In *C. elegans*, the pharmacological treatment is mainly by feeding of drugs or compounds in the form of liquid solution or particulate suspension [19, 20], which reflects the oral administration in mammals to a certain degree. Besides this, with the aid of microinjection, the pharmacological treatment with drugs or compounds can be also performed directly in certain organs, such as the gonad or the intestine. The direct pharmacological treatment with drugs or compounds in the intestine can reflect the intraperitoneal injection to a certain degree. Based on the different developmental stages, the sequential exposure to different compounds can be carried out to assess the role of prevention or treatment of examined drugs or compounds.

1.4 Detection of Translocation and Accumulation of Compounds and Their Effects on Target Organs

The body of *C. elegans* is transparent. This property provides an important basis for our directly visualizing the dynamics in GFP fused proteins and the formation of cell lineage. Moreover, this property provides the important basis for our observing the translocation across biological barriers (such as the intestinal barrier) and the accumulation of compounds after certain labeling.

C. elegans is a powerful animal model for the study of response in target organs after exposure to certain compounds [21]. On the one hand, we can examine the alteration in development and function of target organs for certain drugs or compounds. On the other hand, we can directly determine the molecular response and the underlying mechanisms in certain target organs (such as intestine and gonad) in nematodes after exposure to drugs or compounds. In *C. elegans*, the intact intestine and gonad can be isolated.

1.5 Detection of Chemical Transformation of Compounds in the Body

In the body of organisms, we assume that the compounds will undergo the process of transformation to generate the corresponding derivatives. Examination the chemical transformation process in the body will be useful for our understanding the pharmacological mechanism of certain compounds. Due to the small size with approximately 1 mm diameter, it is not convenient to collect the excreted product of *C. elegans*. Nevertheless, we can examine the possible derivatives of compounds after chemical transformation by collecting the tissue homogenate after grinding. Using the collected tissue homogenate, the chemical components and existence of possible derivatives of examined compound will be determined in nematodes. Considering the very small size, the in situ analysis of chemical transformation at certain aspects for some examined compounds will be also be carried out in *C. elegans*.

1.6 Availability of Developmental, Physiological, and Metabolic Endpoints for Pharmacological Assessment

The *C. elegans* has different tissues or organs, such as nervous system, epidermis, reproductive system, digestive system, and muscle. These developmental structures provide us the basis for assessing pharmacological effects on development and functions of certain tissue or organs. The physiological states of *C. elegans* can be reflected by alterations in locomotion behavior, pharyngeal pumping, defecation behavior, and response to different stresses. The biochemical processes of biological molecules and their molecular control in *C. elegans* are highly conserved compared with those in mammals. Paeoniflorin, an active compound from plants, was used as the example. Using different endpoints, the paeoniflorin treatment was found to have the beneficial effects against *Pseudomonas aeruginosa* infection, high glucose toxicity, and reproductive toxicity of nanoplastics in *C. elegans* [22–24].

1.7 Availability of Certain Pathological Models for Pharmacological Study

In *C. elegans*, there are at least three aspects of pathological models available for both the study of both life science and pharmacology. The first form of pathological models is established by induction with chemicals. For example, addition of nematode growth medium (NGM) with high concentrations of glucose to establish the pathological model for high-glucose toxicity. The second form of pathological models is established by mutation of gene whose homologue in humans are associated with occurrence of certain disease(s). For example, *nlg-1* encodes a homologue to Neuroligin, and disruption in Neuroligin is highly associated with occurrence of autism spectrum disorders. Thus, the *nlg-1* mutants can be used as a pathological model to a certain degree. Actually, at most of the cases, the pathological models are established by transforming wild-type and mutant versions of human genes governing occurrence of diseases into *C. elegans*. In *C. elegans*, the frequently used pathological models for the study of both Alzheimer's disease (AD) and Parkinson disease (PD) belong to this form of pathological models [25–28]. Using these different forms of pathological models, an increasing number of pharmacological studies have been performed in *C. elegans*. The related pathological models can be useful not only for pharmacological study but also designing prevention strategies [29, 30]. Nevertheless, more efforts are needed for the establishment of pathological models in *C. elegans*, such as the establishment of pathological models on rare diseases [31].

1.8 Elucidation of Underlying Mechanisms of Certain Pharmacological Effects

Acting as the smallest model animal, *C. elegans* has received the increasing attention for its application in elucidation of pharmacological mechanisms of drugs or compounds [32–34]. Firstly, *C. elegans* is helpful for understanding the underlying chemical mechanism of examined drugs or compounds, including their chemical transformation. Secondly, *C. elegans* is useful for examining the underlying cellular mechanism of examined drugs or compounds, since *C. elegans* is an important animal model for the study of cell biology. Thirdly, *C. elegans* is useful for determining the underlying metabolic mechanism of examined drugs or compounds. Actually, occurrence of some diseases are associated with the dysfunction in metabolic processes for the animal body or the microbe in the intestine. Fourthly, *C. elegans* is very important for elucidating the underlying molecular mechanism of examined drugs or compounds [25, 35]. Meanwhile, the pharmacological investigations in *C. elegans* can provide the clues for understanding the underlying mechanisms of certain diseases, such as AD [36].

1.9 Identification of Potential Targets of Drugs or Compounds

Due to the role of classic model animal, *C. elegans* has the perfect and well-described molecular and genetic backgrounds. In addition, the rich genetic resources are available in *C. elegans*. These provide the important basis for identification of potential targets of drugs or compounds in *C. elegans*. Using both the human genome sequence and *C. elegans* genome sequence, the pharmacological or therapeutic targets of drugs or compounds have been frequently identified in *C. elegans* [37, 38]. This reflects a form of genetic pharmacology to a great degree with the aim to determine the chemistry-to-gene effects [39, 40]. The molecular targets of drugs or compounds can be identified based on the molecular response and functional analysis of *C. elegans* genes. Besides this, the molecular targets of drugs or compounds can be also identified by phenotypic analysis of *C. elegans* with the transformation of certain human wild-type and/or mutant human gene(s).

1.10 High-Throughput Screen and Identification of Compounds Having Certain Pharmacological Effect

It has been recognized that the small model animals, such as *C. elegans* and zebrafish, are suitable as the initial experimental models for pharmacological study. Due to the small volume of exposure solution and high sensitivity to environmental exposure, *C. elegans* is very suitable for large-scale and high-throughput screen of drugs or compounds for the aims of novel drug discovery, identification of effective derivatives or components in plants, and identification of molecular targets [41–43]. The high-throughput screen of drugs or compounds can be performed based on phenotypic analysis on *C. elegans* [44, 45]. The high-throughput screen of drugs or compounds can also be carried out based on phenotypic analysis on both *C. elegans* and microbes, such as bacterial or fungal pathogens [46, 47]. For the high-throughput screen of molecular targets of drugs or compounds, both transcriptomic sequencing and proteomic data are helpful for providing the candidate targets. Moreover, high-throughput RNAi screening can further provide the systematic data on functional analysis to prove the exact function of candidate targets of drugs or compounds [48].

1.11 Examination of Contributors to Amplify the Pharmacological Effect of Drugs or Compounds

Due to the small volume of exposure solution and high sensitivity to environmental exposure, *C. elegans* is not only suitable for high-throughput screen of drugs or compounds, but also helpful for identification of possible contributors to amply the pharmacological effects of certain drugs or compounds. That is, besides the finding of effective drugs or compounds, we still need to ask which factors can be adjusted to enhance the pharmacological effects of drugs or compounds. The fast experiments with small experimental space in *C. elegans* can provide great convenience for this aim. For example, we can quickly test the effect of different chemical modifications and composites with certain vectors for the examined drugs of compounds in *C. elegans*.

1.12 Safety Assessment of Drugs or Compounds

Besides the lethality analysis, some useful sublethal endpoints have been raised to assess possible adverse effects of examined chemical or pollutants [11]. Some mutation or RNAi backgrounds with susceptible property to chemicals or pollutants have been identified in *C. elegans* [49–55]. These backgrounds can help us to perform the safety analysis under the susceptible conditions. The susceptible conditions also contain the larval stages and the aging process [56–59]. *C. elegans* is not only useful for assessing the effect of drugs or compounds in exposed parents, but also helpful for evaluating their possible effects on multiple generations due to its short life cycle. In *C. elegans*, the transgenerational toxicity of different chemicals has been successfully examined [60–67].

1.13 Limitations of *C. elegans* in Pharmacological Study

As introduced and discussed above, *C. elegans* has multiple aspects of values for the pharmacological study. Nevertheless, meanwhile, some aspects of limitations of *C. elegans* in the pharmacological study still exist. Firstly, *C. elegans* do not have the blood circulation and some important organs, such as heart, liver, lung, and kidney, which exist in mammals and humans. Secondly, *C. elegans* is very small and not convenient to collect the separate tissues and organs for the subsequent study after the pharmacological study. Thirdly, the pharmacological exposure in *C. elegans* mainly reflects the oral administration. Fourthly, not all the genes in humans have the corresponding homologues in *C. elegans*, because the human genome has more complex information and structure than that of *C. elegans*.

References

1. Glazer I, Shapiro-Ilan DI, Sternberg PW (2022) Nematodes as model organisms. CAB International, Oxford
2. Wang D-Y (2018) Nanotoxicology in *Caenorhabditis elegans*. Springer, Singapore
3. Xu R-R, Hua X, Rui Q, Wang D-Y (2022) Alteration in Wnt signaling mediates induction of transgenerational toxicity of polystyrene nanoplastics in *C. elegans*. NanoImpact 28:100425
4. Xu R-R, Hua X, Rui Q, Wang D-Y (2022) Polystyrene nanoparticles caused dynamic alteration in mitochondrial unfolded protein response from parents to the offspring in *C. elegans*. Chemosphere 308:136154
5. Shao Y-T, Wang Y-X, Hua X, Li Y-H, Wang D-Y (2023) Polylactic acid microparticles in the range of μg/L reduce reproductive capacity by affecting the gonad development and the germline apoptosis in *Caenorhabditis elegans*. Chemosphere 336:139193
6. Tang M-F, Ding G-Y, Li L-E, Xiao G-S, Wang D-Y (2023) Exposure to polystyrene nanoparticles at predicted environmental concentrations enhances toxic effects of *Acinetobacter johnsonii* AC15 infection on *Caenorhabditis elegans*. Ecotoxicol Environ Saf 262:115131
7. Wang Y-X, Hua X, Wang D-Y (2023) Exposure to 6-PPD quinone enhances lipid accumulation through activating metabolic sensors of SBP-1 and MDT-15 in *Caenorhabditis elegans*. Environ Pollut 333:121937
8. Wang Y-X, Yuan X-A, Zhou R, Bu Y-Q, Wang D-Y (2023) Combinational exposure to hydroxyatrazine increases neurotoxicity of polystyrene nanoparticles on *Caenorhabditis elegans*. Sci Total Environ 880:163283
9. Hua X, Feng X, Liang G-Y, Chao J, Wang D-Y (2023) Long-term exposure to tire-derived 6-PPD quinone causes intestinal toxicity by affecting functional state of intestinal barrier in *Caenorhabditis elegans*. Sci Total Environ 861:160591
10. Zhao Y-Y, Hua X, Rui Q, Wang D-Y (2023) Exposure to multi-walled carbon nanotubes causes suppression in octopamine signal associated with transgenerational toxicity induction in *C. elegans*. Chemosphere 318:137986
11. Wang D-Y (2020) Exposure toxicology in *Caenorhabditis elegans*. Springer, Singapore
12. Zhao Y, Hua X, Bian Q, Wang D-Y (2022) Nanoplastic exposure at predicted environmental concentrations induces activation of germline Ephrin signal associated with toxicity formation in the *Caenorhabditis elegans* offspring. Toxics 10:699
13. Liu H-L, Zhao Y-L, Hua X, Wang D-Y (2022) Induction of transgenerational toxicity is associated with the activated germline insulin signals in nematodes exposed to nanoplastic at predicted environmental concentrations. Ecotoxicol Environ Saf 243:114022

14. Hua X, Feng X, Liang G-Y, Chao J, Wang D-Y (2023) Exposure to 6-PPD quinone at environmentally relevant concentrations causes abnormal locomotion behaviors and neurodegeneration in *Caenorhabditis elegans*. Environ Sci Technol 57:4940–4950
15. Hua X, Feng X, Liang G-Y, Chao J, Wang D-Y (2023) Long-term exposure to 6-PPD quinone reduces reproductive capacity by enhancing germline apoptosis associated with activation of both DNA damage and cell corpse engulfment in *Caenorhabditis elegans*. J Hazard Mater 454: 131495
16. He W-M, Gu A-H, Wang D-Y (2023) Sulfonate-modified polystyrene nanoparticle at precited environmental concentrations induces transgenerational toxicity associated with increase in germline notch signal of *Caenorhabditis elegans*. Toxics 11:511
17. Hua X, Cao C, Zhang L, Wang D-Y (2023) Activation of FGF signal in germline mediates transgenerational toxicity of polystyrene nanoparticles at predicted environmental concentrations in *Caenorhabditis elegans*. J Hazard Mater 451:131174
18. Wang D-Y (2022) Toxicology at environmentally relevant concentrations in *Caenorhabditis elegans*. Springer, Singapore
19. Zhang L, Wang Y-X, Cao C, Zhu Y-K, Huang W, Yang Y, Qiu H-B, Liu S-Q, Wang D-Y (2022) Beneficial effect of Xuebijing against *Pseudomonas aeruginosa* infection in *Caenorhabditis elegans*. Front Pharmacol 13:949608
20. Zhang W, Lv T, Li M, Wu Q, Yang L, Liu H, Sun D, Sun L, Zhuang Z, Wang D (2013) Beneficial effects of wheat gluten hydrolysate to extend lifespan and induce stress resistance in nematode *Caenorhabditis elegans*. PLoS One 8:e74553
21. Wang D-Y (2019) Target organ toxicology in *Caenorhabditis elegans*. Springer, Singapore
22. Wang Y-X, Zhang L, Yuan X-A, Wang D-Y (2023) Treatment with paeoniflorin increases lifespan of *Pseudomonas aeruginosa* infected *Caenorhabditis elegans* by inhibiting bacterial accumulation in intestinal lumen and biofilm formation. Front Pharmacol 14:1114219
23. Liu T-W, Zhuang Z-H, Wang D-Y (2023) Paeoniflorin mitigates high glucose-induced lifespan reduction by inhibiting insulin signaling in *Caenorhabditis elegans*. Front Pharmacol 14: 1202379
24. Hua X, Feng X, Hua Y-S, Wang D-Y (2023) Paeoniflorin attenuates polystyrene nanoparticle-induced reduction in reproductive capacity and increase in germline apoptosis through suppressing DNA damage checkpoints in *Caenorhabditis elegans*. Sci Total Environ 871: 162189
25. Griffin EF, Caldwell KA, Caldwell GA (2017) Genetic and pharmacological discovery for Alzheimer's disease using *Caenorhabditis elegans*. ACS Chem Neurosci 8:2596–2606
26. Wu Y, Luo Y (2005) Transgenic *C. elegans* as a model in Alzheimer's research. Curr Alzheimer Res 2:37–45
27. Cooper JF, Van Raamsdonk JM (2018) Modeling Parkinson's disease in *C. elegans*. J Parkinsons Dis 8:17–32
28. Tsai CW, Tsai RT, Liu SP, Chen CS, Tsai MC, Chien SH, Hung HS, Lin SZ, Shyu WC, Fu RH (2017) Neuroprotective effects of betulin in pharmacological and transgenic *Caenorhabditis elegans* models of Parkinson's disease. Cell Transplant 26:1903–1918
29. Kraemer B, Schellenberg GD (2007) Using *Caenorhabditis elegans* models of neurodegenerative disease to identify neuroprotective strategies. Int Rev Neurobiol 77:219–246
30. Markaki M, Tavernarakis N (2020) *Caenorhabditis elegans* as a model system for human diseases. Curr Opin Biotechnol 63:118–125
31. Kropp PA, Bauer R, Zafra I, Graham C, Golden A (2021) *Caenorhabditis elegans* for rare disease modeling and drug discovery: strategies and strengths. Dis Model Mech 14: dmm049010
32. Martinez BA, Caldwell KA, Caldwell GA (2017) *C. elegans* as a model system to accelerate discovery for Parkinson disease. Curr Opin Genet Dev 44:102–109
33. Okoro NO, Odiba AS, Osadebe PO, Omeje EO, Liao G, Fang W, Jin C, Wang B (2021) Bioactive phytochemicals with anti-aging and lifespan extending potentials in *Caenorhabditis elegans*. Molecules 26:7323

34. Collins JJ, Evason K, Kornfeld K (2006) Pharmacology of delayed aging and extended lifespan of *Caenorhabditis elegans*. Exp Gerontol 41:1032–1039
35. Parker JA, Holbert S, Lambert E, Abderrahmane S, Néri C (2004) Genetic and pharmacological suppression of polyglutamine-dependent neuronal dysfunction in *Caenorhabditis elegans*. J Mol Neurosci 23:61–68
36. Natale C, Barzago MM, Diomede L (2020) *Caenorhabditis elegans* models to investigate the mechanisms underlying Tau toxicity in tauopathies. Brain Sci 10:838
37. Gill MS (2006) Endocrine targets for pharmacological intervention in aging in *Caenorhabditis elegans*. Aging Cell 5:23–30
38. Link EM, Hardiman G, Sluder AE, Johnson CD, Liu LX (2000) Therapeutic target discovery using *Caenorhabditis elegans*. Pharmacogenomics 1:203–217
39. Rand JB, Johnson CD (1995) Genetic pharmacology: interactions between drugs and gene products in *Caenorhabditis elegans*. Methods Cell Biol 48:187–204
40. Jones AK, Buckingham SD, Sattelle DB (2005) Chemistry-to-gene screens in *Caenorhabditis elegans*. Nat Rev Drug Discov 4:321–330
41. Giunti S, Andersen N, Rayes D, De Rosa MJ (2021) Drug discovery: insights from the invertebrate *Caenorhabditis elegans*. Pharmacol Res Perspect 9:e00721
42. Carretero M, Solis GM, Petrascheck M (2017) *C. elegans* as model for drug discovery. Curr Top Med Chem 17:2067–2076
43. Kim W, Hendricks GL, Lee K, Mylonakis E (2017) An update on the use of *C. elegans* for preclinical drug discovery: screening and identifying anti-infective drugs. Expert Opin Drug Discov 12:625–633
44. Paul D, Chipurupalli S, Justin A, Raja K, Mohankumar SK (2020) *Caenorhabditis elegans* as a possible model to screen anti-Alzheimer's therapeutics. J Pharmacol Toxicol Methods 106: 106932
45. Bulterijs S, Braeckman BP (2020) Phenotypic screening in *C. elegans* as a tool for the discovery of new geroprotective drugs. Pharmaceuticals 13:1
46. Peterson ND, Pukkila-Worley R (2018) *Caenorhabditis elegans* in high-throughput screens for anti-infective compounds. Curr Opin Immunol 54:59–65
47. Kang D, Zhang L, Kirienko NV (2021) High-throughput approaches for the identification of *Pseudomonas aeruginosa* antivirulents. MBio 12:e02240-20
48. Struwe WB, Warren CE (2010) High-throughput RNAi screening for N-glycosylation dependent loci in *Caenorhabditis elegans*. Methods Enzymol 480:477–493
49. Wang D-Y (2019) Molecular toxicology in *Caenorhabditis elegans*. Springer, Singapore
50. Wu Q, Zhao Y, Li Y, Wang D (2014) Susceptible genes regulate the adverse effects of TiO_2-NPs at predicted environmental relevant concentrations on nematode *Caenorhabditis elegans*. Nanomed Nanotechnol Biol Med 10:1263–1271
51. Sun L, Wu Q, Liao K, Yu P, Cui Q, Rui Q, Wang D (2016) Contribution of heavy metals to toxicity of coal combustion related fine particulate matter ($PM_{2.5}$) in *Caenorhabditis elegans* with wild-type or susceptible genetic background. Chemosphere 144:2392–2400
52. Qu M, Xu K, Li Y, Wong G, Wang D (2018) Using *acs-22* mutant *Caenorhabditis elegans* to detect the toxicity of nanopolystyrene particles. Sci Total Environ 643:119–126
53. Zhao L, Kong J, Krasteva N, Wang D (2018) Deficit in epidermal barrier induces toxicity and translocation of PEG modified graphene oxide in nematodes. Toxicol Res 7:1061–1070
54. Ding X, Rui Q, Wang D (2018) Functional disruption in epidermal barrier enhances toxicity and accumulation of graphene oxide. Ecotoxicol Environ Saf 163:456–464
55. Ren M, Zhao L, Ding X, Krasteva N, Rui Q, Wang D (2018) Developmental basis for intestinal barrier against the toxicity of graphene oxide. Part Fibre Toxicol 15:26
56. Qiu Y, Luo L, Yang Y, Kong Y, Li Y, Wang D (2020) Potential toxicity of nanopolystyrene on lifespan and aging process of nematode *Caenorhabditis elegans*. Sci Total Environ 705:135918
57. Qiu Y, Liu Y, Li Y, Li G, Wang D (2020) Effect of chronic exposure to nanopolystyrene on nematode *Caenorhabditis elegans*. Chemosphere 256:127172

58. Xing X, Guo Y, Wang D (2009) Using the larvae nematode *Caenorhabditis elegans* to evaluate neurobehavioral toxicity to metallic salts. Ecotoxicol Environ Saf 72:1819–1823
59. Guo Y, Yang Y, Wang D (2009) Induction of reproductive deficits in nematode *Caenorhabditis elegans* exposed to metals at different developmental stages. Reprod Toxicol 28:90–95
60. Zhao Y, Lin Z, Jia R, Li G, Xi Z, Wang D (2014) Transgenerational effects of traffic-related fine particulate matter ($PM_{2.5}$) on nematode *Caenorhabditis elegans*. J Hazard Mater 274:106–114
61. Zhao Y, Liu Q, Shakoor S, Gong JR, Wang D (2015) Transgenerational safe property of nitrogen-doped graphene quantum dots and the underlying cellular mechanism in *Caenorhabditis elegans*. Toxicol Res 4:270–280
62. Zhao L, Qu M, Wong G, Wang D (2017) Transgenerational toxicity of nanopolystyrene particles in the range of μg/L in nematode *Caenorhabditis elegans*. Environ Sci Nano 4: 2356–2366
63. Sun L, Liao K, Wang D (2021) Comparison of transgenerational reproductive toxicity induced by pristine and amino modified nanoplastics in *Caenorhabditis elegans*. Sci Total Environ 768: 144362
64. Liu H, Tian L, Wang S, Wang D (2021) Size-dependent transgenerational toxicity induced by nanoplastics in nematode *Caenorhabditis elegans*. Sci Total Environ 790:148217
65. Hua X, Zhao Y, Yuan Y, Zhang L, Bian Q, Wang D (2022) Nanoplastics cause transgenerational toxicity through inhibiting germline microRNA *mir-38* in *C. elegans*. J Hazard Mater 437:129302
66. Zhang L, Wang S, Zhao Y, Bi K, Wang D (2022) Increase in germline methyltransferases governing methylation of histone H3K9 is associated with transgenerational nanoplastic toxicity in *Caenorhabditis elegans*. Environ Sci Nano 9:265–274
67. Zhang L, Wang S, Zhao Y, Nurdebek B, Bu Y, Wang D (2022) Long-term exposure to polystyrene nanoparticles causes transgenerational toxicity by affecting the function and expression of MEV-1 and DAF-2 signals in *Caenorhabditis elegans*. NanoImpact 26:100403

Patterns of Pharmacological Treatment 2

Abstract

Due to the existence of certain influencing factors and the different research aims, different patterns of pharmacological treatment with compounds and drugs are formed in *Caenorhabditis elegans*. We here first introduced and discussed pharmacological treatment stages, pharmacological treatment durations, pharmacological treatment routes, and treatment with different forms of compounds or drugs. Some complex systems for pharmacological treatment of compounds and drugs were also introduced and discussed. Finally, pharmacological prevention and pharmacological therapy were further introduced.

Keywords

Treatment patterns · Research aims · Compounds · *Caenorhabditis elegans*

2.1 Introduction

In *Caenorhabditis elegans*, different patterns of pharmacological treatment with compounds and drugs are available. Formation of different patterns of pharmacological treatment with compounds and drugs is due to their different pharmacological treatment stages, pharmacological treatment durations, and pharmacological treatment routes, and the existence of different forms of compounds or drugs (Fig. 2.1). Formation of different patterns of pharmacological treatment with compounds and drugs are also associated with research aims for future clinical interventions, such as pharmacological prevention and pharmacological therapy (Fig. 2.1). In addition, sometimes some complex systems were also raised for the pharmacological treatment of compounds and drugs (Fig. 2.1).

D. Wang, *Pharmacology in Caenorhabditis elegans*,
https://doi.org/10.1007/978-981-92-5166-7_2

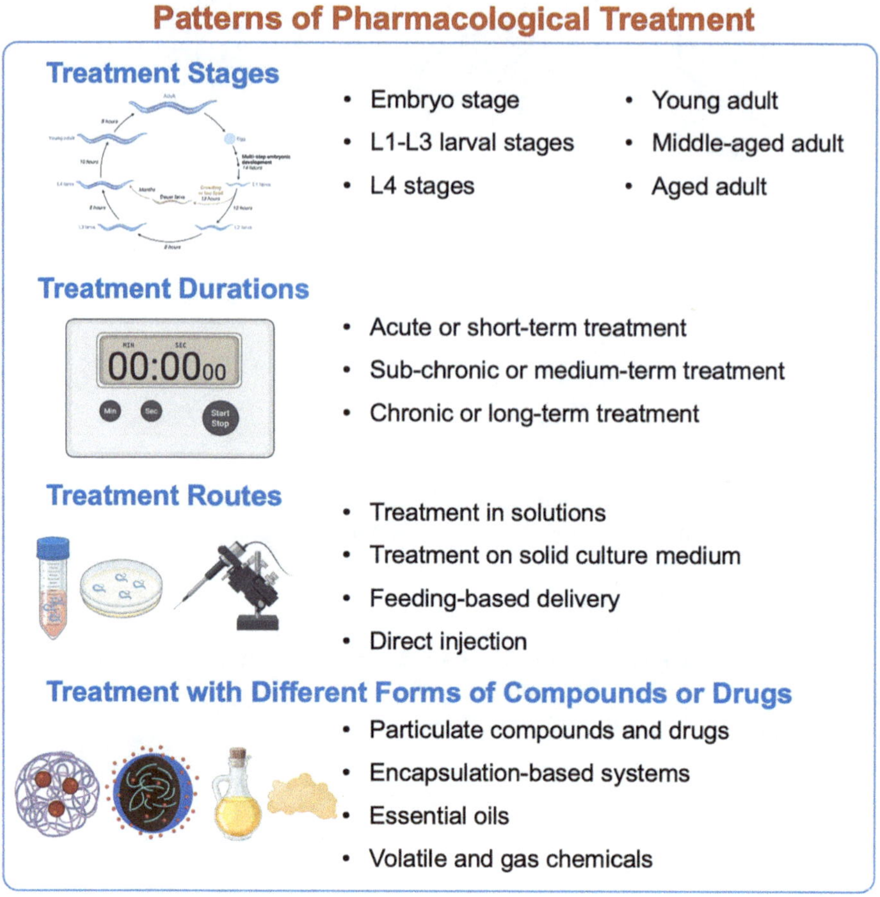

Fig. 2.1 Patterns of pharmacological treatment in *Caenorhabditis elegans*

2.2 Pharmacological Treatment Stages

Pharmacological treatment stages refer to the specific developmental time points or windows during which the *C. elegans* will be treated with compounds and drugs. During the *C. elegans* life cycle, the developmental stages include egg, L1–L4 larvae, young adult, gravid adult, and aged adult. Due to these precisely defined developmental stages, the pharmacological treatment stages for compounds and drugs can contain:

1. Embryo Stage: Pharmacological treatment at this stage can be used to test the effects of compounds and drugs on embryogenesis.

2. L1–L3 Larval Stages: This developmental timing reflects the early to mid-larval development. Pharmacological treatment at this stage can be used to assess the effects of compounds and drugs on larval development and function and the possible effects in causing developmental delay or larval arrest.
3. L4 Stage: This developmental timing reflects the late larval (just before the adulthood). Pharmacological treatment at this stage can be used to evaluate the effects of compounds and drugs on neuronal or gonadal maturation, which can also be considered as the pre-adult priming for adult assays.
4. Young Adult (Day 1–2 Adult): This developmental timing reflects the early reproductive period. For pharmacological treatment with compounds and drugs, especially for the assessment on pharmacological effects on longevity and health span, this stage is normally used as the standard start point. At this stage, almost all aspects of endpoints can be analyzed, including those for assessing development and functions of neurons, reproductive system, and intestine, as well as the metabolisms.
5. Middle-Aged Adult (Day 5–7 Adult): This developmental timing reflects the post-reproductive stage and the early aging stage. Pharmacological treatment at this stage can be used to evaluate the effects of compounds and drugs in rescuing age-related decline, in reducing late-life neurodegeneration (e.g., Aβ or tau models), and in modulating innate immunity.
6. Aged Adult (from the Day 8): This developmental timing reflects the late aging, and the sharp decline of survival curve will occur at this stage. Pharmacological treatment at this stage can be used to assess the possible effects of compounds and drugs for frailty or late-life resilience and in inhibiting immunosuppression.

2.3 Pharmacological Treatment Durations

Pharmacological treatment duration refers to the prescribed length of time for treatment with compounds and drugs in order to achieve specific health outcomes while minimizing the possible risks. The pharmacological treatment durations are selected based on some factors, including the used physiological and pathological models, the treatment goals, and the pharmacokinetic and pharmacodynamic properties of examined compounds and drugs.

The pharmacological treatment durations can be classified into the following several groups:

1. Acute or Short-Term Treatment: This form of treatment duration aims to assess the relatively rapid response of *C. elegans*. This form of duration for treatment with compounds and drugs usually last for 2–48 h. For the aim of future clinical use, the treatments with compounds and drugs are normally suggested to be performed with this form of treatment duration. In the clinical, the drug treatments are usually strictly defined. That is, the treatment course of a drug must be clearly defined to minimize the possible clinical risks.

2. Sub-chronic or Medium-Term Treatment: This form of duration for treatment with compounds and drugs usually last 3–12 days. For certain research aims, such as inhibition in progressive neurodegeneration, this form of treatment duration can be selected. Nevertheless, during the treatment in the clinical, the treatment doses of compounds and drugs need to be strictly designed and defined, and high doses of treatment with compounds and drugs are not suggested.
3. Chronic or Long-Term Treatment: This form of treatment with compounds and drugs are usually performed from a specific stage (e.g., L4-larvae or Day 1 adult) until the death. This form of duration is suggested to be used to assess the long-term effects of treatment with compounds on aging, healthspan, or chronic disease progression. To a great degree, this form of duration is more suitable for evaluating the effects of nutritional intervention, but normally not suitable assessing the effects of drugs for the clinical aims.

2.4 Pharmacological Treatment Routes

Pharmacological treatment route refers to the path by which a compound or a drug is taken into the body or delivered to its site of action. During the design of pharmacological treatment route, the possible influences of efficiency (bioavailability), properties of compounds and drugs, onset of drug's therapeutic effects, and formation of potential side effects are suggested to carefully considered. In *C. elegans*, the pharmacological treatment routes for compounds and drugs include the following several forms:

1. Treatment in Solutions: Normally, we first suggest the use of this form of pharmacological treatment route [1, 2]. In the compound or drug solutions, the adequate uptake of compounds and drugs to nematodes can be guaranteed. For particulate compounds and drugs, the nematodes can be treated in corresponding compound suspensions or drug suspensions. If the treatment with compounds and drugs is from the larval stage, the *E. coli* OP50 needs to be added to satisfy the larval development. Nevertheless, if the treatment with compounds and drugs is from the young adult or adult day-1, the *E. coli* OP50 can also be considered to be added in the treatment solutions to enhance the uptake of compounds and drugs by the nematodes. In addition, this form of pharmacological treatment route is more suitable for performing high throughput screening of compounds and drugs in *C. elegans*.
2. Treatment on solid culture medium: Initially, this form of pharmacological treatment route was the most commonly used treatment route for compounds and drugs. No doubt, it is relatively easy to perform this form of treatment route by adding solutions of compounds or drugs on the surface of solid culture medium (such as the NGM plates) or mixing the compounds or drugs into the molten agar before the plates are poured. However, this form of treatment usually means the not adequate uptake of compounds and drugs to the nematodes. In addition, due to the avoidance behavior, most of the particulate compounds and

drugs will be not bioavailable to the nematodes. For the water-soluble compounds and drugs, this form of treatment route can be considered.

3. Feeding-Based Delivery: This is an improved version of treatment on solid culture medium. That is, the compounds or drugs are mixed with a concentrated pellet of *E. coli* OP50 bacteria, and then placed on the NGM plates. On the surface of NGM plates, the nematodes will ingest the compounds or drugs along with their food.
4. Direct Injection: This means that, with the aid of microinjection, a very fine glass needle is used to inject a compound or drug solution directly into the body cavity (pseudocoelom), gut, or even individual cells of the worms. This is the only direct treatment route, but it is somewhat technically challenging.

2.5 Pharmacological Treatment with Different Forms of Compounds or Drugs

Besides the water-soluble and the lipid-soluble compounds and drugs, there are still some other forms of compounds and drugs (such as the particulate compounds and drugs, the essential oils, and the gas chemicals), which can be used for certain different pharmacological intervention goals.

1. Particulate Compounds and Drugs: As described above, the nematodes can be treated directly in the suspensions of particulate compounds and drugs. Besides this, for certain range of particulate compounds and drugs at the nanosize, the microinjection can also be considered to make them bioavailable to nematodes [3].
2. Encapsulation-Based Systems: The liposomes, a kind of tiny, fat-soluble vesicles, can encapsulate both water-soluble and fat-soluble drugs and deliver them to targets organs. It is suggested that the liposomes encapsulating the compounds and the drugs can be fused with intestinal cells and then efficiently delivered the compounds and the drug into the worm's body [4]. This form of pharmacological treatment may offer the high drug loading capacity and stability.
3. Essential Oils: The essential oils are concentrated and hydrophobic liquids containing volatile aromatic compounds, which usually contain the secondary metabolites in plants. The *C. elegans* can be directed treated in the liquid medium containing the essential oils or treated with essential oils by mixing them into the NGM before pouring the plates or applying them on the surface of NGM plates [5, 6].
4. Volatile and Gas Chemicals: The gas chemical (such as H_2 and H_2S) are different from the volatile molecules (such as the diacetyl). During the treatment of *C. elegans* with gas chemicals, the Petri dish need to be sealed, and the precise control for addition and stopping of gas chemicals should be carefully considered [7]. For the treatment of *C. elegans* with volatile molecules, such as the anesthetic treatment, the Petri dishes are also suggested to be sealed [8].

2.6 Complex Systems for Pharmacological Treatment

In *C. elegans*, sometimes some complex systems for pharmacological treatment can be further raised for certain research or application aims.

1. Combinational Treatment of Compounds and Drugs: This form of pharmacological treatment has been widely performed in different assay systems, including in *C. elegans*. This approach means treating the *C. elegans* with two or more drugs simultaneously. For the aims of future clinical use, usually the synergistic effects between or among compounds and drugs are anticipated [9, 10]. Besides this, the possible additive or antagonistic effects between compounds and drugs also need to be carefully examined [11].
2. Genetic-Pharmacological Combinations: The genetic-pharmacological combination refers to the approach where genetic manipulations (such as mutations, gene knockouts, RNA interference (RNAi), or transgene) are combined with pharmacological treatments to determine the mechanisms of action after the observation of pharmacological effects for certain compounds and drugs [12, 13]. For this aspect, the further high throughput screening to identify anticipated genetic backgrounds and drugs are suggested to be performed.
3. Intermittent Pharmacological Treatment: Normally, the treatment with compounds and drugs is performed as the continuous treatment. For certain research aims, sometimes the intermittent pharmacological treatment can also be performed. Intermittent pharmacological treatment means an experimental regimen in which a compound or a drug is administered in discrete and spaced exposure periods. This approach can be used to evaluate the possible reversibility of drug effects and drug tolerance.

2.7 Pharmacological Prevention

For the clinical intervention, there are mainly two aspects of considerations, the pharmacological prevention and especially the pharmacological therapy. Pharmacological prevention refers to pharmacological effects of compounds and drugs to prevent the severity of pathological process, age-related decline, or pollutants-induced damage before their onset. For this reason, the pharmacological prevention of compounds is usually performed at the larval stage or young adult stage [14, 15]. For example, to determine the potential prevention effect of vitamin E on toxicity of nanomaterials, L2-larvae were first treated with vitamin E, and after that the nematodes were further exposed to Al_2O_3 nanoparticles from the L4-larvae stage [16].

2.8 Pharmacological Therapy

Pharmacological therapy refers to the research aim for administration of compounds and drugs to treat, to reverse, or to ameliorate existing pathological damage or toxicity induced by pollutants or stresses, which contrasts with the pharmacological prevention. Most of the performed pharmacological studies on compounds and drugs in *C. elegans* are about this aspect or aim to achieve this aspect of goals.

References

1. Shu C-J, Wang W, Cao S-Q, Hou B-W, Gu S-L, Fu L, Zhang F-L, Wang D-Y (2026) 6-PPD quinone induces lifespan reduction by causing immunosuppression via DAF-16/PMK-1 signaling in *Caenorhabditis elegans*. Environ Chem Ecotoxicol 8:1176–1184
2. Shu C-J, Hu D-Y, Cao S-Q, Hou B-W, Gu S-L, Fu L, Zhang F-L, Wang D-Y (2026) Exposure to 77PD quinone inhibits longevity and healthspan via affecting mitochondrial signals in *Caenorhabditis elegans*. Ecotoxicol Environ Saf 310:119767
3. Mohan N, Chen C, Hsieh H, Wu Y, Chang H (2010) *In vivo* imaging and toxicity assessment of fluorescent nanodiamonds in *Caenorhabditis elegans*. Nano Lett 10:3692–3699
4. Shibamura A, Ikeda T, Nishikawa Y (2009) A method for oral administration of hydrophilic substances to *Caenorhabditis elegans*: effects of oral supplementation with antioxidants on the nematode lifespan. Mech Ageing Dev 130:652–655
5. Zhu S, Li H, Dong J, Yang W, Liu T, Wang Y, Wang X, Wang M, Zhi D (2017) Rose essential oil delayed Alzheimer's disease-like symptoms by SKN-1 pathway in *C. elegans*. J Agric Food Chem 65:8855–8865
6. Xu R, Li AP, Tan X, Tang X, He XP, Wang LX, Kang JJ, Li SH, Liu Y (2025) Patchouli essential oil extends the lifespan and healthspan of *Caenorhabditis elegans* through JNK-1/DAF-16. Life Sci 360:123270
7. Blackwood EA, Glembotski CC (2022) Hydrogen sulfide: the gas that fuels longevity. J Cardiovasc Aging 2:26
8. Steele LM, Sedensky MM (2018) Approaches to anesthetic mechanisms: the *C. elegans* model. Methods Enzymol 602:133–151
9. Guha S, Mathew ND, Konkwo C, Ostrovsky J, Kwon YJ, Polyak E, Seiler C, Bennett M, Xiao R, Zhang Z, Nakamaru-Ogiso E, Falk MJ (2021) Combinatorial glucose, nicotinic acid and N-acetylcysteine therapy has synergistic effect in preclinical *C. elegans* and zebrafish models of mitochondrial complex I disease. Hum Mol Genet 30:536–551
10. Jarne-Ferrer J, Pallàs M, Griñán-Ferré C, Bellver-Sanchis A (2025) Investigating the synergistic neuroprotective effects of plant-derived antioxidants and the psychedelic N, N-dimethyltryptamine in Alzheimer's disease therapy. Cells 14:934
11. Hernando G, Bouzat C (2025) Drug combination assays using *Caenorhabditis elegans* as a model system. J Pharmacol Toxicol Methods 131:107583
12. Wang Y-X, Zhang L, Yuan X-A, Wang D-Y (2023) Treatment with paeoniflorin increases lifespan of *Pseudomonas aeruginosa* infected *Caenorhabditis elegans* by inhibiting bacterial accumulation in intestinal lumen and biofilm formation. Front Pharmacol 14:1114219
13. Liu T-W, Zhuang Z-H, Wang D-Y (2023) Paeoniflorin mitigates high glucose-induced lifespan reduction by inhibiting insulin signaling in *Caenorhabditis elegans*. Front Pharmacol 14: 1202379
14. Rui Q, Lu Q, Wang D-Y (2009) Administration of *Bushenkangshuai Tang* alleviates the UV irradiation- and oxidative stress-induced lifespan defects in nematode *Caenorhabditis elegans*. Front Med China 3:76–90

15. Ye H-Y, Ye B-P, Wang D-Y (2008) Trace administration of vitamin E can retrieve and prevent UV-irradiation- and metal exposure-induced memory deficits in nematode *Caenorhabditis elegans*. Neurobiol Learn Mem 90:10–18
16. Yu X-M, Guan X-M, Wu Q-L, Zhao Y-L, Wang D-Y (2015) Vitamin E ameliorates the neurodegeneration related phenotypes caused by neurotoxicity of Al_2O_3-nanoparticles in *C. elegans*. Toxicol Res 4:1269–1281
17. Okoro NO, Odiba AS, Osadebe PO, Omeje EO, Liao G, Fang W, Jin C, Wang B (2021) Bioactive phytochemicals with anti-aging and lifespan extending potentials in *Caenorhabditis elegans*. Molecules 26:7323
18. Giunti S, Andersen N, Rayes D, De Rosa MJ (2021) Drug discovery: insights from the invertebrate *Caenorhabditis elegans*. Pharmacol Res Perspect 9:e00721
19. Zarroug SHO, Bajaman JS, Hamza FN, Saleem RA, Abdalla HK (2023) *Caenorhabditis elegans* as an in vivo model for the discovery and development of natural plant-based antimicrobial compounds. Pharmaceuticals 16:1070
20. Kirchweger B, Zwirchmayr J, Grienke U, Rollinger JM (2023) The role of *Caenorhabditis elegans* in the discovery of natural products for healthy aging. Nat Prod Rep 40:1849–1873

Classification of Examined Compounds and Drugs in *Caenorhabditis elegans* 3

Abstract

Caenorhabditis elegans can be used for the pharmacological study of various drugs and compounds. We here introduced and discussed the potential of *C. elegans* for the pharmacological evaluation of chemical compounds, clinical drugs, nutrition, different physical forms of drugs and compounds, active compounds in plants, drugs at the whole organisms level, Traditional Chinese Medicines, and artificial and synthesized drugs.

Keywords

Classification · Drugs · Compounds · *Caenorhabditis elegans*

3.1 Introduction

After the pharmacological exposure, the drugs or the compounds will be bioavailable to, translocated, accumulated, and metabolized in tissues or organs of *Caenorhabditis elegans*. Based on these biological processes, the pharmacological study can be successfully performed in this classic model animal. Initially, the pharmacological treatment is used as an experimental tool to determine the functions of certain examined genes. Later, the pharmacological study has gradually become a relatively independent branch of research in *C. elegans*. The values of *C. elegans* for pharmacological study have been discussed in Chap. 1. In Chap. 2, we introduced how the pharmacological exposure can be carried out in *C. elegans*. So far, an increasing number of literatures have been reported on the application of *C. elegans* in pharmacological study of various drugs or compounds. In this chapter, we further tried to discuss the classification of examined drugs and compounds in the animal model of *C. elegans*. Nevertheless, it is difficult to make such a classification exactly. What we raised here is a glossary classification to a great degree. In addition, one of

D. Wang, *Pharmacology in Caenorhabditis elegans*,
https://doi.org/10.1007/978-981-92-5166-7_3

the main aims of this chapter is to indicate the possible application scope for *C. elegans* in pharmacological study.

C. elegans can be used for the pharmacological study of the following drugs and compounds:

1. Chemical compounds
2. Clinical drugs
3. Nutrition
4. Different physical forms
5. Active compounds in plants
6. Whole organisms level
7. Traditional Chinese Medicines
8. Artificial and synthesized drugs

3.2 Chemical Compounds

In *C. elegans*, one of the attractions for the field of pharmacology is the identification and pharmacological evaluation of chemical compounds, especially the large-scale screen of small molecule library to identify candidate compounds. Theoretically, the pharmacological study of most of the chemical compounds can be performed in *C. elegans*. Nevertheless, there are some exceptions. The corrosive compounds and very large particulate compounds are not suggested to be perform the corresponding pharmacological study in this animal model. The corrosive compounds may cause damage to the epidermal barrier and/or intestinal damage. If the sizes of particulate compounds are larger than the mouth, it is difficult to be bioavailable to nematodes by feeding. For the chemicals, we mainly introduced the small molecules and the screen based on small molecule library. For the chemicals, the biological activities of them can be reflected by their roles as activators and antagonists at certain aspects. Besides these, some metals with beneficial effects were also discussed.

3.2.1 Small Molecule Compounds

Small molecule compounds are organic chemical compounds with the low molecular weight and the therapeutic effect by modulating activity of specific biological targets, such as receptors or enzymes. Some small molecule compounds designed de novo or obtained from natural products can show high affinity to a specific target site by providing hydrophobic pocket or active site to activate/block the receptors, to affect the enzymes, and to modulate protein-protein interactions. These molecule compounds with small size allow bioavailability, cell membrane permeability, and druggable targeting.

For the small molecule compounds, normally, there are three aspects of possibilities to carry out related pharmacological study. First of all, the examined small molecule compound has well-known or other aspects of beneficial effects or exhibits special chemical structure and properties. Tomatidine is the aglycone of α-tomatine. In mice, tomatidine extracted from unripe tomatoes could suppress age-related skeletal muscle atrophy [1]. In *C. elegans*, it was further found that tomatidine administration could extend both lifespan and health span by maintaining mitochondrial homeostasis [2]. For the small molecule compounds with special chemical structure and properties, besides the native compounds, at least the synthesized compounds by chemical integration of different chemicals and the compounds with different surface chemical modifications need to be paid attention to. A series of novel benzofuran-tetrazole derivatives could be synthesized, and some of them exhibited the possible activity against Alzheimer's disease [3].

The high binding potential to targeted proteins can be predicted by molecular docking or confirmed by experiments for certain small molecule compounds. This possibility is on the idea of screening, and associated with screening of chemical binding. The availability of exact or predicted structures for both small molecule compounds and proteins provides the guarantee for this form of screening. Nevertheless, the limitation still exists for this form of screening. At least so far, the key genes contributing to the occurrence of some diseases and underlying molecular mechanisms are still unclear or uncertain. Based on molecular docking screen, 6 examined representative caffeoylquinic acids all showed binding potential to insulin receptor DAF-2, and the lifespan extension effect of these 6 compounds was confirmed in nematodes [4]. Molecular docking can also be used to examine the underlying molecular mechanisms for the observed pharmacological effect of certain small molecule compound. One of the derivatives of xyloketal B was found to have beneficial against Huntington's disease in *C. elegans* transgenic model [5]. Molecular docking analysis indicated that this derivative could bind to GLN369 and GLN393 residues of mutant Htt to form stable trimeric complex, which will be helpful for preventing occurrence of mutant Htt aggregates [5]. The 2-butoxytetrahydrofuran (2-BTHF) isolated from *Holothuria scabra* had beneficial effect against paralysis caused by Aβ toxicity [Tang]. Molecular docking analysis indicated that the 2-BTHF potentially binds to heat shock factor 1 (HSF-1) [6], which provides indirect evidence for the observed beneficial function of 2-BTHF. Besides the finding of compounds with anticipated therapeutic effect, the designed small molecule compounds can help us to perform pharmacological screen for target identification and hit discovery.

Based on phenotypical analysis using some endpoints, certain small molecule compounds can be screened out from small molecule compound library. This possibility is also relevant on the idea of screening, and associated with phenotypical screening. For the animal model of *C. elegans*, there are several aspects of properties which make it suitable and convenient for performing large-scale pharmacological screen in small molecule compound library. The related properties include biological and genetic relevance to human diseases, practical convenience due to small size, rapid life cycle, ease of culture, quantitative phenotypic readouts, available genetic and molecular tools for mechanism elucidation, low amount of compound requirement, and simple systemic exposure route.

For the pharmacological screen in small molecule compound library, we raise several aspects of suggestions. Firstly, this pharmacological screen can be performed in a library with relatively small number of compounds, and certain proportion of small molecule compounds are drugs already approved. Fardghassemi et al. (2021) performed a pharmacological screen of 3942 compounds (many FDA approved) using transgenic ATXN3 strain, and identified 5 small molecule compounds (alfacalcidol, chenodiol, cyclophosphamide, fenbufen, and sulfaphenazole) with the potential to rescue ATXN3 toxicity [7]. The transgene of ATXN3 reflected the polyglutamine expansion disease arising from CAG repeat expansion in exon 10 of ATXN3 [7]. Secondly, high-throughput screen can be performed using small molecule compound library based on phenotypic analysis. A high-throughput screen was carried out to identify small molecules with the function in reprogramming germline fate [8]. Thirdly, using small molecule compound library, high-throughput screen can be performed based on both molecular docking to candidate targets and phenotypic analysis.

3.2.2 Activators

Activators are substances (typically small molecule compounds) with the function of elevating activity of certain targets (e.g., a receptor, enzyme, ion channel, or transcription factor) via positive modulation and specific targeting upon the binding. Thus, the activators contain receptor activator, enzyme activator, channel opener, and transcriptional activator. The activators may exhibit their therapeutic effect by compensating for the deficiency, enhancing protective response, and overriding the pathological inhibition.

In *C. elegans*, activators can be identified based on the following screen:

1. Exaggerates certain phenotype (e.g., hypermotility).
2. Enhances reporter expression (e.g., increases GFP intensity).
3. Suppresses certain mutant phenotype by modulating the compensatory or parallel signals.

In this animal model, 18α-glycyrrhetinic acid treatment elevated proteasome activity and extended proteasome activation-dependent lifespan, suggesting the role of 18α-glycyrrhetinic acid as activator of proteasome [9]. Moreover, in AD model, 18α-glycyrrhetinic acid suppressed the AD progression and related pathological phenotypes, such as the paralysis rate [9]. These observations suggested that administration with proteasome activator is helpful for intervening aggregation-related diseases, such as AD. In addition, 7-ethoxy-4-methylcoumarin was identified as the activator of dopamine D2 receptor, and exhibited the function in ameliorating Parkinson's disease related phenotypes [10]. Based on high-throughput screen, three activators (digoxin, ikarugamycin, and alexidine dihydrochloride) for transcription factor EB were identified to show the function in extending lifespan and promoting autophagolysosomal activity [11].

3.2.3 Antagonists

Antagonists are substances with the function to prevent, reduce, or block the effect of agonists after binding to targeted receptor or enzyme. The pharmacological aims for using antagonists are to prevent agonist binding, to block pathological overactivity, and to maintain homeostasis. The antagonists contain competitive antagonists, non-competitive antagonists, inverse agonists, allosteric antagonists, and physiological antagonists.

In *C. elegans*, some antagonists were identified to extend lifespan, to increase stress resistance, and to protect neurodegeneration-linked proteotoxicity with the functions to inhibit mitochondrial ATP synthesis, DNA synthesis, glycolytic process, protein aggregation, trypsin activity, histone deacetylase activity, angiotensin-converting enzyme activity, prototypic cyclooxygenase activity, and insulin signaling function [12–20]. Certain antagonists were identified as inhibitors of dihydrolipoamide dehydrogenase or dopamine D2 receptor to suppress the tau or Aβ-mediated pathology [21, 22]. Treatment with antagonist for trypsin could reduce lipid accumulation, which required the involvement of insulin signaling [23]. Octopamine inhibits the 5-HT-induced pumping, which could be reversed by antagonist for octopamine [24]. The screen and identification strategies for antagonists are somewhat similar to those for activators in *C. elegans*.

3.2.4 Metals

The metals for pharmacological aims refer to the elements and their corresponding compounds with the function in preventing or treating disease based on their unique physicochemical properties. These metals normally have redox activity, ability as essential cofactors, and oxidation/reduction potential through certain biological interactions. In the clinic, the used therapeutic metals include platinum (Pt), lithium (Li^{+}), gold (Au), silver (Ag^{+}), gadolinium (Gd^{3+}), etc. Moreover, in *C. elegans*, some metals (such as manganese, lithium, selenite, selenium, organoselenium compounds, and organotellurium compounds) were found to have the beneficial effect in enhancing stress resistance and anti-oxidant defense systems, increasing immune response, extending lifespan, and/or inhibiting age-related decline [25–33]. Additionally, in inclusion body myositis model, neuromuscular synaptic defects could be inhibited, and Aβ oligomeric species were reduced by copper treatment [34].

3.3 Clinical Drugs

3.3.1 Hormones

Hormones (typically peptides, steroids, or amines) are endogenous signaling molecules, which can be used for some pharmacological aims, such as metabolic modulation, immunosuppression, and cancer therapy by activating or blocking receptors or causing negative feedback suppression. Hormones have properties of high potency, pleiotropic effects, and complex pharmacokinetics. In *C. elegans*, germline removal extended lifespan partially by stimulating synthesis of 3-β-hydroxy-pregn-5-en-20-one, a steroid hormone [35]. The melatonin regulated locomotion by functioning as a neuromodulator and controlled homeostatic state by acting as a ligand to target distinct receptor pathways [36]. The dafachronic acids, cholesterol-derived hormones, could control *C. elegans* life cycle and initiate the dauer formation [37]. Moreover, *C. elegans* was used to examine the effect of cytokinins, a plant hormone, in enhancing stress resistance [38].

3.3.2 Sterols

Sterols are the steroids with hydroxyl group (-OH) at C-3 position of steroid core and aliphatic side chain at C-17, which can be used to modulate membrane integrity, metabolic pathways, and hormonal signaling by affecting biosynthesis, absorption/reabsorption, and receptors/transporters. Sterols have functional properties of amphipathic nature, role as metabolic precursors for steroid hormones, bile acids, and vitamin D, and membrane modulators. In *C. elegans*, dafachronic acids, 3-keto bile acid-like steroids, regulated longevity. The dafachronic acids extended lifespan in germ-line ablated *daf-9/daf-36* mutants [39]. However, in response to dauer-related signals, dafachronic acids reduced lifespan and inhibited stress resistance [39].

3.3.3 Neurotransmitters

Neurotransmitters are endogenous chemical signals transmitted from presynaptic modulate nervous system function by targeting to their receptors at postsynaptic. Neurotransmitters include serotonin (5-HT), dopamine (DA), norepinephrine (NE), histamine, glutamate, GABA, acetylcholine (ACh), and neuropeptides. In *C. elegans*, serotonin allows for rapid cycle of contraction-relaxation by activating pharyngeal M3 motor neurons and reducing duration of pharyngeal action potential [40]. Moreover, serotonin and octopamine acted together to organize a reciprocal regulation network [40].

3.3.4 Other Clinical Drugs

The identified clinical drugs showing beneficial effect in extending lifespan at least contain (1) clioquinol (anti-neurodegeneration drug) [41], (2) reserpine (anti-hypertensive drug) [42, 43], (3) hydralazine (anti-hypertension drug) [44], (4) valproic acid (mood stabilizer and anti-convulsant) [45], (5) convallatoxin (drug to treat chronic congestive heart failure) [46], (6) mianserin and methiothepin (antidepressant) [47], (7) celecoxib (anti-inflammatory drug) [48], (8) ethosuximide (anticonvulsant) [49], (9) 4,4′-diaminodiphenylsulfone (drug to treat Hansen disease) [50], and (10) ethosuximide, trimethadione, and 3,3-diethyl-2-pyrrolidinone (anticonvulsants) [51]. The identified clinical drugs having beneficial effect in enhancing stress resistance at least contain (1) reserpine (antihypertensive drug) [43], (2) hydralazine (anti-hypertension drug) [44], and (3) convallatoxin (drug to treat chronic congestive heart failure) [46]. The identified clinical drugs having beneficial effect in alleviating neurodegeneration diseases related pathological phenotypes at least contain (1) reserpine (antihypertensive drug) [42], (2) ibuprofen and naproxen (anti-inflammatory drugs) [52], (3) ethosuximide (anti-epileptic drug) [53], (4) bexarotene (anti-cancer drug) [54], (5) clozapine (anti-psychotic drug) [55], and (6) liraglutide and linagliptin (drugs to treat diabetes) [56]. Additionally, the social feeding could be modulated by the antipsychotic drugs: (1) anti-psychotics, chlorpromazine, trifluoperazine, fluphenazine, and haloperidol (the first-generation drugs), and (2) olanzapine (the second-generation drug) [57].

3.4 Nutrition

3.4.1 Anti-oxidants and SOD Mimetics

Antioxidants are endogenous or exogenous substances with the functions to prevent, delay, or remove oxidative damage induced by reactive oxygen/nitrogen species. That is, the pharmacological aims of anti-oxidants include prevention, intervention, and adjuvant therapy to reduce side effects of other treatments. The corresponding underlying mechanisms for functions of anti-oxidants include (1) direct scavenging the reactive oxygen/nitrogen species (such as the function of vitamin C), (2) activating endogenous antio-xidant defenses, and (3) binding pro-oxidant transition metals to inhibit Fenton-like reactions (such as the function of deferoxamine). Anti-oxidants can be classified into two groups: small-molecule anti-oxidants and synthetic mimics of endogenous enzymes. The small-molecule anti-oxidants include both natural anti-oxidants (such as vitamin C/E, flavonoids, and polyphenols) and synthetic anti-oxidants.

In *C. elegans*, at least the anti-oxidants of vitamin C, vitamin E, CoQ10, N-acetyl-cysteine, reduced glutathione, thioproline, and curcumin were shown to have beneficial effects in extending lifespan, and enhancing stress resistance [58–65]. Treatment with ascorbate and N-acetyl-cysteine suppressed TiO_2 nanoparticle toxicity in causing oxidative stress and locomotion inhibition (Fig. 3.1) [61]. Pretreatment with

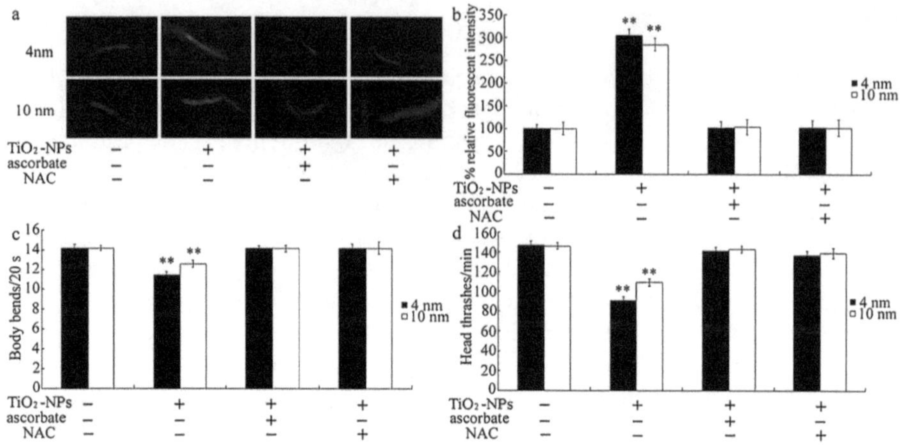

Fig. 3.1 Effects of antioxidant treatment on ROS production and locomotion behavior in nematodes chronically exposed to 4 nm and 10 nm TiO_2-NPs at the concentration of 0.1 g/L [61]. (**a**) Pictures showing the ROS production in nematodes. (**b**) Effects of antioxidant treatment on ROS production in nematodes chronically exposed to 4 nm and 10 nm TiO_2-NPs at the concentration of 0.1 g/L. (**c**) Effects of antioxidant treatment on body bends of nematodes chronically exposed to 4 nm and 10 nm TiO_2-NPs at the concentration of 0.1 g/L. (**d**) Effects of antioxidant treatment on head thrashes of nematodes chronically exposed to 4 and 10 nm TiO_2-NPs at the concentration of 0.1 g/L. The day-8 TiO_2-NPs (4 and 8 nm) exposed (0.1 g/L) adults were treated with 10 mM ascorbate for 24 h or 5 mM N-acetyl-l-cysteine (NAC) for 24 h. Bars represent mean ± S.E.M. $^{**}P < 0.01$

vitamin could prevent Al_2O_3 nanoparticle induced neurotoxicity in causing the neurodegeneration (Fig. 3.2) [63]. With the aid of *gas-1* mutants, anti-oxidants of vitamin E and N-acetylcysteine could rescue the defect of longevity in mitochondrial complex I disease model [66]. The dimethyl sulfoxide (DMSO) had anti-oxidation property, which depended on DAF-16 and SIR-2.1 signals [67]. EGCG and α-lipoic acid are natural anti-oxidants to show beneficial effect against the aging process [68]. MitoQ is the synthesized mitochondria-targeted anti-oxidant to extend lifespan and to improve health span [69]. Nevertheless, the role of anti-oxidants may be complex. Treatment with anti-oxidants could both extend and reduce the lifespan, which depended on their concentrations and experimental conditions [70]. Besides these, Euk-134 and Euk-8, two mimetics of superoxide dismutase/catalase, were found to extend the lifespan and enhance the stress resistance [71, 72].

3.4.2 Vitamins

Vitamins function as essential organic micronutrients by acting as anti-oxidants or have drug-like therapeutic effects by acting as coenzyme precursors. Vitamins contain water-soluble vitamins and lipid-soluble vitamins. In *C. elegans*, besides vitamin C and vitamin E introduced in “3.4.1,” vitamin D could extend lifespan and

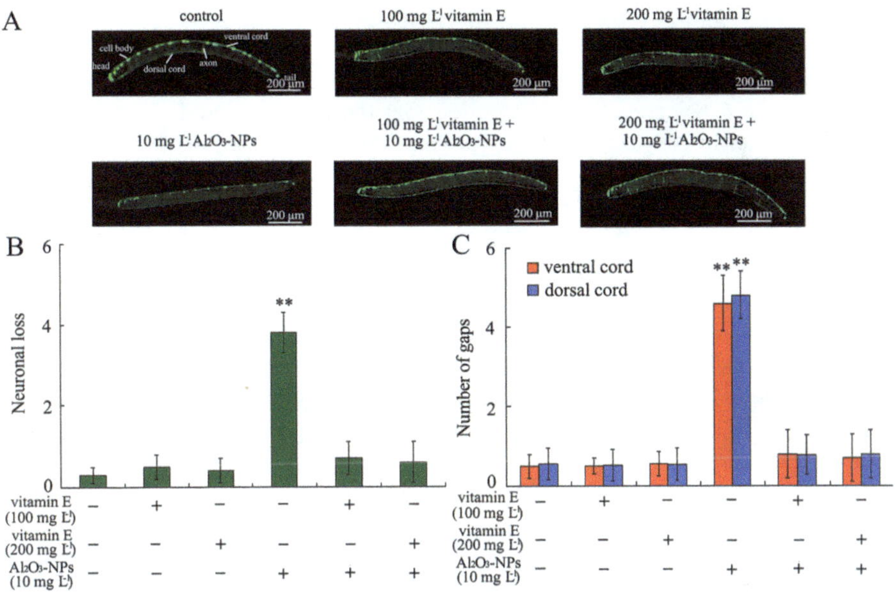

Fig. 3.2 Effects of pretreatment with vitamin E on the neurotoxicity of Al_2O_3-NPs in inducing neurodegeneration in *oxIs12* transgenic nematodes [63]. (**a**) Effects of vitamin E pretreatment on the development of D-type GABAergic neurons in Al_2O_3-NPs exposed *oxIs12* transgenic nematodes. (**b**) Effects of vitamin E pretreatment on neuronal loss of D-type GABAergic neurons in Al_2O_3-NPs exposed *oxIs12* transgenic nematodes. (**c**) Effects of vitamin E pretreatment on gap formation on nerve cords for D-type GABAergic nervous system in Al_2O_3-NPs exposed *oxIs12* transgenic nematodes. Vitamin E pretreatment was performed at the L2-larvae stage for 24 h. Al_2O_3-NPs exposure was performed from L4-larvae for 24 h. Five replicates were performed. Bars represent means ± SEM. $^{**}P < 0.01$

elevate protein homeostasis [73]. The folic acid could extend lifespan and enhance stress resistance [74]. The vitamin B12 exhibited neuroprotective function in Parkinson's disease model [75]. Moreover, some vitamins, such as vitamin C and vitamin B12, were also involved in developmental control, such as the control of neuronal development [76, 77].

3.4.3 Drinking

The drinking refers the liquid substances with beneficial effects, which are associated with the related administration volume, rate, and timing. In *C. elegans*, the effect of lifespan extension was observed after intake with electrolyzed-reduced water [78, 79]. Ingestion with traditional Chinese herbal tea could improve the health span [80]. Administration with certain red wines possibly attenuated virulence of bacterial pathogen in the host of *C. elegans* [81]. Additionally, lifespan could possibly be extended by low doses of ethanol treatment [82].

3.4.4 Food Supplements

Food supplements are concentrated sources of nutrients, botanicals, and other substances. In *C. elegans*, it was reported that the lifespan could be extended by acetic acid, Reishi polysaccharide, royal jelly, and D-glucosamine [83–85]. Moreover, stress resistance could be enhanced by royal jelly and soy sauce [84, 86].

3.4.5 Fatty Acids

Fatty acids can also be considered as the bioactive molecules with the functions to act as precursors to modulate physiological processes, to interact with receptors, and to affect gene expressions. The normally examined fatty acids include: (1) omega-3 polyunsaturated fatty acids (PUFAs) (such as eicosapentaenoic acid (EPA) and docosahexaenoic acid (DHA)), (2) omega-6 fatty acids (such as linoleic acid (LA) and gamma-linolenic acid (GLA)), (3) saturated & monounsaturated fatty acids (such as palmitic acid and oleic acid), and (4) conjugated linoleic acids (CLAs) (such as isomers of linoleic acid). In *C. elegans*, treatment with oleanolic acid and ω-3 fatty acid were shown the beneficial effect to extend the lifespan [87, 88]. The treatment with oleic acid, and omega-3 and -6 fatty acids could induce stress resistance, such as the resistance to oxidative stress and bacterial metabolite [89–91]. The 6-OHDA induced neurodegeneration could be inhibited by administration with α-linolenic acid [92]. Eicosapentaenoic acid long chain polyunsaturated fatty acids were required for alcohol response or neurotransmission process [93, 94].

3.5 Different Physical Forms

3.5.1 Gas Molecules

Gas molecules are those diffusible gases across cell membranes with the role in acting as key signaling molecules. The normally examined gas molecules include nitric oxide (NO), carbon monoxide (CO), hydrogen sulfide (H_2S), hydrogen (H_2), and sulfur dioxide (SO_2). In *C. elegans*, hydrogen sulfide and hydrogen treatment could extend the lifespan [95–97]. In addition, the oxidative damage, bacterial pathogen infection, heat stress, and low oxygen stress were inhibited by hydrogen sulfide, hydrogen cyanide, and hydrogen [95, 97–100]. Moreover, extended lifespan was also caused by certain elevated CO_2 level [101].

3.5.2 Odors

Odors refer to the olfactory stimuli to directly modulate physiological and psychological states usually for the pharmacological aims of stress response attenuation and neuroendocrine modulation. In *C. elegans*, a-pinene odor and isoamyl alcohol odor were helpful to enhance the stress resistance [102, 103]. In addition, the isoamyl alcohol odor possibly extended the lifespan [102].

3.5.3 Oils

Oils refer to the lipid-based bioactive extracts or formulations with the roles in acting as the active drugs, penetration enhancers, or drug delivery systems. In *C. elegans*, lifespan could be extended by clove essential oil and essential oil alloaromadendrene [104, 105]. The clove essential oil and essential oils from leaves of *Cinnamomum osmophloeum* exhibited anti-oxidant property [104, 106, 107]. The essential oil of *Acorus tatarinowii* schott ameliorated Aβ-induced neurodegeneration related phenotypes [108]. Moreover, essential oils from *Cinnamomum cassia* and *Syzygium aromaticum* suppressed pathogen colonization after the infection in *C. elegans* [109].

3.5.4 Powders

Powders refer to finely divided solid particles ranging from nanometers to several 100 μm. Besides acting as the active drugs, powders can be employed as pharmacological agents to modulate drug release, control administration dose, and enable route-specific delivery. In *C. elegans*, protein rich pearl powder showed the anti-oxidant property [110]. In addition, lipid accumulation was reduced and heathspan was improved after feeding the nematodes with oat [111].

3.5.5 Nanomaterials

Besides as drug delivery systems, some nanomaterials themselves have beneficial effects. In *C. elegans*, administration with pristine and TAT conjugated platinum nanoparticles could extend the lifespan [112, 113]. The neurodegeneration diseases-related phenotypes could be relieved by selenium nanoparticles and phytol loaded PLGA nanoparticles [114, 115]. Additionally, oligoguanidine, gold nanoparticles-based nanocomposites, and Ga nanoparticles could increase the survival of bacterial pathogen infected *C. elegans*, and clean the bacterial pathogen from the body [116–118].

3.6 Active Compounds in Plants

Actually, most of the performed pharmacological studies in *C. elegans* are relevant to this aspect. Active compounds in plants are secondary metabolites with certain therapeutic effects. Besides the validated bioactivity, the active compounds in plants can provide structural diversity and multi-target potential. For these reasons, active compounds in plants can be used as direct drugs or lead compounds, or employed in synergistic complexes. The active compounds in plants at least contain (1) alkaloids compounds, (2) phenolics compounds, (3) terpenoids compounds, (4) glycosides compounds, and (5) polysaccharides compounds and peptides.

3.6.1 Active Alkaloids Compounds

1. *Indole Alkaloids*

With the reserpine as the example of indole alkaloids, its treatment could confer the stress resistance in nematodes [43]. Besides this, reserpine treatment induced the lifespan extension, which required the involvement of D2-type receptor, exoribonuclease, and neurotransmitter release [43, 119, 120]. In addition, the Aβ toxicity could also be ameliorated by reserpine treatment [121].

2. *Isoquinoline Alkaloids*

As one of the examples of isoquinoline alkaloids, the berberine treatment could extend the lifespan, increase innate immunity, and confer protective response to neurodegenerative diseases related phenotypes [122, 123]. Among isoquinoline alkaloids, sanguinarine was identified to show the most pronounced effect in reducing lipid accumulation [124]. *C. elegans* was even employed to examine the morphine-induced addiction-like behavior or altered feeding behavior [125–127].

3. *Purine Alkaloids*

As a member of purine alkaloids, treatment with caffeine and its analogs could extend the lifespan by modulating insulin signaling [128, 129]. Caffeine and coffee extract could enhance stress resistance by increasing heat shock response and promoting proteostasis [130]. The Parkinson's disease-related phenotypes could be inhibited by caffeine treatment [131]. Treatment with caffeine was shown to be helpful to reduce lipid accumulation induced by high-glucose feeding [132, 133]. Treatment with caffeine exhibited protective effect in suppressing intestinal aging and enhancing intestinal integrity during the aging process [134, 135]. Moreover, in GNAO1-related disorder model, caffeine was shown to be useful for the control of dyskinesia [136].

Besides the caffeine, theophylline treatment could also promote longevity [128]. Treatment with theophylline could further confer the resistance of nematodes to *Candida albicans* infection [137].

4. *Pyridine/Piperidine*

Treatment with certain amount of nicotine was found to exhibit the protective effect against Parkinson's disease-related phenotypes by modulating oxidative and mitochondrial signals [138–140]. The pilocarpine could protect the *C. elegans* against the hyperthermia by acting as agonist of mAChRs [141].

3.6.2 Active Phenolics Compounds

1. *Simple Phenolics*

With salicylic acid as the first example, 5-octanoyl salicylic acid (C8-SA), a salicylic acid derivative, could extend the lifespan by activating mitochondrial unfolded protein response and autophagy [142]. In addition, treatment with capsaicin and its analogs was shown to have the anti-nociceptive effect [143].

2. *Flavonoids*

With quercetin as the first example of flavonoids, its treatment could extend the lifespan and improve the health span [144, 145]. Administration with quercetin could confer resistance to oxidative stress, UV stress, glucotoxicity, and toxicity of pollutants (such as nanoplastics and bisphenol S) [146–150]. The bisphenol S exposure caused mitochondrial damage and oxidative stress could be inhibited by treatment with quercetin (Fig. 3.3) [150]. The quercetin treatment further exhibited the neuroprotective effect against Alzheimer and Parkinson's pathology [151–153].

Besides the quercetin, genistein treatment also enhanced the resistance of *C. elegans* to oxidative stress, heat stress, and glucose stress [154, 155]. Additionally, amyloid-beta induced neurodegeneration related phenotypes could also be inhibited by genistein treatment [155, 156].

3. *Lignans*

The sesamin and sonneralignan A were shown to have anti-aging effect [157–159]. Alzheimer's disease related pathology could be alleviated by magnolol, sesamin, and sesamolin [159, 160]. Treatment with sesamin, sesamolin, and schisandrin A could increase the survival after bacterial pathogen infection [161, 162]. Moreover, treatment with schisandrin C could alleviate the intestinal barrier dysfunction [163, 164]. Treatment with cubebin could even be helpful to obtain the anti-parasitic activity against *Angiostrongylus cantonensis* [165].

4. *Stilbenes*

With resveratrol as the example of stilbenes, its treatment could extend the lifespan, which required the involvement of SIR-2.1 and MPK-1 regulation [166, 167]. Resveratrol and its derivatives could enhance the stress resistance,

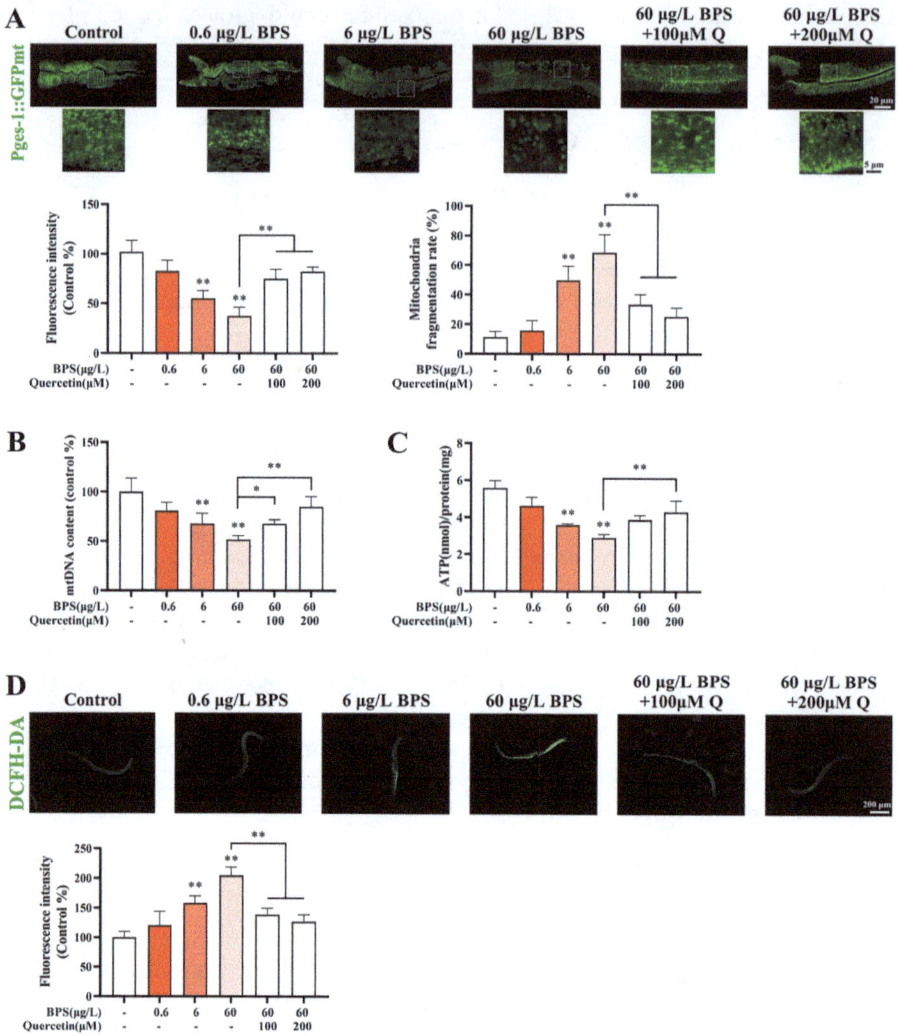

Fig. 3.3 Effects of BPS and quercetin on mitochondrial injury and oxidative stress [150]. (**a**) Representative images of day-8 P*ges-1*::GFPmt nematodes, which mark intestinal mitochondria. The fluorescence intensity, and mitochondrial fragmentation were analyzed. (**b**, **c**) mtDNA content, and ATP production in day-8 nematodes. (**d**) DCFH-DA fluorescence intensity and the representative images of day-8 nematodes. Q quercetin. The data were expressed as the mean ± SEM ($n = 30$). $^{*}P < 0.05$, $^{**}P < 0.01$

such as the resistance to oxidative stress and iron-induced toxicity [168, 169]. Resveratrol treatment also reduced amyloid-beta toxicity in Alzheimer disease model [170]. In addition, the lipid accumulation could be inhibited by trans-trismethoxy resveratrol by modulating FAT-6 and FAT-7 [171].

5. *Tannins*

Among the tannins, treatment with ellagic acid and tannic acid enhanced the stress resistance [172–174]. For example, ellagic acid inhibited deoxyaconitine neurotoxicity by suppressing the induced mitochondrial dysfunction [173]. In addition, treatment with ellagic acid could increase the survival of fungal pathogen infected nematodes [175].

3.6.3 Active Terpenoids Compounds

1. *Monoterpenes*

With thymol as the example, its treatment could reduce lipid accumulation caused by high glucose [176]. Moreover, thymol treatment could enhance innate immunity against fungal infection by modulating p38 MAPK signaling [177].

2. *Sesquiterpenes*

The *C. elegans* lifespan could be extended by β-dihydroagarofuran-type sesquiterpenes [178]. Alzheimer's disease and parkinsonism related phenotypes could be mitigated by treatment with sesquiterpenoids from davana and cadinane-type sesquiterpenes [179, 180]. Santalol isomers administration suppressed transthyretin amyloidogenesis and related pathologies [181]. Moreover, Brevilin A enhanced resistance to oxidative stress and innate immune response by modulating p38 MAPK signaling [182].

3. *Diterpenes*

Among the diterpenes, crassifolin A and carnosol treatments could extend lifespan and/or improve health span [183, 184]. Carnosic acid and diterpenoid caesalmin C had neuroprotective function, such as the inhibition in Abeta-induced paralysis [185, 186]. Treatment with cafestol was shown to be able to elevate the fat oxidation [187]. Additionally, ginkgolide A could confer the resistance to pathogen infection by activating mitochondrial unfolded protein response [188].

4. *Triterpenes*

Among the triterpenes, ginsenosides could extend the lifespan by modulating insulin and stress response-related signals [189, 190]. Both Alzheimer's and Parkinson's-related phenotypes could be reduced by treatment with araloside A, beta-amyrin, hederagenin, and ginsenosides [191–196]. Moreover, lipid accumulation could be reduced by ginsenosides Rg1 [197].

5. *Tetraterpenes*

Among the tetraterpenes, astaxanthin, carotenoids, and beta-carotene enriched lyophilized orange juice could confer stress resistance, such as resistance to oxidative stress and UV-irradiation stress [198–201]. Astaxanthin could extend lifespan by modulating autophagy and redox status [202]. Amyloid-beta induced AD phenotypes could be reduced by astaxanthin [203]. In addition, pretreatment with Z-Astaxanthin could reduce lipid accumulation [11].

3.6.4 Active Glycosides Compounds

1. *Cardiac Glycosides*

Both digoxin and convallatoxin showed the anti-aging effect, and digoxin extended lifespan by activating TFEB [46, 204]. Treatment with convallatoxin also increased the stress resistance [204]. Treatment with ouabain could reduce the sterile and infertile phenotypes of some mutants, such as the *glp-1* mutants [205].

2. *Glucosinolates*

Treatment with sulforaphane could promote the longevity and improve the health span by modulating insulin signaling [206, 207]. Glucotoxic neurodegeneration could also be suppressed by sulforaphane treatment [207]. Sulforaphane treatment suppressed Cd-induced mitochondrial dysfunction by modulating insulin signaling pathway [208]. Additionally, sulforaphane could reduce the polyglutamine toxicity by affecting insulin and AMPK signals [209].

3. *Anthraquinone Glycosides*

Sennosides were selected as the example of anthraquinone glycosides. Sennoside A could inhibit the quorum sensing system of bacterial pathogens [210]. Moreover, sennoside A treatment increased the survival of bacterial pathogen-infected nematodes [210].

4. *Steroidal Glycosides*

Treatment with withanone could extend the lifespan by regulating serotonin receptors [211]. Not only in wild-type nematodes, the lifespan in human EGFR-driven cancerous nematodes and mutants for serotonin receptor genes were also extended by withanone [212, 213]. Diosgenin and shatavarin IV could also extend the lifespan by modulating insulin signaling and autophagy [214, 215]. Treatment with withanone, steroidal saponin, and diosgenin could confer the resistance to oxidative stress, heat stress, glucose toxicity, and osmotic stress [211, 213, 214, 216]. Besides the effect in extending longevity, treatment with diosgenin also reduced lipid accumulation [217]. Treatment with withaferin A, withanolide A, and 18-norspirostane steroidal saponins further showed neuroprotective function in reducing the neurodegenerative diseases related phenotypes [218–220]. Moreover, dioscin conferred the resistance to pathogen infection by activating endoplasmic reticulum unfolded protein response [221].

3.6.5 Active Polysaccharides Compounds and Peptides

1. *Pectin*

The *C. elegans* lifespan could be extended by both ginseng-derived rhamnogalacturonan I (RG-I) pectin and low methyl-esterified ginseng homogalacturonan pectins by modulating insulin signaling and TOR signaling [222, 223]. Besides this aspect, feruloylated sugar chain in sugar beet pectin also showed anti-oxidant activity [224].

2. *β-Glucans*

The paramylon and lentinan were reported to have the potential in extending lifespan, and these compounds extended lifespan by at least modulating insulin signaling [225, 226]. Administration with lentinan and barley beta-glucan also exhibited anti-oxidation effect by affecting insulin signaling pathway [226–228]. Besides these, fermented barley beta-glucan could further reduce the lipid accumulation [229].

3. *Lectins*

In *C. elegans*, overexpression of C-type lectin domain-containing protein CLEC-126 extended the lifespan, suggesting the role of this protein in improving longevity [230]. In addition, agrocybe aegerita lectin could induce apoptosis, implying its potential to treat cancer [231].

4. *Cyclotides*

Both oxidative stress and beta-amyloid toxicity could be inhibited by disulfide-rich cyclic peptides [232]. Additionally, a plant cyclotide has been shown to have the potential in inhibiting protein aggregation, suggesting its possible function against the neurodegenerative diseases [233].

3.7 Whole Organisms Level

3.7.1 Plant Extracts

Plant extracts can be prepared by subjecting plant leaves, roots, bark, flowers, or whole plants to selective extraction process using certain solvents to obtain the concentrated bioactive compounds which means the complex mixture of phytochemicals. The prepared plants extracts can be prepared for a defined biological activity, or used as an active ingredient in herbal medicines. The plant extracts can be divided into four groups, aqueous extracts, alcoholic extracts, hydro-alcoholic extracts, and lipophilic extracts.

1. *Aqueous Extracts*

The aqueous extracts of *Damnacanthus officinarum*, *Curcumae Rhizoma*, *Polygalae Radix*, Red ginseng, *Warburgia salutaris* bark, and garlic had the lifespan-extension activity [234–239]. The aqueous extracts of red ginseng extended the lifespan by modulating the insulin signaling [237]. Accelerated aging caused by oxidative agents could be inhibited by coffee silverskin extract [240]. The aqueous extracts of *Curcumae Rhizoma*, *Ceiba speciosa* bark, and *Uncaria tomentosa* leaves could confer the stress resistance, such as the resistance to oxidative stress, and glucose stress [235, 236, 239, 241, 242]. The aqueous extracts of *Damnacanthus officinarum*, *Uncaria tomentosa*, *Achyrocline satureioides*, and *Decalepis hamiltonii* further showed neuroprotective effects against neurodegenerative diseases [234, 243–245]. Moreover, aqueous extracts of *Allium ursinum* leaves could alleviate the bacterial pathogen infection [246].

2. *Alcoholic Extracts*

The alcoholic extract of sake lees had anti-aging activity [247]. Moreover, this anti-aging activity was under the control of SKN-1/Nrf2 and HSF-1 signals [247].

3. *Hydro-Alcoholic Extracts*

Hydro-alcoholic extracts could be prepared from some *Acacia nilotica* fruits [248]. The prepared hydro-alcoholic extracts further showed the anthelmintic properties [248].

4. *Lipophilic Extracts*

After the preparation of lipophilic extract for *Bixa orellana*, its treatment could extend the lifespan [249]. Moreover, this lipophilic extract could confer the stress resistance [249].

3.7.2 Plant Hydrolysates

Wheat gluten hydrolysate could extend the lifespan [250]. Besides this, treatment with wheat gluten hydrolysate, hydrolysates of mung bean proteins, acid hydrolysates from *Auricularia auricular* polysaccharides, and enzymatic hydrolysates from *Chlorella vulgaris* proteins could enhance stress resistance, such as the resistance to oxidative stress [250–253].

3.7.3 Probiotic Bacteria

The *C. elegans* lifespan could be extended by probiotic bacteria of *Enterococcus faecium* from chicken cecum, *Butyricicoccus pullicaecorum*, *Megasphaera elsdenii*, *Lactobacillus rhamnosus* CNCM I-3690, and lactic acid bacteria from cheese [254–257]. In addition, treatment with probiotic bacteria of *Lactobacillus rhamnosus* CNCM I-3690, *Lactobacillus* fermentum U-21, and lactic acid bacteria enhanced the resistance to oxidative stress and pollutants toxicity [257–259]. Lactic acid bacteria conferred resistance to graphene oxide toxicity largely by maintaining intestinal barrier function (Fig. 3.4) [259]. Treatment with probiotic bacteria of *Enterococcus faecium* from chicken cecum and lactic acid bacteria further increased innate immunity and protected against bacterial pathogen infection

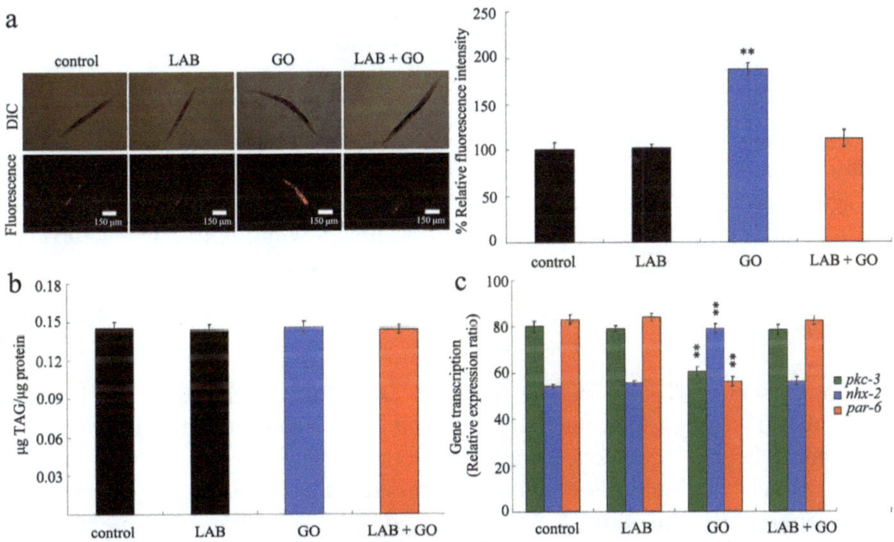

Fig. 3.4 LAB administration maintained the normal intestinal permeability in GO exposed wild-type nematodes [259]. (**a**) Nile red staining results. (**b**) Comparison of triglyceride content. (**c**) Comparison of gene expression patterns of *pkc-3*, *nhx-2*, and *par-6* genes. GO exposure concentration was 100 mg/L. The used LAB strain was L. bulgaricus. L4-larvae were pre-treated with LAB for 12 h, and then exposed to GO for 24 h at 20 °C. Bars represent means± S.E.M. $^{**}P < 0.01$ vs control

[254, 260]. Moreover, treatment with probiotic *Bifidobacterium animalis* could reduce lipid accumulation [261].

3.8 Traditional Chinese Medicines

3.8.1 Chinese Medicines

In Traditional Chinese Medicine (TCM), herbs are directly used to modulate imbalances in body's vital substances (Qi, Blood, Yin, Yang) and organ systems. The TCMs have pharmacological effects of at least anti-inflammatory & immuno-modulatory, cardiovascular, neuro-psychiatric, gastrointestinal, anti-tumor, and anti-diabetic effects.

1. *Astragalus membranaceus*

Astragalus membranaceus extract has several pharmacological effects. In *C. elegans*, the lifespan could be extended by *Astragalus membranaceus* extract [262].

2. *Salvia miltiorrhiza*

The *Salvia miltiorrhiza* extract had a neuroprotective effect for nematodes [263]. Besides this, treatment with *Salvia miltiorrhiza* extract could help the nematodes against *Staphylococcus aureus* infection [264].

3. *Ligusticum chuanxiong*

Ligusticum chuanxiong extract was first found to be able to extend the lifespan [265]. Besides this, it was further found that treatment with *Ligusticum chuanxiong* extract could exhibit the neuroprotective function in nematodes [266].

4. *Ginkgo biloba*

The extracts of *Ginkgo biloba* leaves and seed all could extend the lifespan [267, 268]. Treatment with the *Ginkgo biloba* extract could inhibit the process of sarcopenia, the age-dependent muscle degeneration [269]. Treatment with *Ginkgo biloba* extracts was helpful for enhancing stress resistance, such as the resistance to oxidative stress [267, 268, 270]. *Ginkgo biloba* extracts also provided neuroprotective effect against the amyloid-beta toxicity [235, 271].

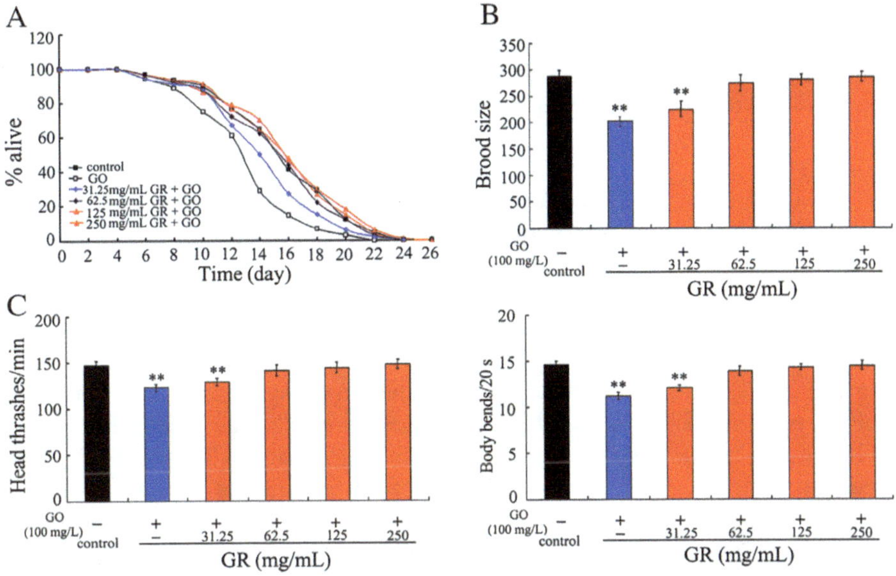

Fig. 3.5 Effects of pretreatment with GR on graphene oxide (GO) toxicity in exposed wild-type nematodes [275]. (**a**) Effects of pretreatment with GR on lifespan in GO (100 mg/L) exposed nematodes. (**b**) Effects of pretreatment with GR on brood size in GO exposed nematodes. (**c**) Effects of pretreatment with GR on head thrash and body bend in GO exposed nematodes. GR, Glycyrrhizae radix. Nematodes were pretreated with GR from L1-larvae to L4-larvae in the presence of food, and then exposed to GO from L4-laevae for 24 h. Bars represent means ± S.E.M. $^{**}P < 0.01$ vs control

5. *Polygalae Radix*

Treatment with *Polygalae Radix* extract showed an anti-aging effect in nematodes [272]. Moreover, *Polygalae Radix* extract could alleviate oxidative stress induced by D-galactose [272].

6. *Schisandra chinensis*

The *Schisandra chinensis* extract could confer the stress resistance [273]. For example, the resistance to ethanol neurotoxicity was suppressed by *Schisandra chinensis* extract [273].

7. *Glycyrrhiza uralensis and Glycyrrhizae radix*

The *Glycyrrhizae radix* extract could extend the lifespan [274]. Besides this, treatment with *Glycyrrhizae radix* extract enhanced stress resistance [274, 275]. For example, treatment with *Glycyrrhizae radix* inhibited the toxicity of graphene oxide in reducing lifespan, reducing reproductive capacity, and in causing locomotion inhibition (Fig. 3.5) [275]. Treatment with *Glycyrrhiza uralensis* extract further showed neuroprotective effect against amyloid-beta toxicity [276].

8. *Trichosanthes kirilowii and T. laceribractea*

In *C. elegans*, the neuroprotective effect of seed oils from *Trichosanthes kirilowii* and *T. laceribractea* could be observed [277]. After the treatment, Aβ aggregation and tau phosphorylation and polymerization were alleviated [277].

3.8.2 Traditional Chinese Medicine Formula

The Traditional Chinese Medicine formula refers to the synergistic combination of multiple medicinal substances (primarily the herbs), which is designed according to TCM theory to achieve certain therapeutic effects. Among the Traditional Chinese Medicine formulas, the *C. elegans* lifespan could be extended by Jianpi-yangwei formula, Liuwei Dihuang formula, and Liu Jun Zi decoction [278–280]. Treatment with Aiweixin formula and Bushenkangshuai Tang could confer stress resistance, such as resistance to chromium toxicity, oxidative stress, and UV irradiation stress [281, 282]. Treatment with Kai-Xin-San, Shengmai formula, Dianxianning formula, Liuwei Dihuang, and Tian-Si-Yin decoction exhibited a neuroprotective effect against neurodegenerative diseases [283–287]. Moreover, treatment with Xuebijing formula showed a beneficial effect against bacterial pathogen infection by inhibiting pathogen accumulation in the body and reducing bacterial pathogen toxicity (Fig. 3.6) [288].

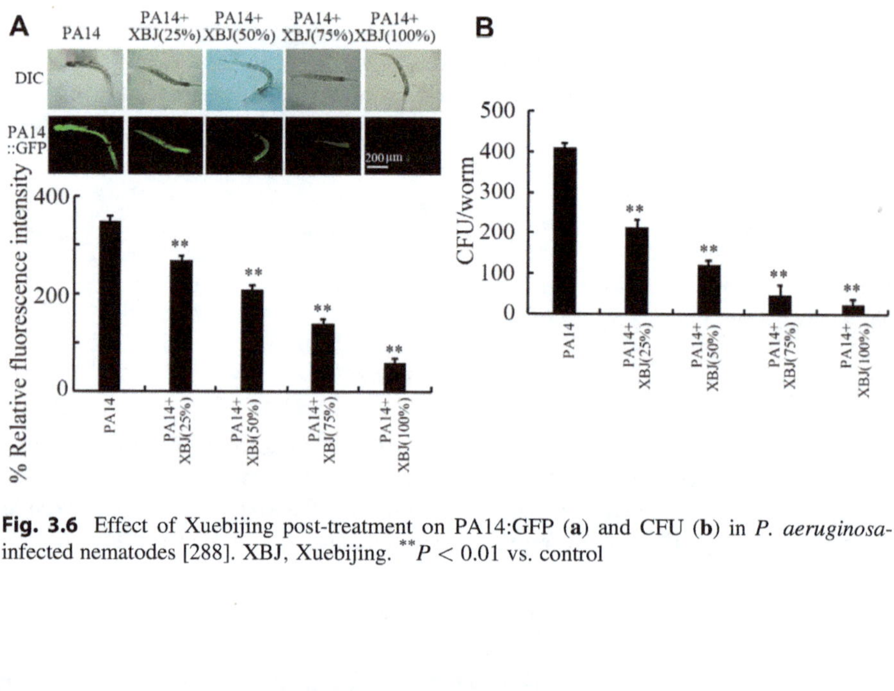

Fig. 3.6 Effect of Xuebijing post-treatment on PA14:GFP (**a**) and CFU (**b**) in *P. aeruginosa*-infected nematodes [288]. XBJ, Xuebijing. $^{**}P < 0.01$ vs. control

3.9 Artificial and Synthesized Drugs

Artificial and synthesized drugs refer to the drugs generated by chemical synthesis. The artificial and synthesized drugs include (1) novel chemical entities, (2) synthetic analogues/mimetics, (3) prodrugs synthetically by modification, (4) chiral/stereoisomeric drugs, and (5) engineered large-molecule drugs.

3.9.1 Synthesis of Conjugates

The anti-oxidative potential could be enhanced by synthesized conjugates of polymeric chitosan-betaxanthin, 4-organylsulfenyl-7-chloro quinolines, and casein-maltodextrin [288–290]. In addition, the lifespan could be extended by synthesized conjugate of polymeric chitosan-betaxanthin [288]. The prepared synthesized conjugates of benzofuran-chalcone, benzofuran-tetrazole, pyrazolo-pyridine, and divalent caffeic amides had the neuroprotective effect against Alzheimer's disease and Parkinson's disease [3, 291–293]. In addition, besides the lifespan extension, treatment with the synthesized conjugate of leucine–metformin could further modulate the glucose and lipid metabolisms [294].

3.9.2 Synthesized Derivatives

The derivatives of curcumin, pyridoperimidine, 1,2,4-triazolo[1,5-a]-pyridine, 1,2,4-triazolo[1,5-a]pyridine, and pinocembrin amino acid could extend the *C. elegans* lifespan [295–299]. The stress resistance, such as the resistance to oxidative stress, could be further induced by treatment with derivative of 1,2,4-triazolo[1,5-a]-pyridine [299].

3.9.3 Synthesized Analogues, Activators, and Antagonists

The fluorinated analogues were synthesized [300]. Among the prepared analogues, 24-difluoromethyl showed activator activity for DAF-12 receptor, whereas the difluoromethylidene exhibited the antagonist activity for DAF-12 receptor [300]. 27-Nor-D4-dafachronic acid was synthesized as the ligand of DAF-12 receptor to activate its activity [301]. In addition, AM251 was identified as the antagonist of cannabinoid receptor to influence the dauer formation [302].

3.9.4 Synthesized Peptides and Enzymes

The synthesized hybrid polyketide-nonribosomal peptide had the function to increase the larval survival [303]. The synthesized caenopore-5 could act as the intestinal anti-microbial protein in nematodes [304]. In addition, recombinant buckwheat trypsin inhibitor could be prepared to improve mitochondrial and protein homeostasis [17].

3.10 Perspectives

In this chapter, the potential pharmacological evaluation of chemical compounds, clinical drugs, nutrition, different physical forms of drugs and compounds, active compounds in plants, drugs at the whole organisms level, Traditional Chinese Medicines, and artificial and synthesized drugs in *C. elegans* was introduced and discussed. Nevertheless, not all the drugs and compounds can be suitably classified into these groups. For example, some biochemical metabolites have also been proven to show beneficial effects against the stresses or pollutants toxicity. Carnitine could protect *C. elegans* from glucose toxicity [305]. Treatment with cholesterol, pyruvate, and vitamin D3 could inhibit the toxicity of pollutants, such as 6-PPD quinone [306–308].

Besides this aspect, even for the introduced eight groups of drugs and compounds, only limited drugs and compounds have been examined in *C. elegans*. With the plant extracts as the example, pharmacological effects of many more herb extracts have not been assessed in *C. elegans*. Moreover, in most cases, the examined pharmacological effect of drugs and compounds is on anti-aging, stress resistance, neuroprotective function, reduction in lipid accumulation, and anti-infection. Actually, more aspects of biological, physiological, and metabolic effects of drugs and compounds are suggested to be assessed after the treatment in nematodes.

References

1. Dyle MC, Ebert SM, Cook DP, Kunkel SD, Fox DK, Bongers KS, Bullard SA, Dierdorff JM, Adams CM (2014) Systems-based discovery of tomatidine as a natural small molecule inhibitor of skeletal muscle atrophy. J Biol Chem 289:14913–14924
2. Fang EF, Waltz TB, Kassahun H, Lu Q, Kerr JS, Morevati M, Fivenson EM, Wollman BN, Marosi K, Wilson MA, Iser WB, Eckley DM, Zhang Y, Lehrmann E, Goldberg IG, Scheibye-Knudsen M, Mattson MP, Nilsen H, Bohr VA, Becker KG (2017) Tomatidine enhances lifespan and healthspan in *C. elegans* through mitophagy induction via the SKN-1/Nrf2 pathway. Sci Rep 7:46208
3. Kushwaha P, Fatima S, Upadhyay A, Gupta S, Bhagwati S, Baghel T, Siddiqi MI, Nazir A, Sashidhara KV (2019) Synthesis, biological evaluation and molecular dynamic simulations of novel Benzofuran-tetrazole derivatives as potential agents against Alzheimer's disease. Bioorg Med Chem Lett 29:66–72
4. Li R, Tao M, Wu T, Zhuo Z, Xu T, Pan S, Xu X (2021) A promising strategy for investigating the anti-aging effect of natural compounds: a case study of caffeoylquinic acids. Food Funct 12:8583–8593
5. Zeng Y, Guo W, Xu G, Wang Q, Feng L, Long S, Liang F, Huang Y, Lu X, Li S, Zhou J, Burgunder JM, Pang J, Pei Z (2016) Xyloketal-derived small molecules show protective effect by decreasing mutant Huntingtin protein aggregates in Caenorhabditis elegans model of Huntington's disease. Drug Des Devel Ther 10:1443–1451

6. Tangrodchanapong T, Sornkaew N, Yurasakpong L, Niamnont N, Nantasenamat C, Sobhon P, Meemon K (2021) Beneficial effects of cyclic ether 2-butoxytetrahydrofuran from sea cucumber *Holothuria scabra* against Aβ aggregate toxicity in transgenic *Caenorhabditis elegans* and potential chemical interaction. Molecules 26:2195
7. Fardghassemi Y, Maios C, Parker JA (2021) Small molecule rescue of ATXN3 toxicity in *C. elegans* via TFEB/HLH-30. Neurotherapeutics 18:1151–1165
8. Benson JA, Cummings EE, O'Reilly LP, Lee M, Pak SC (2014) A high-content assay for identifying small molecules that reprogram *C. elegans* germ cell fate. Methods 68:529–535
9. Papaevgeniou N, Sakellari M, Jha S, Tavernarakis N, Holmberg CI, Gonos ES, Chondrogianni N (2016) 18a-Glycyrrhetinic acid proteasome activator decelerates aging and Alzheimer's disease progression in *Caenorhabditis elegans* and neuronal cultures. Antioxid Redox Signal 25:855–869
10. Li H, Feng Y, Chen Z, Jiang X, Zhou Z, Yuan J, Li F, Zhang Y, Huang X, Fan S, Wu X, Huang C (2021) Pepper component 7-ethoxy-4-methylcoumarin, a novel dopamine D2 receptor agonist, ameliorates experimental Parkinson's disease in mice and *Caenorhabditis elegans*. Pharmacol Res 163:105220
11. Wang C, Niederstrasser H, Douglas PM, Lin R, Jaramillo J, Li Y, Oswald NW, Zhou A, McMillan EA, Mendiratta S, Wang Z, Zhao T, Lin Z, Luo M, Huang G, Brekken RA, Posner BA, MacMillan JB, Gao J, White MA (2017) Small-molecule TFEB pathway agonists that ameliorate metabolic syndrome in mice and extend *C. elegans* lifespan. Nat Commun 8:2270
12. Cuanalo-Contreras K, Park K, Mukherjee A, Pena LM, Soto C (2017) Delaying aging in *Caenorhabditis elegans* with protein aggregation inhibitors. Biochem Biophys Res Commun 482:62–67
13. Ikedaa T, Kishikawaa J, Hayashidaa Y, Fujikawab M, Yokoyama K (2020) Identification of chemical compounds as an inhibitor of mitochondrial ATP synthesis, leading to an increased stress resistance and an extended lifespan in *C. elegans*. Biochim Biophys Acta Bioenerg 1861:148281
14. El-Ami T, Moll L, Marques FC, Volovik Y, Reuveni H, Cohen E (2014) A novel inhibitor of the insulin/IGF signaling pathway protects from age-onset, neurodegeneration-linked proteotoxicity. Aging Cell 13:165–174
15. Angeli S, Klang I, Sivapatham R, Mark K, Zucker D, Bhaumik D, Lithgow GJ, Andersen JK (2013) A DNA synthesis inhibitor is protective against proteotoxic stressors via modulation of fertility pathways in *Caenorhabditis elegans*. Aging 5:759–769
16. Edwards C, Canfield J, Copes N, Rehan M, Lipps D, Bradshaw PC (2014) D-beta-hydroxybutyrate extends lifespan in *C. elegans*. Aging 6:621–644
17. Li J, Cui X, Ma X, Li C, Wang Z (2019) Recombinant buckwheat trypsin inhibitor improves the protein and mitochondria homeostasis in *Caenorhabditis elegans* model of aging and age-related disease. Gerontology 65:513–523
18. Kumar S, Dietrich N, Kornfeld K (2016) Angiotensin converting enzyme (ACE) inhibitor extends *Caenorhabditis elegans* life span. PLoS Genet 12:e1005866
19. Huang X-B, Mu X-H, Wan Q-L, He X-M, Wu G-S, Luo H-R (2017) Aspirin increases metabolism through germline signalling to extend the lifespan of *Caenorhabditis elegans*. PLoS One 12:e0184027
20. Schmeisser S, Zarse K, Ristow M (2011) Lonidamine extends lifespan of adult *Caenorhabditis elegans* by increasing the formation of mitochondrial reactive oxygen species. Horm Metab Res 43:687–692
21. Ahmad W, Ebert PR (2018) 5-Methoxyindole-2-carboxylic acid (MICA) suppresses A-β-mediated pathology in *C. elegans*. Exp Gerontol 108:215–225
22. McCormick AV, Wheeler JM, Guthrie CR, Liachko NF, Kraemer BC (2013) Dopamine D2 receptor antagonism suppresses tau aggregation and neurotoxicity. Biol Psychiatry 73:464–471

23. Li C, Ning L, Cui X, Ma X, Li J, Wang Z (2019) Recombinant buckwheat trypsin inhibitor decreases fat accumulation via the IIS pathway in *Caenorhabditis elegans*. Exp Gerontol 128: 110753
24. Packham R, Walker RJ, Holden-Dye L (2010) The effect of a selective octopamine antagonist, epinastine, on pharyngeal pumping in *Caenorhabditis elegans*. Invertebr Neurosci 10:47–52
25. Lin Y, Hoang H, Hsieh SI, Rangel N, Foster AL, Sampayo JN, Lithgow GJ, Srinivasan C (2006) Manganous ion supplementation accelerates wild type development, enhances stress resistance, and rescues the life span of a short-lived *Caenorhabditis elegans* mutant. Free Radic Biol Med 40:1185–1193
26. Coutts F, Palmo AB, Duarte RRR, de Jong S, Lewis CM, Dima D, Powel TR (2019) The polygenic nature of telomere length and the anti-ageing properties of lithium. Neuropsychopharmacology 44:757–765
27. Tam ZY, Gruber J, Ng LF, Halliwell B, Gunawan R (2014) Effects of lithium on age-related decline in mitochondrial turnover and function in *Caenorhabditis elegans*. J Gerontol A Biol Sci Med Sci 69:810–820
28. Stefanello ST, Gubert P, Puntel B, Mizdal CR, de Campos MMA, Salman SM, Dornelles L, Avila DS, Aschner M, Soares FAA (2015) Protective effects of novel organic selenium compounds against oxidative stress in the nematode *Caenorhabditis elegans*. Toxicol Rep 2: 961–967
29. Li W, Shi Y, Chang C, Huang C, Liao VH (2014) Selenite protects *Caenorhabditis elegans* from oxidative stress via DAF-16 and TRXR-1. Mol Nutr Food Res 58:863–874
30. Avila DS, Benedetto A, Au C, Manarin F, Erikson K, Soares FA, Rocha JBT, Aschner M (2012) Organotellurium and organoselenium compounds attenuate Mn-induced toxicity in *C. elegans* by preventing oxidative stress. Free Radic Biol Med 52:1903–1910
31. Rohn I, Raschke S, Aschner M, Tuck S, Kuehnelt D, Kipp A, Schwerdtle T, Bornhorst J (2019) Treatment of *Caenorhabditis elegans* with small selenium species enhances antioxidant defense systems. Mol Nutr Food Res 63:e1801304
32. Li W-H, Shi Y-C, Tseng I-L, Liao VH-C (2013) Protective efficacy of selenite against lead-induced neurotoxicity in *Caenorhabditis elegans*. PLoS One 8:e62387
33. Li W-H, Chang C-H, Huang C-W, Wei C-C, Liao VH-C (2014) Selenite enhances immune response against *Pseudomonas aeruginosa* PA14 via SKN-1 in *Caenorhabditis elegans*. PLoS One 9:e105810
34. Rebolledo DL, Aldunate R, Kohn R, Neira I, Minniti AN, Inestrosa NC (2011) Copper reduces Aβ oligomeric species and ameliorates neuromuscular synaptic defects in a *C. elegans* model of inclusion body myositis. J Neurosci 31:10149–10158
35. Broué F, Liere P, Kenyon C, Baulieu E (2007) A steroid hormone that extends the lifespan of *Caenorhabditis elegans*. Aging Cell 6:87–94
36. Tanaka D, Furusawa K, Kameyama K, Okamoto H, Doi M (2007) Melatonin signaling regulates locomotion behavior and homeostatic states through distinct receptor pathways in *Caenorhabditis elegans*. Neuropharmacology 53:157–168
37. Martin R, Entchev EV, Kurzchalia TV, Knolker H (2010) Steroid hormones controlling the life cycle of the nematode *Caenorhabditis elegans*: stereoselective synthesis and biology. Org Biomol Chem 8:739–750
38. Kadlecová A, Jirsa T, Novák O, Kammenga J, Strnad M, Voller J (2018) Natural plant hormones increase stress resistance and longevity of *Caenorhabditis elegans*. Biogerontology 19:109–120
39. Gerisch B, Rottiers V, Li D, Motola DL, Cummins CL, Lehrach H, Mangelsdorf DJ, Antebi A (2007) A bile acid-like steroid modulates *Caenorhabditis elegans* lifespan through nuclear receptor signaling. Proc Natl Acad Sci USA 104:5014–5019
40. Niacaris T, Avery L (2003) Serotonin regulates repolarization of the *C. elegans* pharyngeal muscle. J Exp Biol 206:223–231

41. Wang Y, Branicky R, Stepanyan Z, Carroll M, Guimond M, Hihi A, Hayes S, McBride K, Hekimi S (2009) The anti-neurodegeneration drug clioquinol inhibits the aging-associated protein CLK-1. J Biol Chem 284:314–323
42. Saharia K, Arya U, Kumar R, Sahu R, Das CK, Gupta K, Dwivedi H, Subramaniam JR (2012) Reserpine modulates neurotransmitter release to extend lifespan and alleviate age-dependent Aβ proteotoxicity in *Caenorhabditis elegans*. Exp Gerontol 47:188–197
43. Srivastava D, Arya U, SoundaraRajan T, Dwivedi H, Kumar S, Subramaniam JR (2008) Reserpine can confer stress tolerance and lifespan extension in the nematode *C. elegans*. Biogerontology 9:309–316
44. Dehghan E, Zhang Y, Saremi B, Yadavali S, Hakimi A, Dehghani M, Goodarzi M, Tu X, Robertson S, Lin R, Chudhuri A, Mirzaei H (2017) Hydralazine induces stress resistance and extends *C. elegans* lifespan by activating the NRF2/SKN-1 signalling pathway. Nat Commun 8:2223
45. Evason K, Collins JJ, Huang C, Hughes S, Kornfeld K (2008) Valproic acid extends *Caenorhabditis elegans* lifespan. Aging Cell 7:305–317
46. Xu J, Guo Y, Sui T, Wang Q, Zhang Y, Zhang R, Wang M, Guan S, Wang L (2017) Molecular mechanisms of anti-oxidant and anti-aging effects induced by convallatoxin in *Caenorhabditis elegans*. Free Radic Res 51:529–544
47. Petrascheck M, Ye X, Buck LB (2007) An antidepressant that extends lifespan in adult *Caenorhabditis elegans*. Nature 450:553–556
48. Ching T, Chiang W, Chen C, Hsu A (2011) Celecoxib extends *C. elegans* lifespan via inhibition of insulinlike signaling but not cyclooxygenase-2 activity. Aging Cell 10:506–519
49. Collins JJ, Evason K, Pickett CL, Schneider DL, Kornfeld K (2008) The anticonvulsant ethosuximide disrupts sensory function to extend *C. elegans* lifespan. PLoS Genet 4:e1000230
50. Cho SC, Park MC, Keam B, Choi KM, Cho Y, Hyun S, Parka SC, Lee J (2010) DDS, 4,4′-diaminodiphenylsulfone, extends organismic lifespan. Proc Natl Acad Sci USA 107: 19236–19331
51. Evason K, Huang C, Yamben I, Covey DF, Kornfeld K (2005) Anticonvulsant medications extend worm life-span. Science 307:258–262
52. Morita M, Osoda K, Yamazaki M, Shirai F, Matsuoka N, Arakawa H, Nishimura S (2009) Effects of non-steroidal anti-inflammatory drugs on Aβ deposition in Aβ1-42 transgenic *C. elegans*. Brain Res 1295:186–191
53. Chen X, McCueF HV, Wong SQ, Kashyap SS, Kraemer BC, Barclay JW, Burgoyne RD, Morgan A (2015) Ethosuximide ameliorates neurodegenerative disease phenotypes by modulating DAF-16/FOXO target gene expression. Mol Neurodegener 10:51
54. Habchi J, Arosio P, Perni M, Costa AR, Yagi-Utsumi M, Joshi P, Chia S, Cohen SIA, Müller MBD, Linse S, Nollen EAA, Dobson CM, Knowles TPJ, Vendruscolo M (2016) An anticancer drug suppresses the primary nucleation reaction that initiates the production of the toxic Aβ42 aggregates linked with Alzheimer's disease. Sci Adv 2:e1501244
55. Hao L, Ben-David O, Babb SM, Futerman AH, Cohen BM, Buttner EA (2017) Clozapine modulates glucosylceramide, clears aggregated proteins, and enhances ATG8/LC3 in *Caenorhabditis elegans*. Neuropsychopharmacology 42:951–962
56. Wongchai K, Schlotterer A, Lin J, Humpert PM, Klein T, Hammes HP, Morcos M (2016) Protective effects of liraglutide and linagliptin in *C. elegans* as a new model for glucose-induced neurodegeneration. Horm Metab Res 48:70–75
57. Dwyer DS, Awatramani P, Thakur R, Seeni R, Aamodt EJ (2015) Social feeding in *Caenorhabditis elegans* is modulated by antipsychotic drugs and calmodulin and may serve as a protophenotype for asociality. Neuropharmacology 92:56–62
58. Sonane M, Moin N, Satish A (2017) The role of antioxidants in attenuation of *Caenorhabditis elegans* lethality on exposure to TiO_2 and ZnO nanoparticles. Chemosphere 187:240–247
59. Ishii N, Senoo-Matsuda N, Miyake K, Yasuda K, Ishii T, Hartman PS, Furukawa S (2004) Coenzyme Q_{10} can prolong *C. elegans* lifespan by lowering oxidative stress. Mech Ageing Dev 125:41–46

60. Shibamura A, Ikeda T, Nishikawa Y (2009) A method for oral administration of hydrophilic substances to *Caenorhabditis elegans*: effects of oral supplementation with antioxidants on the nematode lifespan. Mech Ageing Dev 130:652–655
61. Wu Q-L, Wang W, Li Y-X, Li Y-P, Ye B-P, Tang M, Wang D-Y (2012) Small sizes of TiO_2-NPs exhibit adverse effects at predicted environmental relevant concentrations on nematodes in a modified chronic toxicity assay. J Hazard Mater 243:161–168
62. Wu Q-L, Liu P-D, Li Y-X, Du M, Xing X-J, Wang D-Y (2012) Inhibition of ROS elevation and damage to mitochondrial function prevents lead-induced neurotoxic effects on structures and functions of AFD neurons in *Caenorhabditis elegans*. J Environ Sci 24:1–10
63. Yu X-M, Guan X-M, Wu Q-L, Zhao Y-L, Wang D-Y (2015) Vitamin E ameliorates neurodegeneration related phenotypes caused by neurotoxicity of Al_2O_3-nanoparticles in *C. elegans*. Toxicol Res 4:1269–1281
64. Adachi H, Ishii N (2000) Effects of tocotrienols on life span and protein carbonylation in *Caenorhabditis elegans*. J Gerontol Biol Sci 55A:B280–B285
65. Ye H-Y, Ye B-P, Wang D-Y (2008) Trace administration of vitamin E can retrieve and prevent UV-irradiation- and metal exposure-induced memory deficits in nematode *Caenorhabditis elegans*. Neurobiol Learn Mem 90:10–18
66. Polyak E, Ostrovsky J, Peng M, Dingley SD, Tsukikawa M, Kwon YJ, McCormack S, Bennett M, Xiao R, Seiler C, Zhang Z, Falk MJ (2018) N-acetylcysteine and vitamin E rescue animal longevity and cellular oxidative stress in preclinical models of mitochondrial complex I disease. Mol Genet Metab 123:449–462
67. Wang X, Wang X, Li L, Wang D (2010) Lifespan extension in *Caenorhabditis elegans* by DMSO is dependent on *sir-2.1* and *daf-16*. Biochem Biophys Res Commun 400:613–618
68. Brown MK, Evans JL, Luo Y (2006) Beneficial effects of natural antioxidants EGCG and α-lipoic acid on life span and age-dependent behavioral declines in *Caenorhabditis elegans*. Pharmacol Biochem Behav 85:620–628
69. Ng LF, Gruber J, Cheah IK, Goo CK, Cheong WF, Shui G, Sit KP, Wenk MR, Halliwell B (2014) The mitochondria-targeted antioxidant MitoQ extends lifespan and improves healthspan of a transgenic *Caenorhabditis elegans* model of Alzheimer disease. Free Radic Biol Med 71:390–401
70. Desjardins D, Cacho-Valadez B, Liu J, Wang Y, Yee C, Bernard K, Khaki A, Breton L, Hekimi S (2017) Antioxidants reveal an inverted U-shaped dose-response relationship between reactive oxygen species levels and the rate of aging in *Caenorhabditis elegans*. Aging Cell 16:104–112
71. Sampayo JN, Olsen A, Lithgow GJ (2003) Oxidative stress in *Caenorhabditis elegans*: protective effects of superoxide dismutase/catalase mimetics. Aging Cell 2:319–326
72. Melov S, Ravenscroft J, Malik S, Gill MS, Walker DW, Clayton PE, Wallace DC, Malfroy B, Doctrow SR, Lithgow GJ (2000) Extension of life-span with superoxide dismutase/catalase mimetics. Science 289:1567–1569
73. Mark KA, Dumas KJ, Bhaumik D, Schilling B, Davis S, Oron TR, Sorensen DJ, Lucanic M, Brem RB, Melov S, Ramanathan A, Gibson BW, Lithgow GJ (2016) Vitamin D promotes protein homeostasis and longevity via the stress response pathway genes SKN-1, IRE-1, and XBP-1. Cell Rep 17:1227–1237
74. Rathor L, Akhoon BA, Pandey S, Srivastava S, Pandey R (2015) Folic acid supplementation at lower doses increases oxidative stress resistance and longevity in *Caenorhabditis elegans*. Age 37:113
75. Schaffner A, Li X, Gomez-Llorente Y, Leandrou E, Memou A, Clemente N, Yao C, Afsari F, Zhi L, Pan N, Morohashi K, Hua X, Zhou M, Wang C, Zhang H, Chen SG, Elliott CJ, Rideout H, Ubarretxena-Belandia I, Yue Z (2019) Vitamin B12 modulates Parkinson's disease LRRK2 kinase activity through allosteric regulation and confers neuroprotection. Cell Res 29:313–329

76. Dallaire A, Proulx S, Simard MJ, Lebel M (2014) Expression profile of *Caenorhabditis elegans* mutant for the Werner syndrome gene ortholog reveals the impact of vitamin C on development to increase life span. BMC Genomics 15:940
77. Zhang A, Ackley B, Yan D (2020) Vitamin B12 regulates glial migration and synapse formation through isoform-specific control of PTP-3/LAR PRTP expression. Cell Rep 30: 3981–3988
78. Park S, Park S (2013) Electrolyzed-reduced water increases resistance to oxidative stress, fertility, and lifespan via insulin/IGF-1-like signal in *C. elegans*. Biol Res 46:147–153
79. Yan H, Tian H, Kinjo T, Hamasaki T, Tomimatsu K, Nakamichi N, Teruya K, Kabayama S, Shirahata S (2010) Extension of the lifespan of *Caenorhabditis elegans* by the use of electrolyzed reduced water. Biosci Biotechnol Biochem 74:2011–2015
80. Lin C, Zhang X, Zhuang C, Lin Y, Cao Y, Chen Y (2020) Healthspan improvements in *Caenorhabditis elegans* with traditional Chinese herbal tea. Oxidative Med Cell Longev 2020: 4057841
81. Cho HS, Lee J, Cho MH, Lee J (2015) Red wines and flavonoids diminish *Staphylococcus aureus* virulence with anti-biofilm and anti-hemolytic activities. Biofouling 31:1–11
82. Patananan AN, Budenholzer LM, Eskin A, Torresa ER, Clarke SG (2015) Ethanol-induced differential gene expression and acetyl-CoA metabolism in a longevity model of the nematode *Caenorhabditis elegans*. Exp Gerontol 61:20–30
83. Chuang M, Chiou S, Huang C, Yang W, Wong C (2009) The lifespan-promoting effect of acetic acid and Reishi polysaccharide. Bioorg Med Chem 17:7831–7840
84. Wang X, Cook LF, Grasso LM, Cao M, Dong Y (2015) Royal jelly-mediated prolongevity and stress resistance in *Caenorhabditis elegans* is possibly modulated by the interplays of DAF-16, SIR-2.1, HCF-1, and 14-3-3 proteins. J Gerontol Biol Sci 70:827–838
85. Weimer S, Priebs J, Kuhlow D, Groth M, Priebe S, Mansfeld J, Merry TL, Dubuis S, Laube B, Pfeiffer AF, Schulz TJ, Guthke R, Platzer M, Zamboni N, Zarse K, Ristow M (2014) D-Glucosamine supplementation extends life span of nematodes and of ageing mice. Nat Commun 5:3563
86. Sugawara T, Saraprug D, Sakamoto K (2019) Soy sauce increased the oxidative stress tolerance of nematode via p38 MAPK pathway. Biosci Biotechnol Biochem 83:709–716
87. Zhang J, Lu L, Zhou L (2015) Oleanolic acid activates *daf-16* to increase lifespan in *Caenorhabditis elegans*. Biochem Biophys Res Commun 468:843–849
88. Qi W, Gutierrez GE, Gao X, Dixon H, McDonough JA, Marini AM, Fisher AL (2017) The ω-3 fatty acid a-linolenic acid extends *Caenorhabditis elegans* lifespan via NHR-49/PPARα and oxidation to oxylipins. Aging Cell 16:1125–1135
89. Yoon K, Lee TY, Moon J, Choi SY, Choi YJ, Mitchell RJ, Lee JI (2020) Consumption of oleic acid during matriphagy in free-living nematodes alleviates the toxic effects of the bacterial metabolite violacein. Sci Rep 10:8087
90. Lynn DA, Dalton HM, Sowa JN, Wang MC, Soukas AA, Curran SP (2015) Omega-3 and -6 fatty acids allocate somatic and germline lipids to ensure fitness during nutrient and oxidative stress in *Caenorhabditis elegans*. Proc Natl Acad Sci USA 112:15378–15383
91. Wei C-C, Yen P-L, Chang S-T, Cheng P-L, Lo Y-C, Liao VH-C (2016) Antioxidative activities of both oleic acid and *Camellia tenuifolia* seed oil are regulated by the transcription factor DAF-16/FOXO in *Caenorhabditis elegans*. PLoS One 11:e0157195
92. Shashikumar S, Pradeep H, Chinnu S, Rajini PS, Rajanikant GK (2015) Alpha-linolenic acid suppresses dopaminergic neurodegeneration induced by 6-OHDA in *C. elegans*. Physiol Behav 151:563–569
93. Raabe RC, Mathies LD, Davies AG, Bettinger JC (2014) The omega-3 fatty acid eicosapentaenoic acid is required for normal alcohol response behaviors in *C. elegans*. PLoS One 9:e105999
94. Lesa GM, Palfreyman M, Hall DH, Clandinin MT, Rudolph C, Jorgensen EM, Schiavo G (2003) Long chain polyunsaturated fatty acids are required for efficient neurotransmission in *C. elegans*. J Cell Sci 116:4965–4975

95. Qabazard B, Li L, Gruber J, Peh MT, Ng LF, Kumar SD, Rose P, Tan C, Dymock BW, Wei F, Swain SC, Halliwell B, Sturzenbaum SR, Moore PK (2014) Hydrogen sulfide is an endogenous regulator of aging in *Caenorhabditis elegans*. Antioxid Redox Signal 20:2621–2630
96. Qabazard B, Ahmed S, Li L, Arlt VM, Moore PK, Sturzenbaum SR (2013) *C. elegans* aging is modulated by hydrogen sulfide and the sulfhydrylase/cysteine synthase *cysl-2*. PLoS One 8: e80135
97. Zhang M, Li Z, Gao D, Gong W, Gao Y, Zhang C (2020) Hydrogen extends *Caenorhabditis elegans* longevity by reducing reactive oxygen species. PLoS One 15:e0231972
98. Budde MW, Roth MB (2011) The response of *Caenorhabditis elegans* to hydrogen sulfide and hydrogen cyanide. Genetics 189:521–532
99. Budde MW, Roth MB (2010) Hydrogen sulfide increases hypoxia-inducible factor-1 activity independently of von Hippel–Lindau tumor suppressor-1 in *C. elegans*. Mol Biol Cell 21:212–217
100. Miller DL, Roth MB (2007) Hydrogen sulfide increases thermotolerance and lifespan in *Caenorhabditis elegans*. Proc Natl Acad Sci USA 104:20618–20622
101. Sharabi K, Hurwitz A, Simon AJ, Beitel GJ, Morimoto RI, Rechavi G, Sznajder JI, Gruenbaum Y (2009) Elevated CO_2 levels affect development, motility, and fertility and extend life span in *Caenorhabditis elegans*. Proc Natl Acad Sci USA 106:4024–4029
102. Kurino C, Furuhashi T, Sudoh K, Sakamoto K (2017) Isoamyl alcohol odor promotes longevity and stress tolerance via DAF-16 in *Caenorhabditis elegans*. Biochem Biophys Res Commun 485:395–399
103. Ensaka N, Sakamoto K (2020) a-Pinene odor exposure enhances heat stress tolerance through Daf-16 in *Caenorhabditis elegans*. Biochem Biophys Res Commun 528:726–731
104. Zhang L, Gu B, Wang Y (2021) Clove essential oil confers antioxidant activity and lifespan extension in *C. elegans* via the DAF-16/FOXO transcription factor. Comparat Biochem Physiol Part C 242:108938
105. Yu C, Li W, Hsu F, Yen P, Chang S, Liao VH (2014) Essential oil alloaromadendrene from mixed-type *Cinnamomum osmophloeum* leaves prolongs the lifespan in *Caenorhabditis elegans*. J Agric Food Chem 62:6159–6165
106. Hsu F, Li W, Yu C, Hsieh Y, Yang Y, Liu J, Shih J, Chu Y, Yen P, Chang S, Liao VH (2012) In vivo antioxidant activities of essential oils and their constituents from leaves of the Taiwanese *Cinnamomum osmophloeum*. J Agric Food Chem 60:3092–3097
107. Rossi M, Caruso F, Kwok L, Lee G, Caruso A, Gionfra F, Candelotti E, Belli SL, Molasky N, RaleySusman KM, Leone S, Filipsky T, Tofani D, Pedersen J, Incerpi S (2017) Protection by extra virgin olive oil against oxidative stress in vitro and in vivo chemical and biological studies on the health benefits due to a major component of the Mediterranean diet. PLoS One 12:e0189341
108. Chen X, Liao D, Sun M, Cui X, Wang H (2020) Essential oil of *Acorus tatarinowii* schott ameliorates Aβ-induced toxicity in *Caenorhabditis elegans* through an autophagy pathway. Oxid Med Cell Longev 2020:3515609
109. Lang M, Montjarret A, Duteil E, Bedoux G (2021) *Cinnamomum cassia* and *Syzygium aromaticum* essential oils reduce the colonization of *Salmonella* typhimurium in an *in vivo* infection model using *Caenorhabditis elegans*. Molecules 26:5598
110. Chiu H, Hsiao S, Lu Y, Han Y, Shen Y, Venkatakrishnan K, Wang C (2018) Efficacy of protein rich pearl powder on antioxidant status in a randomized placebo-controlled trial. J Food Drug Anal 26:309–312
111. Gao C, Gao Z, Greenway FL, Burton JH, Johnson WD, Keenan MJ, Enright FM, Martin RJ, Chu Y, Zheng J (2015) Oat consumption reduced intestinal fat deposition and improved health span in *Caenorhabditis elegans* model. Nutr Res 35:834–843
112. Kim J, Shirasawa T, Miyamoto Y (2010) The effect of TAT conjugated platinum nanoparticles on lifespan in a nematode *Caenorhabditis elegans* model. Biomaterials 31:5849–5854

113. Kim J, Takahashi M, Shimizu T, Shirasawa T, Kajita M, Kanayama A, Miyamoto Y (2008) Effects of a potent antioxidant, platinum nanoparticle, on the lifespan of *Caenorhabditis elegans*. Mech Ageing Dev 129:322–331
114. Cong W, Bai R, Li Y, Wang L, Chen C (2019) Selenium nanoparticles as an efficient nanomedicine for the therapy of Huntington's disease. ACS Appl Mater Interfaces 11: 34725–34735
115. Sathya S, Shanmuganathan B, Balasubramaniam B, Balamurugan K, Devi KP (2020) Phytol loaded PLGA nanoparticles regulate the expression of Alzheimer's related genes and neuronal apoptosis against amyloid-β induced toxicity in Neuro-2a cells and transgenic *Caenorhabditis elegans*. Food Chem Toxicol 136:110962
116. Chen Z, Zhou C, Xu Y, Wen K, Song J, Bai S, Wu C, Huang W, Cai Q, Zhou K, Wang H, Wang Y, Feng X, Bai Y (2021) An alternatingly amphiphilic, resistance-resistant antimicrobial oligoguanidine with dual mechanisms of action. Biomaterials 275:120858
117. Ramasamy M, Lee J, Lee J (2017) Development of gold nanoparticles coated with silica containing the antibiofilm drug cinnamaldehyde and their effects on pathogenic bacteria. Int J Nanomedicine 12:2813–2828
118. Choi S, Britigan BE, Narayanasamy P (2019) Iron/heme metabolism-targeted gallium(III) nanoparticles are active against extracellular and intracellular *Pseudomonas aeruginosa* and *Acinetobacter baumannii*. Antimicrob Agents Chemother 63:e02643–e02618
119. Saharia K, Kumar R, Gupta K, Mishra S, Subramaniam JR (2016) Reserpine requires the D2-type receptor, *dop-3*, and the exoribonuclease, *eri-1*, to extend the lifespan in *C. elegans*. J Biosci 41:689–695
120. Saharia K, Arya U, Kumar R, Sahu R, Das CK, Gupta K, Dwivedi H, Subramaniam JR (2012) Reserpine modulates neurotransmitter release to extend lifespan and alleviate age-dependent Abeta proteotoxicity in *Caenorhabditis elegans*. Exp Gerontol 47:188–197
121. Arya U, Dwivedi H, Subramaniam JR (2009) Reserpine ameliorates Abeta toxicity in the Alzheimer's disease model in *Caenorhabditis elegans*. Exp Gerontol 44:462–466
122. Xiao Y, Zhang L, Zhou H, Cui Y, Chen K, Zhang H, Wu Q, Liu F (2025) Berberine extends healthspan and delays neurodegenerative diseases in *Caenorhabditis elegans* through ROS-dependent PMK-1/SKN-1 activation. Arch Gerontol Geriatr 128:105644
123. Xiao Y, Cui Y, Zhang Y, Fu W, Liu Y, Liu F (2025) Berberine hydrochloride enhances innate immunity to protect against pathogen infection via p38 MAPK pathway. Front Immunol 16: 1536143
124. Chow YL, Sato F (2013) Screening of isoquinoline alkaloids for potent lipid metabolism modulation with *Caenorhabditis elegans*. Biosci Biotechnol Biochem 77:2405–2412
125. Ide S, Kunitomo H, Iino Y, Ikeda K (2022) *Caenorhabditis elegans* exhibits morphine addiction-like behavior via the opioid-like receptor NPR-17. Front Pharmacol 12:802701
126. Cheong MC, Artyukhin AB, You YJ, Avery L (2015) An opioid-like system regulating feeding behavior in *C. elegans*. elife 4:e06683
127. Li H, Roxo M, Cheng X, Zhang S, Cheng H, Wink M (2019) Pro-oxidant and lifespan extension effects of caffeine and related methylxanthines in *Caenorhabditis elegans*. Food Chem X 1:100005
128. Du X, Guan Y, Huang Q, Lv M, He X, Yan L, Hayashi S, Fang C, Wang X, Sheng J (2018) Low concentrations of caffeine and its analogs extend the lifespan of *Caenorhabditis elegans* by modulating IGF-1-like pathway. Front Aging Neurosci 10:211
129. Brunquell J, Morris S, Snyder A, Westerheide SD (2018) Coffee extract and caffeine enhance the heat shock response and promote proteostasis in an HSF-1-dependent manner in *Caenorhabditis elegans*. Cell Stress Chaperones 23:65–75
130. Manalo RVM, Medina PMB (2020) Caffeine reduces deficits in mechanosensation and locomotion induced by L-DOPA and protects dopaminergic neurons in a transgenic *Caenorhabditis elegans* model of Parkinson's disease. Pharm Biol 58:721–731

131. Guo C, Shen W, Jin W, Jia X, Ji Z, Jinling L, Bin L (2023) Effects of epigallocatechin gallate, caffeine, and their combination on fat accumulation in high-glucose diet-fed *Caenorhabditis elegans*. Biosci Biotechnol Biochem 87:898–906
132. Wang Y, Xiang YF, Liu AL (2022) Comparative and combined effects of epigallocatechin-3-gallate and caffeine in reducing lipid accumulation in *Caenorhabditis elegans*. Plant Foods Hum Nutr 77:279–285
133. Min H, Youn E, Shim YH (2021) Long-term caffeine intake exerts protective effects on intestinal aging by regulating vitellogenesis and mitochondrial function in an aged *Caenorhabditis elegans* model. Nutrients 13:2517
134. Lee M, Lee J, Kim D, Min H, Shim YH (2024) Caffeine-induced upregulation of *pas-1* and *pas-3* enhances intestinal integrity by reducing vitellogenin in aged *Caenorhabditis elegans* model. Nutrients 16:4298
135. Di Rocco M, Galosi S, Lanza E, Tosato F, Caprini D, Folli V, Friedman J, Bocchinfuso G, Martire A, Di Schiavi E, Leuzzi V, Martinelli S (2022) *Caenorhabditis elegans* provides an efficient drug screening platform for GNAO1-related disorders and highlights the potential role of caffeine in controlling dyskinesia. Hum Mol Genet 31:929–941
136. Singh S, Fatima Z, Ahmad K, Hameed S (2020) Repurposing of respiratory drug theophylline against *Candida albicans*: mechanistic insights unveil alterations in membrane properties and metabolic fitness. J Appl Microbiol 129:860–8875
137. Ullah I, Zhao L, Uddin S, Zhou Y, Wang X, Li H (2024) Nicotine-mediated therapy for Parkinson's disease in transgenic *Caenorhabditis elegans* model. Front Aging Neurosci 16: 1358141
138. Lu X, Zhang Y, Li H, Jin Y, Zhao L, Wang X (2021) Nicotine prevents in vivo Abeta toxicity in *Caenorhabditis elegans* via SKN-1. Neurosci Lett 761:136114
139. Ullah I, Uddin S, Xin W, Li H (2025) Therapeutic potential of nicotine in modulating oxidative and mitochondrial pathways in a *C. elegans* Parkinson's disease model. Free Radic Res 59:716–733
140. Kalinnikova TB, Kolsanova RR, Belova EB, Shagidullin RR, Gainutdinov MK (2016) Opposite effects of moderate heat stress and hyperthermia on cholinergic system of soil nematodes *Caenorhabditis elegans* and *Caenorhabditis briggsae*. J Therm Biol 62:37–49
141. Shamalnasab M, Gravel SP, St-Pierre J, Breton L, Jäger S, Aguilaniu H (2018) A salicylic acid derivative extends the lifespan of *Caenorhabditis elegans* by activating autophagy and the mitochondrial unfolded protein response. Aging Cell 17:e12830
142. Nkambeu B, Salem JB, Beaudry F (2020) Capsaicin and its analogues impede nocifensive response of *Caenorhabditis elegans* to noxious heat. Neurochem Res 45:1851–1859
143. Li J, Wang J, Huang Z, Cui X, Li C (2023) Oxidized quercetin has stronger anti-amyloid activity and anti-aging effect than native form. Comp Biochem Physiol C Toxicol Pharmacol 271:109676
144. Sugawara T, Sakamoto K (2020) Quercetin enhances motility in aged and heat-stressed *Caenorhabditis elegans* nematodes by modulating both HSF-1 activity, and insulin-like and p38-MAPK signalling. PLoS One 15:e0238528
145. Ayuda-Durán B, González-Manzano S, Miranda-Vizuete A, Sánchez-Hernández E, Romero M, Dueñas M, Santos-Buelga C, González-Paramás AM (2019) Exploring target genes involved in the effect of quercetin on the response to oxidative stress in *Caenorhabditis elegans*. Antioxidants 8:585
146. Li SM, Liu D, Liu YL, Liu B, Chen XH (2020) Quercetin and its mixture increase the stress resistance of *Caenorhabditis elegans* to UV-B. Int J Environ Res Public Health 17:1572
147. Civelek M, Flory S, Meloh H, Fitzenberger E, Wenzel U (2020) The polyphenol quercetin protects from glucotoxicity depending on the aggresome in *Caenorhabditis elegans*. Eur J Nutr 59:485–491
148. Lei S, Hu Z, Liu H (2024) Treatment with quercetin mitigates polystyrene nanoparticle-induced reduction in neuron capacity by inhibiting dopaminergic neurodegeneration and facilitating dopamine metabolism in *Caenorhabditis elegans*. Chemosphere 364:143303

149. He W, Liu Z, Zhang H, Liu Q, Weng Z, Wang D, Guo W, Xu J, Wang D, Jiang Z, Gu A (2024) Bisphenol S decreased lifespan and healthspan via insulin/IGF-1-like signaling-against mitochondrial stress in *Caenorhabditis elegans*. Ecotoxicol Environ Saf 285:117136
150. Schiavi A, Cirotti C, Gerber LS, Di Lauro G, Maglioni S, Shibao PYT, Montresor S, Kirstein J, Petzsch P, Köhrer K, Schins RPF, Wahle T, Barilà D, Ventura N (2023) Abl depletion via autophagy mediates the beneficial effects of quercetin against Alzheimer pathology across species. Cell Death Discov 9:376
151. Das SS, Sarkar A, Chabattula SC, Verma PRP, Nazir A, Gupta PK, Ruokolainen J, Kesari KK, Singh SK (2022) Food-grade quercetin-loaded nanoemulsion ameliorates effects associated with Parkinson's disease and cancer: studies Employing a transgenic *C. elegans* model and human cancer cell lines. Antioxidants 11:1378
152. Garzón-García L, Ayuda-Durán B, González-Manzano S, Santos-Buelga C, González-Paramás AM (2025) Neuroprotective potential of the flavonoids quercetin and epicatechin in a *C. elegans* tauopathy model. Mol Nutr Food Res 69:e70108
153. Zhang SY, Qin ZC, Sun YY, Chen YS, Chen WB, Wang HG, An D, Sun D, Liu YQ (2023) Genistein promotes anti-heat stress and antioxidant effects via the coordinated regulation of IIS, HSP, MAPK, DR, and mitochondrial pathways in *Caenorhabditis elegans*. Antioxidants 12:125
154. Wang J, Jiang B, Lin X, Zhang J, Xiong L, Feng Y, Cheang WS, Xu B (2026) Genistein ameliorates glucose-induced beta-amyloid toxicity, oxidative stress, and aging in the *C. elegans* model of Alzheimer's disease. Free Radic Biol Med 242:167–187
155. Li K, Zhang S, Sun Y, Chen Y, Chen W, Ruan W, Liu Y (2022) Antiamyloid beta toxicity effect of genistein via activation of DAF-16 and HSP-16.2 signal pathways in *Caenorhabditis elegans*. J Biochem Mol Toxicol 36:e23055
156. Nakatani Y, Yaguchi Y, Komura T, Nakadai M, Terao K, Kage-Nakadai E, Nishikawa Y (2018) Sesamin extends lifespan through pathways related to dietary restriction in *Caenorhabditis elegans*. Eur J Nutr 57:1137–1146
157. Li YY, Xia JL, Bai M, Liu K, Wei JH, Lv ZR, Liu YH, Gao CH, Yi XX (2025) Sonneralignan A, a new lignan from the fruit of mangrove Sonneratia apetala. Nat Prod Res 39:1983–1988
158. Keowkase R, Shoomarom N, Bunargin W, Sitthithaworn W, Weerapreeyakul N (2018) Sesamin and sesamolin reduce amyloid-beta toxicity in a transgenic *Caenorhabditis elegans*. Biomed Pharmacother 107:656–664
159. Xie Z, Zhao J, Wang H, Jiang Y, Yang Q, Fu Y, Zeng H, Hölscher C, Xu J, Zhang Z (2020) Magnolol alleviates Alzheimer's disease-like pathology in transgenic *C. elegans* by promoting microglia phagocytosis and the degradation of beta-amyloid through activation of PPAR-gamma. Biomed Pharmacother 124:109886
160. Xiao Y, Zhou H, Cui Y, Zhu X, Li S, Yu C, Jiang N, Liu L, Liu F (2024) Schisandrin A enhances pathogens resistance by targeting a conserved p38 MAPK pathway. Int Immunopharmacol 128:111472
161. Anju VT, Busi S, Ranganathan S, Ampasala DR, Kumar S, Suchiang K, Kumavath R, Dyavaiah M (2021) Sesamin and sesamolin rescues *Caenorhabditis elegans* from *Pseudomonas aeruginosa* infection through the attenuation of quorum sensing regulated virulence factors. Microb Pathog 155:104912
162. Kim MR, Cho SY, Lee HJ, Kim JY, Nguyen UTT, Ha NM, Choi KY, Cha KH, Kim JH, Kim WK, Kang K (2022) Schisandrin C improves leaky gut conditions in intestinal cell monolayer, organoid, and nematode models by increasing tight junction protein expression. Phytomedicine 103:154209
163. Tran SH, Zhang S, Lee HS, Hitayezu E, Nguyen UTT, Lee S, Kang SW, Park K, Kim SH, Kwon HC, Cha K, Kwon J, Kim J, Kang K (2026) Schisandrin C alleviates depressive-like behaviors by modulating the AKT/CREB/BDNF pathway, the serotonin pathway of tryptophan metabolism, and the gut microbiota composition. Phytomedicine 150:157581

164. Teixeira TR, Schmidt B, Sperlich E, Lemes BL, Amaro MC, Pérez R, Céspedes-Méndez C, Villegas C, Burgos V, de Moraes J, Paz C (2025) Cubebin, a lignan isolated from *Drimys andina*, exhibits potent and selective antiparasitic activity against *Angiostrongylus cantonensis*. ACS Omega 10:31161–31169
165. Lee J, Kwon G, Park J, Kim JK, Lim YH (2016) Brief communication: SIR-2.1-dependent lifespan extension of *Caenorhabditis elegans* by oxyresveratrol and resveratrol. Exp Biol Med 241:1757–1763
166. Yoon DS, Cha DS, Choi Y, Lee JW, Lee MH (2019) MPK-1/ERK is required for the full activity of resveratrol in extended lifespan and reproduction. Aging Cell 18:e12867
167. Fischer N, Büchter C, Koch K, Albert S, Csuk R, Wätjen W (2017) The resveratrol derivatives trans-3,5-dimethoxy-4-fluoro-4′-hydroxystilbene and trans-2,4′,5-trihydroxystilbene decrease oxidative stress and prolong lifespan in *Caenorhabditis elegans*. J Pharm Pharmacol 69:73–81
168. Soares MV, Puntel RL, Ávila DS (2018) Resveratrol attenuates iron-induced toxicity in a chronic post-treatment paradigm in *Caenorhabditis elegans*. Free Radic Res 52:939–951
169. Regitz C, Fitzenberger E, Mahn FL, Dußling LM, Wenzel U (2016) Resveratrol reduces amyloid-beta (Abeta1-42)-induced paralysis through targeting proteostasis in an Alzheimer model of *Caenorhabditis elegans*. Eur J Nutr 55:741–747
170. Yue Y, Shen P, Chang AL, Qi W, Kim KH, Kim D, Park Y (2019) trans-Trismethoxy resveratrol decreased fat accumulation dependent on *fat-6* and *fat-7* in *Caenorhabditis elegans*. Food Funct 10:4966–4974
171. Bai S, Yu Y, An L, Wang W, Fu X, Chen J, Ma J (2022) Ellagic acid increases stress resistance via insulin/IGF-1 signaling pathway in *Caenorhabditis elegans*. Molecules 27:6168
172. Wang W, Guo J, Ye D, Wu R, Zhang Y, Liu Y (2025) Ellagic acid mitigates deoxyaconitine-induced neurotoxicity by improving mitochondrial function. Life Sci 381:124007
173. Petry R, de Almeida JM, Côa F, Crasto de Lima F, Martinez DST, Fazzio A (2024) Interaction of graphene oxide with tannic acid: computational modeling and toxicity mitigation in *C. elegans*. Beilstein J Nanotechnol 15:1297–1311
174. Possamai Rossatto FC, Tharmalingam N, Escobar IE, d'Azevedo PA, Zimmer KR, Mylonakis E (2021) Antifungal activity of the phenolic compounds ellagic acid (EA) and caffeic acid phenethyl ester (CAPE) against drug-resistant *Candida auris*. J Fungi 7:763
175. Ban F, Hu L, Zhou XH, Zhao Y, Mo H, Li H, Zhou W (2021) Inverse molecular docking reveals a novel function of thymol: Inhibition of fat deposition induced by high-dose glucose in *Caenorhabditis elegans*. Food Sci Nutr 9:4243–4253
176. Shu C, Sun L, Zhang W (2016) Thymol has antifungal activity against *Candida albicans* during infection and maintains the innate immune response required for function of the p38 MAPK signaling pathway in *Caenorhabditis elegans*. Immunol Res 64:1013–1024
177. Gao L, Zhang R, Lan J, Ning R, Wu D, Chen D, Zhao W (2016) beta-Dihydroagarofuran-type sesquiterpenes from the seeds of *Celastrus monospermus* and their lifespan-extending effects on the nematode *Caenorhabditis elegans*. J Nat Prod 79:3039–3046
178. Yu YF, Liu YH, Chen XH, Zhi DJ, Qi FM, Zhang ZP, Li YQ, Zhang ZX, Fei DQ (2020) Cadinane-type sesquiterpenes from the resinous exudates of *Commiphora myrrha* and their anti-Alzheimer's disease bioactivities. Fitoterapia 142:104536
179. Trivedi M, Singh S, Pandey T, Gupta SK, Verma RS, Pandey R (2022) Sesquiterpenoids isolated from davana (*Artemisia pallens* Wall. ex DC) mitigates parkinsonism in *Caenorhabditis elegans* disease model. Biochem Biophys Res Commun 609:15–22
180. Mohankumar A, Kalaiselvi D, Thiruppathi G, Muthusaravanan S, Vijayakumar S, Suresh R, Tawata S, Sundararaj P (2022) Santalol isomers inhibit transthyretin amyloidogenesis and associated pathologies in *Caenorhabditis elegans*. Front Pharmacol 13:924862
181. Zhu X, Liu F, Wu Q, Li S, Ruan G, Yang J, Yu C, Jiang N, Xiao Y, Liu Y (2022) Brevilin A enhances innate immunity and the resistance of oxidative stress in *Caenorhabditis elegans* via p38 MAPK pathway. Int Immunopharmacol 113:109385

182. Dai W, Deng L, He C, Fu X, Liu J, Wang GC, Yang J, Zhang YB, Xiao F, Wan QL (2025) Crassifolin A prolongs lifespan and healthspan in *Caenorhabditis elegans* via activating autophagy. J Ethnopharmacol 342:119399
183. Lin C, Zhang X, Su Z, Xiao J, Lv M, Cao Y, Chen Y (2019) Carnosol improved lifespan and healthspan by promoting antioxidant capacity in *Caenorhabditis elegans*. Oxidative Med Cell Longev 2019:5958043
184. Chen Y, Wang Y, Qin Q, Zhang Y, Xie L, Xiao J, Cao Y, Su Z, Chen Y (2022) Carnosic acid ameliorated Abeta-mediated (amyloid-beta peptide) toxicity, cholinergic dysfunction and mitochondrial defect in *Caenorhabditis elegans* of Alzheimer's model. Food Funct 13: 4624–4640
185. Zhang ZP, Bai X, Cui WB, Chen XH, Liu X, Zhi DJ, Zhang ZX, Fei DQ, Wang DS (2022) Diterpenoid caesalmin C delays Abeta-induced paralysis symptoms via the DAF-16 pathway in *Caenorhabditis elegans*. Int J Mol Sci 23:6871
186. Farias-Pereira R, Kim E, Park Y (2020) Cafestol increases fat oxidation and energy expenditure in *Caenorhabditis elegans* via DAF-12-dependent pathway. Food Chem 307:125537
187. Cui Y, Wang R, Li X, Bai G, Xiao Y (2025) Ginkgolide a enhances the resistance to pathogen infection through mitochondrial unfolded protein response. Cell Mol Life Sci 82:349
188. Yu X, Li H, Lin D, Guo W, Xu Z, Wang L, Guan S (2021) Ginsenoside prolongs the lifespan of *C. elegans* via lipid metabolism and activating the stress response signaling pathway. Int J Mol Sci 22:9668
189. Chen Q, Chen X, Chen L, Zhang X, Yang Z, Hao J, Ma Z (2025) Ginsenoside Re regulates the insulin/insulin-like growth factor-1 signaling pathway and mediates lipid metabolism to achieve anti-aging effects in *Caenorhabditis elegans*. Molecules 30:3463
190. Zhao C, Yue J, Xie Y, Liu B, Xu S, Zhi D, Wang D (2025) A ginsenoside composition ameliorated Abeta and Tau aggregation via autophagy lysosome pathway. Mol Neurobiol 62: 11822–11833
191. Chen X, Zhou XY, Lan C, Fu HJ, Li ZC, Chen MY, Wen YP, Yu L, Qin DL, Wu AG, Wu JM, Zhou XG (2025) Araloside A induces Raf/MEK/ERK-dependent autophagy to mitigate Alzheimer's and Parkinson's pathology in cellular and *C. elegans* models. Mol Neurobiol 62:16061–16080
192. Xie K, Wang H, Yao X, Lv J, Wang Q, Zhao Y, Yang S, Xu L, Shi Y, Hu J, Shan Y (2025) Hederagenin reduces Aβ-induced oxidative damage, decreases Aβ deposition, and promotes cell survival by the P13K/Akt signaling pathway. J Leukoc Biol 117:qiae124
193. Wei CC, Chang CH, Liao VH (2017) Anti-Parkinsonian effects of beta-amyrin are regulated via LGG-1 involved autophagy pathway in *Caenorhabditis elegans*. Phytomedicine 36:118–125
194. Li X, Hu M, Zhou X, Yu L, Qin D, Wu J, Deng L, Huang L, Ren F, Liao B, Wu A, Fan D (2024) Hederagenin inhibits mitochondrial damage in Parkinson's disease via mitophagy induction. Free Radic Biol Med 224:740–756
195. Zhi D, Yang W, Yue J, Xu S, Ma W, Zhao C, Wang X, Wang D (2022) HSF-1 mediated combined ginsenosides ameliorating Alzheimer's disease like symptoms in *Caernorhabditis elegans*. Nutr Neurosci 25:2136–2148
196. Shi H, Zhao J, Li Y, Li J, Li Y, Zhang J, Qiu Z, Wu C, Qin M, Liu C, Zeng Z, Zhang C, Gao L (2023) Ginsenosides Rg1 regulate lipid metabolism and temperature adaptation in *Caenorhabditis elegans*. J Ginseng Res 47:524–533
197. Paulo IRF, de Oliveira Caland RB, Cadavid COM, Melo GM, Bezerra LSDC, Pons E, Peña L, de Paula Oliveira R (2022) beta-carotene genetically-enriched lyophilized orange juice increases antioxidant capacity and reduces beta-amyloid proteotoxicity and fat accumulation in *Caenorhabditis elegans*. Food Chem 5:100141
198. González-Peña MA, Lozada-Ramírez JD, Ortega-Regules AE (2021) Carotenoids from mamey (*Pouteria sapota*) and carrot (*Daucus carota*) increase the oxidative stress resistance of *Caenorhabditis elegans*. Biochem Biophys Rep 26:100989

199. Martorell P, Llopis S, Gil JV, Genovés S, Ramón D, Zacarías L, Rodrigo MJ (2020) Evaluation of carotenoids protection against oxidative stress in the animal model *Caenorhabditis elegans*. Methods Mol Biol 2083:387–401
200. Lin X, Shao CS, Elsherbiny SM, Huang Q (2025) Astaxanthin attenuates UV-irradiation aging process via activating JNK-1/DAF-16 in *Caenorhabditis elegans*. Photochem Photobiol 101: 133–146
201. Ding F, Zhao Y (2025) Astaxanthin promotes the longevity of *Caenorhabditis elegans* via modulation of the intracellular redox status and PHA-4-mediated autophagy. Food Funct 16: 617–627
202. Li J, Sun W, Wang X, Ding F, Li L, Lin X, Zhao Y (2025) Astaxanthin suppresses amyloid-beta-Induced toxicity in AD transgenic *Caenorhabditis elegans* via promoting Skn-1-dependent proteasomal activity. ACS Chem Neurosci 16:3954–3963
203. Xie J, Cai R, Hou X, Zhao K, Xiao J, Cao Y, Liu X (2025) Z-Astaxanthin exhibits superior anti-obesity effects in *Caenorhabditis elegans*: insights from geometric isomers and signaling pathways. J Sci Food Agric 105:4795–4807
204. Medina PM, Ponce JM, Cruz CA (2021) Revealing the anticancer potential of candidate drugs in vivo using *Caenorhabditis elegans* mutant strains. Transl Oncol 14:100940
205. Qi Z, Ji H, Le M, Li H, Wieland A, Bauer S, Liu L, Wink M, Herr I (2021) Sulforaphane promotes *C. elegans* longevity and healthspan via DAF-16/DAF-2 insulin/IGF-1 signaling. Aging 13:1649–1670
206. Schlotterer A, Masri B, Humpert M, Krämer BK, Hammes HP, Morcos M (2021) Sulforaphane and Vitamin E protect from glucotoxic neurodegeneration and lifespan reduction in *C. elegans*. Exp Clin Endocrinol Diabetes 129:887–894
207. Hernández-Cruz EY, Aparicio-Trejo OE, Eugenio-Pérez D, Juárez-Peredo E, Zurita-León M, Valdés VJ, Pedraza-Chaverri J (2024) Sulforaphane exposure prevents cadmium-induced toxicity and mitochondrial dysfunction in the nematode *Caenorhabditis elegans* by regulating the insulin/insulin-like growth factor signaling (IIS) Pathway. Antioxidants 13:584
208. Balducci M, Pérez JT, Del Río CT, Pérez MC, Carranza ADV, Gomez Escribano AP, Vázquez-Manrique RP, Tarozzi A (2024) Erucin, a natural isothiocyanate, prevents polyglutamine-induced toxicity in *Caenorhabditis elegans* via *aak-2*/AMPK and *daf-16*/FOXO signaling. Int J Mol Sci 25:12220
209. Han X, Nan M, Cai X, Qiao B, Chen L, Shen L (2022) Sennoside A inhibits quorum sensing system to attenuate its regulated virulence and pathogenicity via targeting LasR in *Pseudomonas aeruginosa*. Front Microbiol 13:1042214
210. Naß J, Efferth T (2021) Withanone ameliorates stress symptoms in *Caenorhabditis elegans* by acting through serotonin receptors. Pharmacopsychiatry 54:215–223
211. Akhoon BA, Rathor L, Pandey R (2018) Withanolide A extends the lifespan in human EGFR-driven cancerous *Caenorhabditis elegans*. Exp Gerontol 104:113–117
212. Naß J, Abdelfatah S, Efferth T (2021) Induction of stress resistance and extension of lifespan in *Caenorhabditis elegans* serotonin-receptor knockout strains by withanolide A. Phytomedicine 84:153482
213. Shanmugam G, Mohankumar A, Kalaiselvi D, Nivitha S, Murugesh E, Shanmughavel P, Sundararaj P (2017) Diosgenin a phytosterol substitute for cholesterol, prolongs the lifespan and mitigates glucose toxicity via DAF-16/FOXO and GST-4 in *Caenorhabditis elegans*. Biomed Pharmacother 95:1693–1703
214. Smita SS, Trivedi S, Pandey T, Trivedi M, Pandey R (2020) A bioactive compound shatavarin IV-mediated longevity as revealed by dietary restriction-induced autophagy in *Caenorhabditis elegans*. Biogerontology 21:827–844
215. Wu AG, Teng JF, Wong VK, Zhou XG, Qiu WQ, Tang Y, Wu JM, Xiong R, Pan R, Wang YL, Tang B, Ding TY, Yu L, Zeng W, Qin DL, Law BY (2019) Novel steroidal saponin isolated from Trillium tschonoskii maxim. exhibits anti-oxidative effect via autophagy induction in cellular and *Caenorhabditis elegans* models. Phytomedicine 65:153088

216. Govindhan T, Amirthalingam M, Govindan S, Duraisamy K, Cho JH, Tawata S, Periyakali SB, Palanisamy S (2024) Diosgenin intervention: targeting lipophagy to counter high glucose diet-induced lipid accumulation and lifespan reduction. 3 Biotech 14:171
217. Parrales V, Arcile G, Laserre L, Normant S, Le Goff G, Da Costa Noble C, Ouazzani J, Callizot N, Haïk S, Rabhi C, Bizat N (2025) Neuroprotective effect of withaferin derivatives toward MPP^+ and 6-OHDA toxicity to dopaminergic neurons. ACS Chem Neurosci 16:802–817
218. Akhoon BA, Pandey S, Tiwari S, Pandey R (2016) Withanolide A offers neuroprotection, ameliorates stress resistance and prolongs the life expectancy of *Caenorhabditis elegans*. Exp Gerontol 78:47–56
219. Wu AG, Yong YY, He CL, Li YP, Zhou XY, Yu L, Chen Q, Lan C, Liu J, Yu CL, Qin DL, Wu JM, Zhou XG (2024) Novel 18-norspirostane steroidal saponins: extending lifespan and mitigating neurodegeneration through promotion of mitophagy and mitochondrial biogenesis in *Caenorhabditis elegans*. Mech Ageing Dev 218:111901
220. Xiao Y, Liu F, Wu Q, Zhu X, Yu C, Jiang N, Li S, Liu Y (2024) Dioscin activates endoplasmic reticulum unfolded protein response for defense sgainst pathogenic bacteria in *Caenorhabditis elegans* via IRE-1/XBP-1 pathway. J Infect Dis 229:237–244
221. Wang J, Zhou Y, Yu Y, Wang Y, Xue D, Zhou Y, Li X (2023) A ginseng-derived rhamnogalacturonan I (RG-I) pectin promotes longevity via TOR signalling in *Caenorhabditis elegans*. Carbohydr Polym 312:120818
222. Wang J, Wang Y, Xu X, Song C, Zhou Y, Xue D, Feng Z, Zhou Y, Li X (2024) Low methyl-esterified ginseng homogalacturonan pectins promote longevity of *Caenorhabditis elegans* via impairing insulin/IGF-1 signalling. Carbohydr Polym 346:122600
223. Xiang C, Teng H, Sheng Z, Zhao C, Deng J, Zhao C, He B, Chen L, Ai C (2025) Structural characterization and antioxidant activity mechanism of the ferulic acid-rich subfraction from sugar beet pectin. Carbohydr Polym 347:122691
224. Higashimura Y, Isobe M, Miura K, Desaka N, Nishikawa H, Nishida N, Naito J, Kawano T, Miyada T, Naito Y (2025) Paramylon isolated from *Euglena gracilis* EOD-1 extends lifespan through activation of DAF-16-mediated antioxidant pathway via *clec-196* in *Caenorhabditis elegans*. Sci Rep 15:42202
225. Liu H, Wang Y, Zhang W, Sun W, Ji X, Zhang S, Qiao K (2022) Lentinan extends lifespan and increases oxidative stress resistance through DAF-16 and SKN-1 pathways in *Caenorhabditis elegans*. Int J Biol Macromol 202:286–295
226. Xiao X, Zhou Y, Tan C, Bai J, Zhu Y, Zhang J, Zhou X, Zhao Y (2021) Barley beta-glucan resist oxidative stress of *Caenorhabditis elegans* via *daf-2/daf-16* pathway. Int J Biol Macromol 193:1021–1031
227. Zhang J, Wang P, Tan C, Zhao Y, Zhu Y, Bai J, Xiao X (2022) Integrated transcriptomics and metabolomics unravel the metabolic pathway variations for barley beta-glucan before and after fermentation with L. plantarum DY-1. Food Funct 13:4302–4314
228. Xiao X, Tan C, Sun X, Zhao Y, Zhang J, Zhu Y, Bai J, Dong Y, Zhou X (2020) Fermented barley beta-glucan regulates fat deposition in *Caenorhabditis elegans*. J Sci Food Agric 100: 3408–3417
229. Wang J, Wang D, Setrerrahmane S, Martinez J, Xu HM (2023) The peptide Acein promotes dopamine secretion through *clec-126* to extend the lifespan of elderly *C. elegans*. Aging 15: 14651–14665
230. Liu W, Yu G, Yu W, Ye X, Jin Y, Shrestha A, Yang Q, Sun H (2017) Autophagy inhibits apoptosis induced by agrocybe aegerita lectin in hepatocellular carcinoma. Anti Cancer Agents Med Chem 17:221–229
231. Kalmankar NV, Hari H, Sowdhamini R, Venkatesan R (2021) Disulfide-rich cyclic peptides from *Clitoria ternatea* protect against beta-amyloid toxicity and oxidative stress in transgenic *Caenorhabditis elegans*. J Med Chem 64:7422–7433

232. Kalmankar NV, Gehi BR, Sowdhamini R (2022) Effects of a plant cyclotide on conformational dynamics and destabilization of beta-amyloid fibrils through molecular dynamics simulations. Front Mol Biosci 9:986704
233. Yang X, Zhang P, Wu J, Xiong S, Jin N, Huang Z (2012) The neuroprotective and lifespan-extension activities of *Damnacanthus officinarum* extracts in *Caenorhabditis elegans*. J Ethnopharmacol 141:41–47
234. Jing L, Zhao Y, Jiang L, Song F, An L, Qi E, Fu X, Chen J, Ma J (2025) Unlocking the potential of *Curcumae Rhizoma* aqueous extract in stress resistance and extending lifespan in *Caenorhabditis elegans*. Molecules 30:1668
235. Zhang J, Zheng W, Zhou Z, An Y, Zheng H, Zhang Y, Wei Y, Zhang Q, Zheng J, Wang F (2025) Aqueous *Polygalae Radix* extract (PRE) prolongs the lifespan of *C. elegans* and alleviates D-galactose-induced oxidative stress in the mouse liver and brain by modulating PPARgamma/MAPK. J Ethnopharmacol 348:119878
236. Hu M, Cheng N, Liu R, Shao L, Yang R, Du S, Chu L, Wang W, Wang H (2025) Red ginseng aqueous extract superior to white ginseng exhibited anti-aging property through IIS signaling pathway in *Caenorhabditis elegans*. Chem Biodivers 22:e202500604
237. Huang CH, Hsu FY, Wu YH, Zhong L, Tseng MY, Kuo CJ, Hsu AL, Liang SS, Chiou SH (2015) Analysis of lifespan-promoting effect of garlic extract by an integrated metabolo-proteomics approach. J Nutr Biochem 26:808–817
238. Abdelfattah MAO, Dmirieh M, Ben Bakrim W, Mouhtady O, Ghareeb MA, Wink M, Sobeh M (2022) Antioxidant and anti-aging effects of *Warburgia salutaris* bark aqueous extract: evidences from in silico, in vitro and in vivo studies. J Ethnopharmacol 292:115187
239. Iriondo-DeHond A, Martorell P, Genovés S, Ramón D, Stamatakis K, Fresno M, Molina A, Del Castillo MD (2016) Coffee silverskin extract protects against accelerated aging caused by oxidative agents. Molecules 21:721
240. Dos Santos FB, Quines CB, Pilissão LEB, Dal Forno AHC, Rodrigues CF, Denardin CC, Farias FM, Ávila DS (2020) Aqueous bark extract of *Ceiba speciosa* (A. St.-Hill) Ravenna protects against glucose toxicity in *Caenorhabditis elegans*. Oxid Med Cell Longev 2020: 1321354
241. Azevedo BC, Roxo M, Borges MC, Peixoto H, Crevelin EJ, Bertoni BW, Contini SHT, Lopes AA, França SC, Pereira AMS, Wink M (2019) Antioxidant activity of an aqueous leaf extract from *Uncaria tomentosa* and its major alkaloids mitraphylline and isomitraphylline in *Caenorhabditis elegans*. Molecules 24:3299
242. Shi Z, Lu Z, Zhao Y, Wang Y, Zhao-Wilson X, Guan P, Duan X, Chang YZ, Zhao B (2013) Neuroprotective effects of aqueous extracts of *Uncaria tomentosa*: Insights from 6-OHDA induced cell damage and transgenic *Caenorhabditis elegans* model. Neurochem Int 62:940–947
243. Kamireddy K, Chinnu S, Priyanka PS, Rajini PS, Giridhar P (2018) Neuroprotective effect of *Decalepis hamiltonii* aqueous root extract and purified 2-hydroxy-4-methoxy benzaldehyde on 6-OHDA induced neurotoxicity in *Caenorhabditis elegans*. Biomed Pharmacother 105:997–1005
244. Santos PA, Pflüger P, Rott MB, Gómez-Couso H, Rodrigues I, Pereira P, Fontenla JÁ (2025) *Achyrocline satureioides*: in silico and in vivo evaluation of the potential neuroprotective effect of the aqueous extract. J Ethnopharmacol 353:120335
245. Hangweirer K, Sandner G, König A, Kirchsteiger J, Blank-Landeshammer B, Winkler S, Schurr J, Weghuber J (2025) Allium ursinum leaves aqueous extract alleviates *Pseudomonas aeruginosa* infection in *Caenorhabditis elegans*. Phytomedicine 146:157100
246. Zou X, Xiao Y, Ke L, Nie Y, Xiao J, Yang J, Guo C, Liu X (2025) Sake lees extract obtained using a novel continuous phase-transition extraction method: evaluation of its bioactive composition, anti-aging efficacy and mechanism. Food Funct 16:3862–3878
247. Dikti Vildina J, Kalmobe J, Djafsia B, Schmidt TJ, Liebau E, Ndjonka D (2017) Anti-onchocerca and anti-*Caenorhabditis* activity of a hydro-alcoholic extract from the fruits of *Acacia nilotica* and some proanthocyanidin derivatives. Molecules 22:748

248. Gómez-Linton DR, Alavez S, Navarro-Ocaña A, Román-Guerrero A, Pinzón-López L, Pérez-Flores LJ (2021) Achiote (*Bixa orellana*) lipophilic extract, bixin, and delta-tocotrienol effects on lifespan and stress resistance in *Caenorhabditis elegans*. Planta Med 87:368–374
249. Zhang W-M, Lv T, Li M, Wu Q-L, Yang L-S, Liu H, Sun D-F, Sun L-M, Zhuang Z-H, Wang D-Y (2013) Beneficial effects of wheat gluten hydrolysate to extend lifespan and induce stress resistance in nematode *Caenorhabditis elegans*. PLoS One 8:e74553
250. Fang Z, Chen Y, Wang G, Feng T, Shen M, Xiao B, Gu J, Wang W, Li J, Zhang Y (2019) Evaluation of the antioxidant effects of acid hydrolysates from *Auricularia auricular* polysaccharides using a *Caenorhabditis elegans* model. Food Funct 10:5531–5543
251. Zhang Y, Jiang W, Hao X, Tan J, Wang W, Yu M, Zhang G, Zhang Y (2021) Preparation of the enzymatic hydrolysates from *Chlorella vulgaris* protein and assessment of their antioxidant potential using *Caenorhabditis elegans*. Mol Biotechnol 63:1040–1048
252. Zhang Y, Li W, Hou P, Yang T, Xie J (2025) Physicochemical and stability analysis of mung bean protein hydrolysates with lipid peroxidation inhibition. Food Chem 463:141135
253. Sim I, Park K, Kwon G, Koh J, Lim Y (2018) Probiotic potential of *Enterococcus faecium* isolated from chicken cecum with immunomodulating activity and promoting longevity in *Caenorhabditis elegans*. J Microbiol Biotechnol 28:883–892
254. Kwon G, Lee J, Koh J, Lim Y (2018) Lifespan extension of *Caenorhabditis elegans* by *Butyricicoccus pullicaecorum* and *Megasphaera elsdenii* with probiotic potential. Curr Microbiol 75:557–564
255. Azat R, Liu Y, Kayir A, Lin D, Zhou W, Zheng X (2016) Probiotic properties of lactic acid bacteria isolated from traditionally fermented Xinjiang cheese. J Zhejiang University-Sci B Biomed Biotechnol 17:597–609
256. Grompone G, Martorell P, Llopis S, Gonzalez N, Genoves S, Mulet AP, Fernandez-Calero T, Tiscornia I, Bollati-Fogolın M, Chambaud I, Foligne B, Montserrat A, Ramon D (2012) Anti-inflammatory *Lactobacillus rhamnosus* CNCM I-3690 strain protects against oxidative stress and Increases lifespan in *Caenorhabditis elegans*. PLoS One 7:e52493
257. Marsova M, Poluektova E, Odorskaya M, Ambaryan A, Revishchin A, Pavlova G, Danilenko V (2020) Protective effects of *Lactobacillus* fermentum U-21 against paraquat-induced oxidative stress in *Caenorhabditis elegans* and mouse models. World J Microbiol Biotechnol 36:104
258. Zhao Y-L, Yu X-M, Jia R-H, Yang R-L, Rui Q, Wang D-Y (2015) Lactic acid bacteria protects *Caenorhabditis elegans* from toxicity of graphene oxide by maintaining normal intestinal permeability under different genetic backgrounds. Sci Rep 5:17233
259. Chelliah R, Choi J, Hwang S, Park B, Daliri EB, Kim S, Wei S, Ramakrishnan SR, Oh D (2018) In vitro and *in vivo* defensive effect of probiotic LAB against *Pseudomonas aeruginosa* using *Caenorhabditis elegans* model. Virulence 9:1489–1507
260. Martorell P, Llopis S, Gonzalez N, Chenoll E, Lopez-Carreras N, Aleixandre A, Chen Y, Karoly ED, Ramon D, Genoves S (2016) Probiotic strain *Bifidobacterium animalis* subsp. lactis CECT 8145 reduces fat content and modulates lipid metabolism and antioxidant response in *Caenorhabditis elegans*. J Agric Food Chem 64:3462–3472
261. Setel DD, Beker M, Terzioglu-Usak S, Elibol B (2022) *Astragalus membranaceus* treatment combined with caloric restriction may enhance genesis factors and decrease apoptosis in the hippocampus of rats. Arch Gerontol Geriatr 99:104584
262. Yuen CW, Murugaiyah V, Najimudin N, Azzam G (2021) Danshen (*Salvia miltiorrhiza*) water extract shows potential neuroprotective effects in *Caenorhabditis elegans*. J Ethnopharmacol 266:113418
263. Jiang JM, Chen G, Chen YY, Wan SJ, Chen SM, Ren HG, Lin ZX, Feng H, Zhang H, Xu HX (2022) The anti-infective activity of *Salvia miltiorrhiza* against *Staphylococcus aureus* by attenuating accessory gene regulator system-mediated virulence. Food Funct 13:5050–5060
264. Wang X, Wang X, Wang D, Li L, Niu X (2010) Effect of *Ligusticum chuanxiong* extract on lifespan of *Caenorhabditis elegans* and its underlying molecular mechanisms. Zhongguo Zhong Yao Za Zhi 35:1042–1045

265. Qin Y, Chen F, Tang Z, Ren H, Wang Q, Shen N, Lin W, Xiao Y, Yuan M, Chen H, Bu T, Li Q, Huang L (2022) *Ligusticum chuanxiong* Hort as a medicinal and edible plant foods: antioxidant, anti-aging and neuroprotective properties in *Caenorhabditis elegans*. Front Pharmacol 13:1049890
266. Shen N, Zeng W, Leng F, Lu J, Lu Z, Cui J, Wang L, Jin B (2021) Ginkgo seed extract promotes longevity and stress resistance of *Caenorhabditis elegans*. Food Funct 12:12395–12406
267. Wu Z, Smith JV, Paramasivam V, Butko P, Khan I, Cypser JR, Luo Y (2002) *Ginkgo biloba* extract EGb 761 increases stress resistance and extends life span of *Caenorhabditis elegans*. Cell Mol Biol 48:725–731
268. Cao Z, Wu Y, Curry K, Wu Z, Christen Y, Luo Y (2007) *Ginkgo biloba* extract EGb 761 and *Wisconsin Ginseng* delay sarcopenia in *Caenorhabditis elegans*. J Gerontol A Biol Sci Med Sci 62:1337–1345
269. Kampkötter A, Pielarski T, Rohrig R, Timpel C, Chovolou Y, Wätjen W, Kahl R (2007) The *Ginkgo biloba* extract EGb761 reduces stress sensitivity, ROS accumulation and expression of catalase and glutathione S-transferase 4 in *Caenorhabditis elegans*. Pharmacol Res 55:139–147
270. Smith JV, Luo Y (2003) Elevation of oxidative free radicals in Alzheimer's disease models can be attenuated by *Ginkgo biloba* extract EGb 761. J Alzheimer's Dis 5:287–300
271. Wu Y, Wu Z, Butko P, Christen Y, Lambert MP, Klein WL, Link CD, Luo Y (2006) Amyloid-beta-induced pathological behaviors are suppressed by *Ginkgo biloba* extract EGb 761 and ginkgolides in transgenic *Caenorhabditis elegans*. J Neurosci 26:13102–13113
272. Wu ZQ, Li K, Tian X, Zhou MX, Li ZJ (2020) *Schisandra chinensis* water extract protects ethanol-induced neurotoxicity in *Caenorhabditis elegans*. J Food Biochem 44:e13249
273. Ruan Q, Qiao Y, Zhao Y, Xu Y, Wang M, Duan J, Wang D (2016) Beneficial effects of *Glycyrrhizae radix* extract in preventing oxidative damage and extending the lifespan of *Caenorhabditis elegans*. J Ethnopharmacol 177:101–110
274. Zhao Y-L, Jia R-H, Qiao Y, Wang D-Y (2016) Glycyrrhizic acid, active component from *Glycyrrhizae radix*, prevents toxicity of graphene oxide by influencing functions of microRNAs in nematode *Caenorhabditis elegans*. Nanomedicine 12:735–744
275. Link P, Wetterauer B, Fu Y, Wink M (2015) Extracts of *Glycyrrhiza uralensis* and isoliquiritigenin counteract amyloid-beta toxicity in *Caenorhabditis elegans*. Planta Med 81: 357–362
276. Wang W, Li S, Zhu Y, Cui X, Sheng Z, Wang H, Cheng Z (2024) Antioxidant and Neuroprotective effects of seed oils from *Trichosanthes kirilowii* and *T. laceribractea* in *Caenorhabditis elegans*: a comparative analysis and mechanism study. Antioxidants 13:861
277. Zeng L, Yang Z, Yun T, Fan S, Pei Z, Chen Z, Sun C, Xu F (2019) Antiaging effect of a Jianpi-yangwei formula in *Caenorhabditis elegans*. BMC Complement Altern Med 19:313
278. Chen W, Wang J, Shi J, Yang X, Yang P, Wang N, Yang S, Xie T, Yang H, Zhang M, Wang H, Fei, j. (2019) Longevity effect of Liuwei Dihuang in both *Caenorhabditis elegans* and aged mice. Aging Dis 10:578–591
279. Wang X, Zhang L, Gao X, Zhang L, Yu J, Liu Y, Fang M, Yan Y, Chen L, Du J, Guan H, Huang C, Fan S (2025) Liu Jun Zi decoction extends lifespan and healthspan through p16/p21 signaling in *Caenorhabditis elegans*. Biogerontology 26:201
280. Zhu B, Yang P, Mammat N, Ding H, He J, Qian Y, Fei J, Abdukerim K (2015) Aiweixin, a traditional Uyghur medicinal formula, protects against chromium toxicity in *Caenorhabditis elegans*. BMC Complement Altern Med 15:285
281. Rui Q, Lu Q, Wang D-Y (2009) Administration with Bushenkangshuai Tang alleviates UV irradiation- and oxidative stress-induced lifespan defects in nematode *Caenorhabditis elegans*. Front Med China 3:76–90
282. Li D, Xue A, Guan Y, Qiu Q, Ren Y, Zhang N (2025) Investigation of Kai-Xin-San in alleviating cognitive impairment in aβ transgenic *Caenorhabditis elegans* through mitochondrial function regulation. Fitoterapia 187:106921

283. Zhang W, Zhi D, Ren H, Wang D, Wang X, Zhang Z, Fei D, Zhu H, Li H (2016) Shengmai formula ameliorates pathological characteristics in AD *C. elegans*. Cell Mol Neurobiol 36: 1291–1302
284. Zhi D, Wang D, Yang W, Duan Z, Zhu S, Dong J, Wang N, Wang N, Fei D, Zhang Z, Wang X, Wang M, Li H (2017) Dianxianning improved amyloid β-induced pathological characteristics partially through DAF-2/DAF-16 insulin like pathway in transgenic *C. elegans*. Sci Rep 7:11408
285. Sangha JS, Sun X, Wally OSD, Zhang K, Ji X, Wang Z, Wang Y, Zidichouski J, Prithiviraj B, Zhang J (2012) Liuwei Dihuang (LWDH), a Traditional Chinese Medicinal formula, protects against β-amyloid toxicity in transgenic *Caenorhabditis elegans*. PLoS One 7:e43990
286. Zhou L, Yang C, Liu Z, Chen L, Wang P, Zhou Y, Yuan M, Zhou LT, Wang X, Zhu LQ (2024) Neuroprotective effect of the traditional decoction Tian-Si-Yin against Alzheimer's disease via suppression of neuroinflammation. J Ethnopharmacol 321:117569
287. Zhang L, Wang Y-X, Cao C, Zhu Y-K, Huang W, Yang Y, Qiu H-B, Liu S-Q, Wang D-Y (2022) Beneficial effect of Xuebijing against *Pseudomonas aeruginosa* infection in *Caenorhabditis elegans*. Front Pharmacol 13:949608
288. Guerrero-Rubio MA, Hernandez-García S, García-Carmona F, Gandía-Herrero F (2021) Biosynthesis of a novel polymeric chitosan-betaxanthin and characterization of the first sugar-derived betalains and their effects in the in vivo model *Caenorhabditis elegans*. Carbohydr Polym 252:117141
289. Salgueiro WG, Xavier MCDF, Duarte LFB, Câmara DF, Fagundez DA, Soares ATG, Perin G, Alves D, Avila DS (2014) Direct synthesis of 4-organylsulfenyl-7-chloro quinolines and their toxicological and pharmacological activities in *Caenorhabditis elegans*. Eur J Med Chem 75: 448–459
290. Sun X, Wu X, Chen X, Guo R, Kou Y, Li X, Sheng Y, Wu Y (2021) Casein-maltodextrin Maillard conjugates encapsulation enhances the antioxidative potential of proanthocyanidins: An *in vitro* and *in vivo* evaluation. Food Chem 346:128952
291. Sashidhara KV, Modukuri RK, Jadiya P, Dodda RP, Kumar M, Sridhar B, Vikash Kumar V, Haque R, Siddiqi MI, Nazir A (2014) Benzofuran–chalcone hybrids as potential multifunctional agents against Alzheimer's disease: synthesis and in vivo studies with transgenic *Caenorhabditis elegans*. ChemMedChem 9:2671–2684
292. Tu L, Tseng N, Tsai Y, Lin T, Lo Y, Charng J, Hsu H, Chen Y, Chen R, Wu Y, Chan Y, Chen C, Fang J, Chen Y (2018) Rationally designed divalent caffeic amides inhibit amyloid-β fibrillization, induce fibril dissociation, and ameliorate cytotoxicity. Eur J Med Chem 158: 393–404
293. Maqbool M, Rajvansh R, Srividya K, Hoda N (2020) Deciphering the robustness of pyrazolopyridine carboxylate core structure-based compounds for inhibiting α-synuclein in transgenic *C. elegans* model of Synucleinopathy. Bioorg Med Chem 28:115640
294. Banerjee J, Bruckbauer A, Zemel MB (2016) Activation of the AMPK/Sirt1 pathway by a leucine–metformin combination increases insulin sensitivity in skeletal muscle, and stimulates glucose and lipid metabolism and increases life span in *Caenorhabditis elegans*. Metabolism 65:1679–1691
295. Kim B, Kim S, Baek S, Lee EY, Lee ES, Chung CH, Ahn CM, Park S (2019) Cur2004-8, a synthetic curcumin derivative, extends lifespan and modulates age-related physiological changes in *Caenorhabditis elegans*. Drug Discoveries Therapeutics 13:198–206
296. Sayed AAR, El-Shaieb KM, Mourad AE (2012) Life span extension of *Caenorhabditis elegans* by novel pyridoperimidine derivative. Arch Pharm Res 35:69–76
297. Mekheimer RA, Sayed AAR, Ahmed EA, Sadek KU (2015) Synthesis and characterization of new 1,2,4-triazolo[1,5-a]-pyridines that extend the life span of *Caenorhabiditis elegans* via Their anti-inflammatory/antioxidant effects. Arch Pharm Chem Life Sci 348:650–665
298. Wang W, Feng X, Du Y, Liu C, Pang X, Jiang K, Wang X, Liu Y (2021) Synthesis of novel pinocembrin amino acid derivatives and their antiaging effect on *Caenorhabditis elegans* via the modulating DAF-16/FOXO. Drug Design Dev Therapy 15:4177–4193

299. Mekheimer RA, Sayed AAR, Ahmed EA (2012) Novel 1,2,4-triazolo[1,5-a]pyridines and their fused ring systems attenuate oxidative stress and prolong lifespan of *Caenorhabiditis elegans*. J Med Chem. 55(9):4169–4177
300. Rodriguez CR, del Fueyo MC, Santillán VJ, Dansey IMV, Veleiro AS, Castro OA, Burton G (2019) Synthesis and biological activity of fluorinated analogues of the DAF-12 receptor antagonist 24-hydroxy-4-cholen-3-one. Steroids 151:108469
301. Samaja GA, Castro O, Alvarez LD, Dansey MV, Escudero DS, Veleiro AS, Adalí Pecci A, Burton G (2013) 27-Nor-D^4-dafachronic acid is a synthetic ligand of *Caenorhabditis elegans* DAF-12 receptor. Bioorg Med Chem Lett 23:2893–2896
302. Rodrigues PR, Kaul TK, Ho J, Lucanic M, Burkewitz K, Mair WB, Held JM, Bohn LM, Gil MS (2016) Synthetic ligands of cannabinoid receptors affect dauer formation in the nematode *Caenorhabditis elegans*. G3 (Bethesda) 6(6):1695–1705
303. Shou Q, Feng L, Long Y, Han J, Nunnery JK, Powell DH, Butcher RA (2016) A hybrid polyketide-nonribosomal peptide in nematodes that promotes larval survival. Nat Chem Biol 12:770–772
304. Medini K, Harris PWR, Hards K, Dingley AJ, Cook GM, Brimble MA (2015) Chemical synthesis of a pore-forming antimicrobial protein, caenopore-5, by Using native chemical ligation at a glu-cys site. ChemBioChem 16:328–336
305. Deusing DJ, Beyrer M, Fitzenberger E, Wenzel U (2015) Carnitine protects the nematode *Caenorhabditis elegans* from glucose-induced reduction of survival depending on the nuclear hormone receptor DAF-12. Biochem Biophys Res Commun 460:747–752
306. Wang Y-X, Hu G-C, Wang D-Y (2026) 6-PPD quinone reduces lifespan by activating a feedback loop between cholesterol transformation related signal and insulin signaling in *C. elegans*. J Environ Sci 166:700–710. https://doi.org/10.1016/j.jes.2025.10.044
307. Wan X, Liang G-Y, Wang D-Y (2025) 6-PPD quinone at environmentally relevant concentrations disrupts citric acid cycle in *Caenorhabditis elegans*: role of reduction in acetyl CoA and pyruvate contents. Environ Chem Ecotoxicol 7:1119–1129
308. Wu J-W, Li L-E, Hu D-Y, Liu R, Bian Q, Wang D-Y (2025) Environmentally relevant concentrations of 6-PPDQ disrupts vitamin D3 adsorption and receptor function in *Caenorhabditis elegans*. Environ Sci Process Impacts 27:2798–2808

Physiological and Pathological Models for Pharmacological Study

4

Abstract

The introduced physiological models in *Caenorhabditis elegans* contain aging and aging-related changes and stress models. The introduced and discussed pathological models include models for metabolic-related disorders, neurodegeneration-related diseases models, channelopathy models, muscle dystrophy models, seizure-like models, addiction models, anxiety-like models, autism-like models, tumor-like models, wound healing models and regeneration models, mtDNA replicative polymerase (POLG)-related disease models, α1-antitrypsin deficiency (ATD) models, pathogen infection models, and immunosuppression models. These established physiological and pathological models provide important supports for the pharmacological study in *C. elegans*.

Keywords

Physiological model · Pathological model · Pharmacological assessment · *Caenorhabditis elegans*

4.1 Introduction

Caenorhabditis elegans can be employed to examine the pharmacological effects of various compounds and drugs and underlying pharmacological mechanisms. One of the important supports for this is that the genetic and molecular backgrounds have been well described, and molecular biology-related techniques are also well-established in *C. elegans*. Besides this aspect, the established physiological models and pathological models provide another important support for the pharmacological study in this animal model.

In Chap. 2, we have introduced the compounds and drugs suitable to be examined in *C. elegans*. In this chapter, the introduced physiological models include (1) aging

D. Wang, *Pharmacology in Caenorhabditis elegans*,
https://doi.org/10.1007/978-981-92-5166-7_4

and aging-related changes and (2) stress models. The introduced pathological models include (1) models for metabolic-related disorders, (2) neurodegeneration-related diseases models, (3) channelopathy models, (4) muscle dystrophy models, (5) seizure-like models, (6) addiction models, (7) anxiety-like models, (8) autism-like models, (9) tumor-like models, (10) wound healing models and regeneration models, (11) mtDNA replicative polymerase (POLG)-related disease models, (12) α1-antitrypsin deficiency (ATD) models, (13) pathogen infection models, and (14) immunosuppression models.

4.2 Aging and Aging-Related Changes

4.2.1 Longevity

The longevity refers to extension of *C. elegans* median and/or maximum lifespan. The median lifespan indicates the average survival time of examined population, and the maximum lifespan indicates the survival time of the longest-lived individuals for the examined population. During the lifespan assay, the possible influences from slowed growth and reduced reproduction are suggested to be carefully considered, and difference between lifespan curves are needed to be validated statistically (e.g., log-rank tests). The *C. elegans* longevity control is linked to some conserved signaling pathways (insulin, dietary restriction (DR), mitochondrial function, and germline related) [1–3]. A large number of compounds and drugs have been shown to be able to extend the *C. elegans* lifespan [4–9]. For example, treatment with 0.1–1 mg/mL wheat gluten hydrolysate could extend the lifespan (Fig. 4.1) [10].

4.2.2 Healthspan

Healthspan refers to the normal physiological state during the aging process. Maintenance of healthspan is helpful for *C. elegans* to delay the onset of age-related

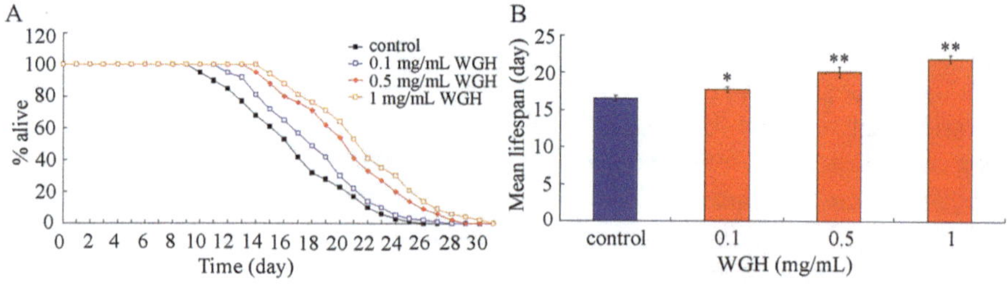

Fig. 4.1 Effects of WGH treatment on lifespan of *C. elegans*. (**a**) Lifespan curves of nematodes treated with WGH [10]. (**b**) Comparison of mean lifespans in nematodes treated with WGH. WGH treatment was performed throughout the lifespan of nematodes from L4-larvae. WGH wheat gluten hydrolysate. Bars represent means ± S.E.M. $^{*}P < 0.05$, $^{**}P < 0.01$

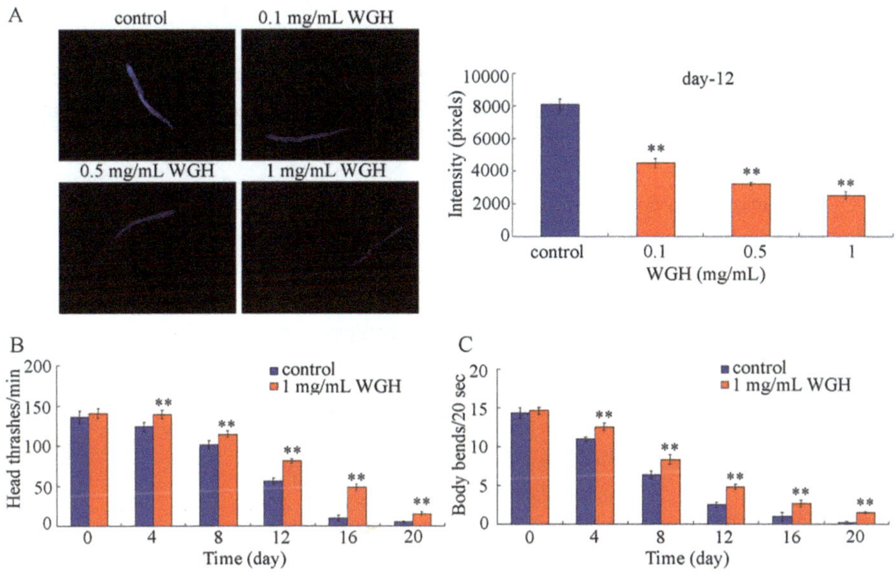

Fig. 4.2 Effects of WGH treatment to decrease intestinal autofluorescence (**a**) and enhance locomotion behavior (**b**, **c**) during the aging process [10]. WGH treatment was performed throughout the lifespan of nematodes from L4-larvae. WGH wheat gluten hydrolysate. Bars represent means ± S.E.M. $^{**}P < 0.01$

functional decline. Compared with quantity of life reflected by lifespan, quality of life can be reflected by healthspan. The functional markers to quantify the healthspan include:

1. Physiological state: Pharyngeal pumping and defecation cycle length.
2. Locomotion: Head thrash, body bend, and swimming or crawling speed
3. Sensory perception (such as chemotaxis and thermotaxis), learning, and memory.
4. Stress response: The response to oxidative stress as indicated by intestinal reactive oxygen species (ROS) and intestinal lipofuscin accumulation, and heat stress.
5. Intestinal barrier integrity to reflect the functional state of intestinal barrier.

The healthspan could be improved by some compounds or drugs, such as ergothioneine, metformin, collagen peptides, nitazoxanide, and *Lactiplantibacillus brownii* strain MH-1 [11–15]. With the wheat gluten hydrolysate as the example, its treatment suppressed the intestinal lipofuscin accumulation and the locomotion inhibition during the aging process (Fig. 4.2) [10].

4.2.3 Immunosenescence

Immunosenescence refers to age-related decline in innate immune response and defense capacity against the pathogen infection. The functional markers to quantify the immunosenescence during the aging process include:

1. Innate immune response as indicated by change in expression of antimicrobial peptide genes.
2. Response of signaling pathway activity reporters, such as DAF-16::GFP and PMK-1::GFP.
3. Response to bacterial or fungal pathogen infection as indicated by change of survival curve and pathogen colonization in body of *C. elegans*.

4.2.4 Reproductive Aging

Reproductive aging refers to functional decline in hermaphrodite's reproductive system, which is a consequence of systemic physiological decline. For *C. elegans*, its peak for reproductive output is at days 1–3 of adulthood. *C. elegans* reproductive span is typically about 5 days, and *C. elegans* will cease progeny productions at adult day 7 or 8. The functional markers to assess the reproductive aging include:

1. Alteration in gonad morphology and loss of somatic gonad integrity.
2. Germline atrophy and apoptosis.
3. Decline in oocyte quality.
4. Sperm depletion reflected by both direct observation of sperm amount and morphology and assessment by mating aged hermaphrodites with young males.
5. Distended and dysfunctional spermatheca where the fertilization will occur.
6. Reduction in reproductive capacity reflected by change of total brood size and dynamic daily brood size.
7. Change of ovulation rate.

With the anticonvulsant medicine ethosuximide as the example of drugs, its treatment could delay the *C. elegans* reproductive aging [16].

4.3 Stress Models

4.3.1 Oxidative Stress

The *C. elegans* oxidative stress models can be employed to screen, validate, and characterize pharmacological effect of compounds or drugs in preventing or mitigating oxidative stress induced damage, thereby identifying candidates with potential therapeutic value for oxidative stress related diseases. The *C. elegans* oxidative stress models can be generated at least through these two aspects:

1. Generation by treatment with chemical pro-oxidants, such as paraquat or juglone to generate superoxide or H_2O_2.
2. Using the mutant strains with altered redox homeostasis, such as *mev-1* and *gas-1* mutants.

Using the prepared oxidative stress models, the following several endpoints are suggested to be assessed after the treatment with pharmacological compounds and drugs:

1. ROS levels measured by fluorescent probes of H_2DCFDA and MitoSOX.
2. Protein oxidation assessed by carbonyl content, and lipid peroxidation reflected by level of malondialdehyde (MDA).
3. Activity of antioxidant enzymes (SOD, catalase, and glutathione peroxidase) and expression of related genes (such as GST-4::GFP, SOD-3::GFP, SKN-1::GFP, and DAF-16::GFP).
4. Survival and lifespan.
5. Healthspan reflected by locomotion and pharyngeal pumping during the aging process.

Using the established *C. elegans* oxidative stress models, the suggested examined pharmacological targeted signaling pathways can include SKN-1/Nrf2 signaling, DAF-16/FOXO signaling, and mitochondrial function related signals (such as component genes of electron transfer chain). A large number of compounds and drugs have the beneficial effect against oxidative stress in nematodes [17–22]. With the wheat gluten hydrolysate as the example, its treatment could inhibit paraquat induced oxidative stress and extend the lifespan of paraquat exposed nematodes (Fig. 4.3) [10].

4.3.2 Heat Stress and Cold Stress

The heat stress model can be generated by shifting the nematodes from 20 °C to 35–37 °C for 2–8 h. The functional markers to assess the heat stress can include:

1. Survival and lifespan.
2. Stress response activation reflected by change of HSF-1::GFP and heat shock proteins (such as HSP-16.2 and HSP-70).
3. Proteostasis assessment.

Using the established *C. elegans* heat stress model, the suggested examined pharmacological targeted signaling pathways can include HSF-1 and proteostasis related signals. The resistance to heat stress could be enhanced by icariin, vanillic acid, deuterohemin peptide, and *Jujubae Fructus* extract [23–26]. With the wheat gluten hydrolysate as the example, its treatment could increase the lifespan of heat stress exposed nematodes (Fig. 4.3) [10].

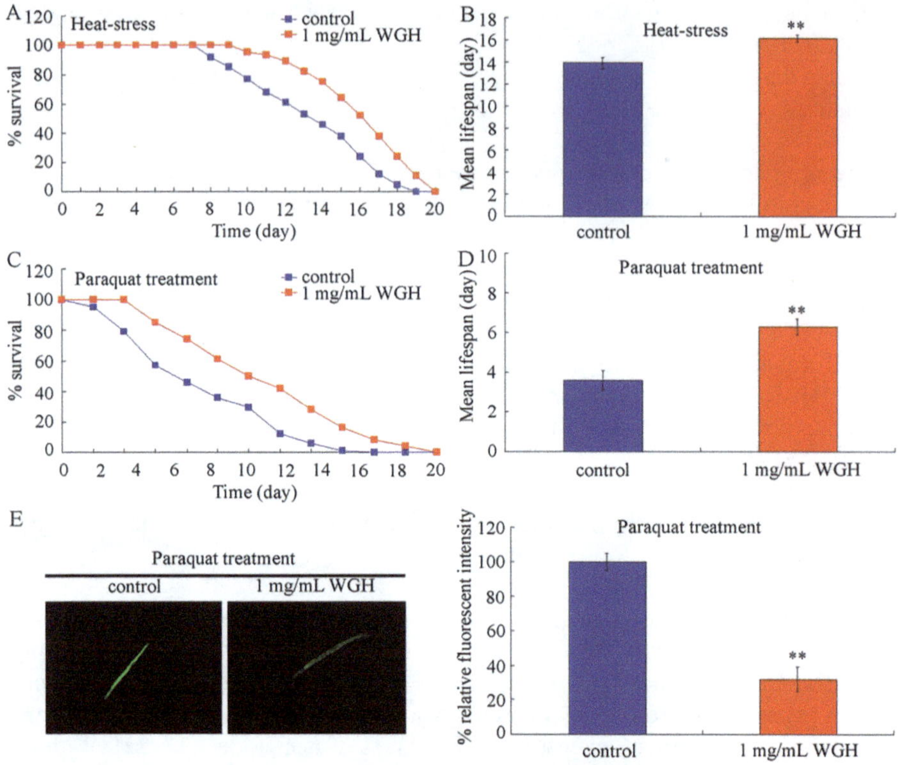

Fig. 4.3 Heat-stress and oxidative stress resistance of WGH-treated *C. elegans* [10]. (**a**) Lifespans of heat-stress exposed nematodes pretreated with or without 1 mg/mL of WGH. (**b**) Mean lifespans of heat-stress exposed nematodes pretreated with or without 1 mg/mL of WGH. (**c**) Lifespans of paraquat exposed nematodes pretreated with or without 1 mg/mL of WGH. (**d**) Mean lifespans of paraquat exposed nematodes pretreated with or without 1 mg/mL of WGH. (**e**) Intestinal ROS production in paraquat exposed nematodes pretreated with or without 1 mg/mL of WGH. Nematodes pre-treated with 1 mg/mL of WGH for 48-hr were transferred to the 35uC condition for 16-hr or 2 mmol/L of paraquat for 6–hr. WGH wheat gluten hydrolysate. Bars represent means ± S.E.M. $^{**}P < 0.01$

The cold stress models can be generated by shifting the nematodes from 20 °C to 4–10 °C for 1–24 h, or allowing the nematodes to be maintained at 10 °C or 15 °C. The suggested functional markers to assess the cold stress can include:

1. Survival and lifespan.
2. Development (such as the time to adulthood).
3. Motility and locomotion.
4. Membrane integrity assessed by lipophilic fluorescent dye staining or sensitivity to detergents.

4.3.3 Hyperosmotic Stress

Hyperosmotic stress models can be generated by transferring the nematodes to 400–500 mM NaCl, 400 mM sorbitol, or 200 mM fructose. The suggested functional markers to assess the hyperosmotic stress can include:

1. Survival and lifespan.
2. Avoidance to high-osmolarity zones in choice assay.
3. Motility and locomotion.
4. Developmental arrest (such as the L1 arrest).
5. Morphological changes (such as the body shrinkage or swelling).

The resistance to osmotic stress could be induced by treatment with brazilin, verbascoside, ursolic acid, and açaí aqueous extract [27–30].

4.3.4 UV Stress

UV stress models can be generated by treating the nematodes on agar plates or in liquid or suspension under the UV irradiation with precise energy dosing (J/m^2). The functional markers to assess the UV stress can include:

1. Survival and lifespan.
2. DNA damage (such as the induction of replication blocks or mutations).
3. DNA repair and genomic stability.
4. Reduction in brood size.
5. Locomotion, sensory perception, and related behaviors.

The resistance to UV stress could be caused by treatment with ferulic acid, triptolide, and *Curcumae Rhizoma* extract [31–33]. With the vitamin E as the example of compounds, UV stress induced deficit in memory for thermosensation could be retrieved by vitamin E treatment (Fig. 4.4) [34].

4.3.5 Hypoxia Stress

Hypoxia stress model can be generated by treating the nematodes with $<0.1\%$ O_2 for 24 h or less to screen strong protective drugs. Hypoxia stress model can also be generated by treating the nematodes with moderate hypoxia, such as 1–5% O_2 for days. HIF-1 (Hypoxia-Inducible Factor-1) is the central regulator for hypoxia stress. These hypoxia stress models are helpful to screen compounds and drugs to alleviate the damage of hypoxia population. The functional markers to assess the hypoxia stress can include:

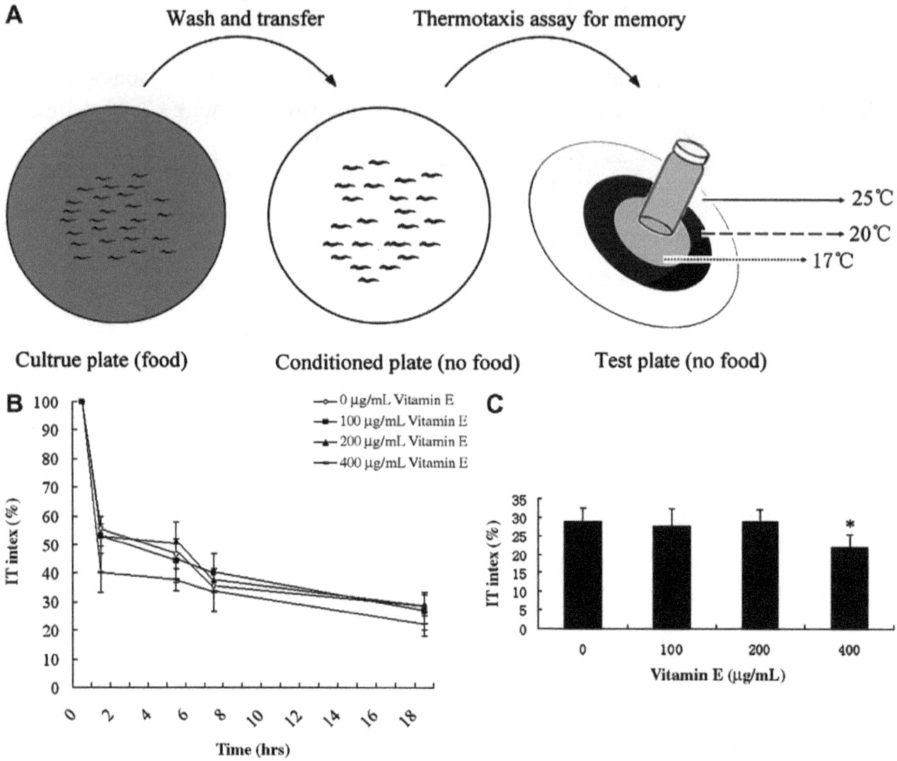

Fig. 4.4 Effects of vitamin E administration on the memory for thermosensation [34]. (**a**) Thermotaxis assay model for memory. (**b**) The extinctions of the association (food at 20 °C) of control and vitamin E treated animals. (**c**) Comparison of the extinctions of control and vitamin E treated animals at the time interval of 18-h. IT isothermotracking. Data are expressed as means ± SD. $^{*}P < 0.05$

1. Survival and lifespan.
2. Locomotion and sensory perception.
3. Growth and reproductive capacity.
4. Expression of marker HIF-1::GFP.

It was reported that the hypoxia damage could be reduced by *Velvet Antler* methanol extracts [35].

4.3.6 Simulated Microgravity Stress and Ionizing Radiation

The simulated microgravity stress model can be generated by treating the nematodes suspended in liquid buffer with rotary system. In *C. elegans*, simulated microgravity stress caused multiple toxicities, such as intestinal damage, oxidative stress, and locomotion inhibition [36–40]. Besides this, the ionizing radiation models can be generated by treating the nematodes with high doses or chronic low doses of certain irradiation, such as gamma rays (^{60}Co or ^{137}Cs sources), heavy ions, and X-rays [41]. The established simulated microgravity stress and ionizing radiation models are helpful to screen compounds and drugs for maintaining the human health during the spaceflight. The suggested functional markers to assess the simulated microgravity stress and ionizing radiation can include:

1. Lifespan and healthspan.
2. Intestinal oxidative damage and mitochondrial function.
3. Germline development and reproductive capacity.
4. Locomotion, sensory perception, and muscle function.

Besides these endpoints, the lethality, DNA damage, and mutagenesis are suggested to be further examined after the ionizing radiation treatment. The ionizing radiation-induced damage could be reduced by resveratrol, n-butanol and ethyl acetate fractions from oolong tea, and *Acanthopanax senticosus* polysaccharide [42–44]. In addition, Hong Huang Tang was reported to mitigate microgravity-induced oxidative stress [45].

4.3.7 Pollutants Toxicity

In *C. elegans*, for a certain pollutant, if its toxicity at different aspects and underlying molecular mechanisms have been well determined and described, the corresponding pollutant toxicity models can be generated. The 6-PPD quinone (6-PPDQ) is a typical emergent contaminant, and was selected as the example of pollutants. In *C. elegans*, exposure to 6-PPDQ caused intestinal damage, germline apoptosis and suppressed reproductive capacity, locomotion inhibition and damage on neuronal development, inhibition in longevity and healthspan, disruption in several aspects of biological metabolisms, and even the induction of transgenerational toxicity, and these aspects of damage induced by 6-PPDQ was under the control at least of insulin and mitochondrial related signals and some nuclear hormone receptors [46–71]. Moreover, the 6–PPDQ caused intestinal oxidative stress, damage on intestinal barrier, reduced reproductive capacity, and inhibited locomotion could be alleviated by paeoniflorin treatment (Fig. 4.5) [72].

4.4 Models for Metabolic-Related Disorders

4.4.1 Glucose Toxicity

The glucose toxicity models can be generated by treating the nematodes on nematode growth medium (NGM) plates added with D-glucose (typically at concentration of 20–100 mM). During this aspect of assay, the osmotic control (e.g., mannitol or sorbitol at equimolar concentration) is suggested to be set up in order to distinguish

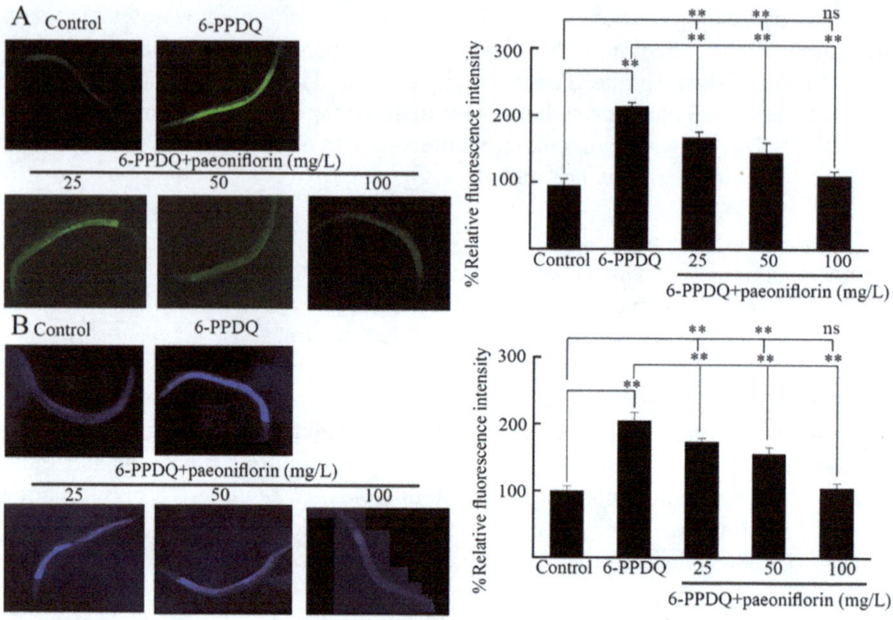

Fig. 4.5 Effect of paeoniflorin treatment on intestinal ROS production (**a**) and intestinal autofluorescence (**b**) in 6-PPDQ exposed nematodes [72]. Exposure concentration of 6-PPDQ was 100 μg/L. $^{**}P < 0.01$. ns no significant difference

the glucose toxicity from the general osmotic stress. The suggested functional markers to assess the glucose stress can include:

1. Lifespan and healthspan.
2. Mitochondrial function.
3. Intestinal oxidative stress and related markers.
4. Locomotion, sensory perception, learning, and memory.

The examined glucose toxicity in *C. elegans* is somewhat relevant to age-related diseases. In *C. elegans*, treatment with carnitine, diosgenin, genistein, green tea extract, and *Ceiba speciosa* bark extract could provide the beneficial effect against the glucose toxicity [73–77]. With the paeoniflorin as the example of compounds, its treatment could reduce high glucose induced lifespan reduction by modulating insulin signaling pathway (Fig. 4.6) [78].

4.4.2 Lipid Accumulation and Obesity-Related Phenotype

The lipid accumulation models can be generated by dietary feeding: (1) feeding with high-glucose diets (≥ 2% glucose in NGM), (2) feeding with high-fat diets, and (3) feeding with bacterial food sources with altered lipid composition. Besides this

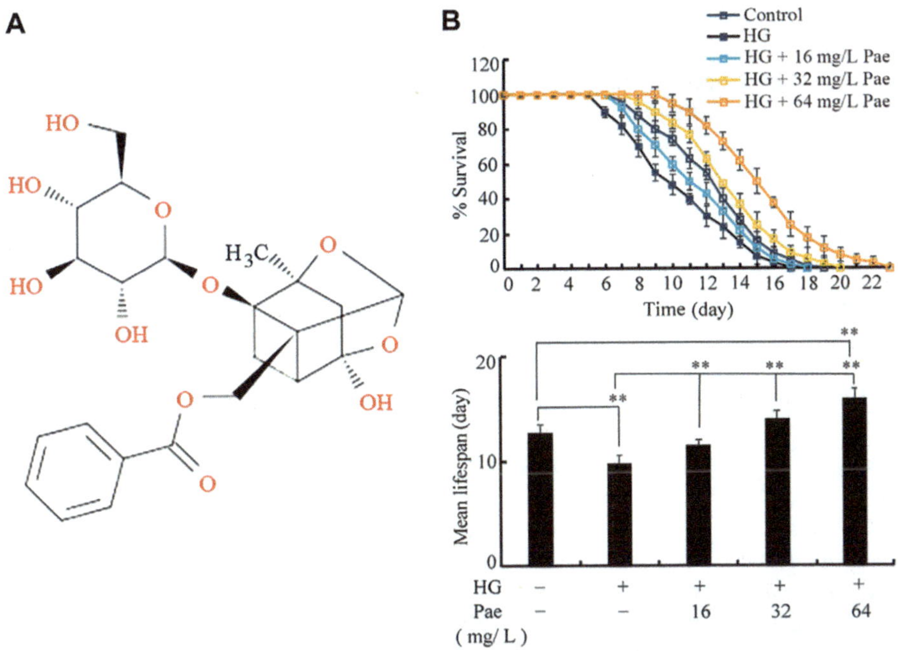

Fig. 4.6 Effect of paeoniflorin administration in reducing glucose toxicity [78]. (**a**) Chemical structure of paeoniflorin. (**b**) Effect of paeoniflorin administration on lifespan in glucose treated nematodes. Lifespan curve of HG was significantly ($p < 0.01$) different from control. Lifespan curves of HG + 16 mg/L Pae, HG + 32 mg/L Pae, and HG + 64 mg/L Pae were significantly ($p < 0.01$) different from HG. Lifespan curve of HG + 64 mg/L Pae was significantly ($p < 0.01$) different from control. HG high glucose (50 mM); Pae paeoniflorin. $^{**}p < 0.01$

aspect of methods, some of the loss-of-function mutants can also be considered as the lipid accumulation and/or the obesity related models. The considered loss-of-function mutants at lease include: (1) *daf-2(e1370)*, insulin receptor mutant with constitutive fat accumulation; (2) *tub-1*, Tubby homolog mutant with obesity related phenotype; and (3) *maoc-1*, enoyl-CoA hydratase gene mutant with increased lipid droplets. The suggested functional markers or methods to assess the lipid accumulation can include:

1. Oil Red O staining to detect neutral lipid.
2. Nile Red staining and Sudan Black staining to detect the lipid droplets.
3. BODIPY staining to detect the fatty acid uptake and storage.
4. Triglyceride quantification.
5. GC-MS assay for detecting the fatty acid compositions.
6. Expression of enzyme genes governing fatty acid synthesis and β-oxidation.

Accompanied with assay of these markers, some secondary phenotypes associated with altered lipid accumulation or related signals are suggested to be further examined. These secondary phenotypes can include: (1) longevity, (2) fecundity, (2) locomotory, (4) pharyngeal pumping rate, and (5) stress resistance. The suggested pharmacological targets associated with altered lipid accumulation can include: (1) insulin signaling, (2) AMPK activation, (3) PPAR agonists (via NHR-49/NHR-80), (4) SREBP/SBP-1, and (5) lipase activation.

It was reported that treatment with betulinic acid, rutin, piceatannol, esculetin, swertiamarin, and fermented barley extracts have the function in reducing lipid accumulation [79–84]. Additionally, the glucose induced lipid accumulation could be further reduced by tetrahydrocurcumin, curcumin, cycloastragenol, marine macroalgae extract, and polyphenols from *Mallotus peltatus* [85–89].

4.4.3 Glycogen Accumulation

The *agl-1* is a homologue to human AGL genes. The *agl-1* mutant with increased glycogen accumulatio in intestine, hyperdermis, and muscle was suggested as the genetic model of glycogen accumulation [90]. This model may be helpful to screen compounds and drugs to treat the Glycogen Storage Disease Type III (GSDIII) caused by human AGL mutations. Using this animal model, 25 candidate compounds were identified to improve the phenotypes in *agl-1* mutants [90]. The suggested functional endpoints to assess the glycogen accumulation can include:

1. Glycogen quantification.
2. Glycogen staining.
3. Lifespan.
4. Motility.
5. Reproductive capacity.

4.4.4 Mitochondrial Disorders

The mutation of genes encoding components of complexes on electron transport chain can be employed as the genetic models to reflect the mitochondrial disorders. The corresponding electron transport chain complex mutants include:

1. *gas-1* (complex I/subunit) has phenotypes of low ATP, slow growth and pharyngeal pumping, reduced fecundity, and hypersensitivity to oxidative stress. This mutant is suggested to be used to reflect the MELAS/Leigh syndrome (complex I deficiency). In this genetic model, N-acetylcysteine and vitamin E were found to be able to rescue the deficit in lifespan and cellular oxidative stress phenotypes [91].
2. *mev-1* (complex II/succinate dehydrogenase cytochrome b subunit) has phenotypes of increased ROS, hypersensitivity to oxidative stress, reduced lifespan, and inhibited locomotion. This mutant is suggested to be used to reflect the Leigh-like syndrome. Using this genetic model, organoselenotriazoles,

organoruthenium(II) complexes, and manganese could attenuate the oxidative damage and/or lifespan reduction in *mev-1* mutants [92–94].

3. *isp-1* (complex III/ Rieske iron-sulfur protein) has phenotypes of reduced oxygen consumption, oxidative stress resistance, slow development, and long lifespan. This mutant is suggested to be used to reflect the GRACILE syndrome, a mitochondrial encephalopathy. Using this genetic model, the compounds or drugs acting as the electron transport chain modulators might be screened. The mitochondrial mutant *isp-1* showed the drug-resistant property to hemiasterlin, an antimitotic drug [95].
4. *clk-1* (Coenzyme Q biosynthesis enzyme) has the phenotypes of oxidative stress resistance, long lifespan, delayed development, and slow rhythmic behaviors. This mutant is suggested to be used to reflect the primary CoQ10 deficiency. Using this genetic model, 190 compounds were identified to show active property on this mutant [96]. In addition, the activity of chlorogenic acid in extending lifespan was also affected by *clk-1* mutation [97].

The suggested functional endpoints to assess the mitochondrial disorders can include:

1. Oxygen consumption rate (OCR).
2. ATP content.
3. Mitochondrial reactive oxygen species.
4. Activity of mitochondrial complexes.
5. Expression of genes encoding components of mitochondrial complexes.
6. Mitochondrial membrane potential.
7. Lethality, lifespan, and healthspan.
8. Locomotion and growth.

4.5 Neurodegeneration-Related Diseases Models

4.5.1 Alzheimer' Disease (AD) Models

In *C. elegans*, there are several widely used AD models. The first group are the Aβ expression models, which can be generated by expression of $A\beta_{1-42}$ in body wall muscle or neurons. The suggested functional endpoints for assessing corresponding strains can include:

1. Aβ deposits.
2. Neurodegeneration related phenotypes.
3. Progressive paralysis.
4. Lethality and lifespan.
5. Sensory perception, learning, and memory.
6. Oxidative stress and mitochondrial damage.

In these Aβ expression models, diphenyl-diselenide, oxoisoaporphine alkaloid derivative 8–1, monascin, dauricine, 2-arylethenylquinoline derivatives, and mitochondria-targeted MitoQ showed the neuroprotective effects by suppressing amyloid-beta peptide [98–104].

The second groups are Tauopathy models, which can be generated by expression of wild-type and mutant Tau in the neurons. The suggested functional endpoints for assessing corresponding strains can include:

1. Tau aggregation.
2. Neurodegeneration-related phenotypes.
3. Progressive paralysis.
4. Lethality and lifespan.
5. Sensory perception, learning, and memory.
6. Oxidative stress and mitochondrial damage.

In these Tauopathy models, curcumin, tetrahydropyrimidine analogue, *Dendrobium officinale* phenolic extract, *Reineckia carnea* ether fraction, 2-thioxo-1,2,3,4-tetrahydropyrimidines, and antagonism of dopamine D2 receptor exhibited the function against the neurodegeneration-related diseases [105–110].

4.5.2 Parkinson Disease (PD) Models

In *C. elegans*, there are also several widely used PD models. The first group are the α-Synuclein aggregation models, which were generated by preparing pan-muscular and pan-neuronal α-synuclein expression strains [111]. The suggested functional endpoints for assessing corresponding strains can include:

1. α-Synuclein inclusion formation.
2. Dopaminergic neurodegeneration related phenotypes.
3. Progressive paralysis.
4. Lethality and lifespan.
5. Synaptic development and function.
6. Sensory perception, learning, and memory.
7. Oxidative stress and mitochondrial damage.

In these α-Synuclein aggregation models, chrysin, policosanol, botulin, palmitic acid, chlorogenic acid, ergothioneine, rose essential oil, and probiotic *Bacillus subtilis* exhibited neuroprotective role by suppressing α-Synuclein aggregation [112–119].

The second group are dopaminergic neurons labeled models, which can be generated by expression of *dat-1* promoter fused with GFP. In these models, the development and neurodegeneration of DA neurons (CEP, ADE, and PDE pairs) can be determined. Using these models, diterpene glycosides and *Sorbus alnifolia* could protect the dopaminergic neurodegeneration [120, 121].

The third group are the chemical induced models, which are prepared by treating the nematodes with neurotoxic 6-hydroxydopamine (6-OHDA), 1-methyl-4-phenyl-pyridinium (MPP^+), and rotenone to kill the dopaminergic neurons [122]. The suggested functional endpoints for the assessment can include:

1. Dopaminergic neurodegeneration related phenotypes.
2. Progressive paralysis.
3. Lethality and lifespan.
4. Sensory perception, learning, and memory.
5. Oxidative stress and mitochondrial damage.

Using these models, galangin, damaurone D, phenothiazines, tricetin, neurogrit gold, N-gamma-(L-glutamyl)-L-selenomethionine, xyloketal B, bassiamide A, withaferin derivatives, maackiain, and carpesii fructus extract were proven to have the potential function in treating the PD [123–134].

The fourth group are the genetic models with the help of genetic mutants, such as *pink-1* and *pdr-1*. The suggested functional endpoints for the assessment can include:

1. Dopaminergic neurodegeneration-related phenotypes.
2. Mitochondrial function.
3. Mitophagy.
4. Progressive paralysis.
5. Lethality and lifespan.
6. Sensory perception, learning, and memory.

Using these genetic models, hederagenin, and peiminine were shown to have the neuroprotective functions [135, 136].

4.5.3 Huntington's Disease (HD) Models

There are several HD models in *C. elegans*. The first group is the body-wall muscle aggregation models by expressing polyQ::YFP/GFP fusion proteins under muscle-specific promoters (such as P*myo-3*). The corresponding strains have phenotypes of discrete, fluorescent aggregates in muscle and progressive age-dependent paralysis. Using these strains, quinic acid, rutin, perillaldehyde, meclizine, and garana extract may have the beneficial effect to counteract the HD phenotypes [137–141].

The second group are the neuronal-specific toxicity models, which were generated by expressing mHTT (exon1 or longer fragments) in whole nervous system or certain neuronal subtypes, such as the GABAergic motor neurons under P*unc-47* or sensory neurons under P*osm-10*. The third group are the transgenic strains expressing wild-type or mutant human HTT exon1 or longer N-terminal fragment of HTT. Using these strains, the clofazimine, selenium nanoparticles, and xyloketal-derived small molecules were proven to exhibit the neuroprotective effects [142–144].

The suggested functional endpoints for the assessment can include:

1. Protein aggregation as indicated by the formation of insoluble, fluorescently tagged mHTT inclusions.
2. Neuronal vulnerability, such as the toxicity on GABAergic neurons and sensory neurons.
3. Expression of genes involved in autophagy, ubiquitin-proteasome, chaperones.
4. Progressive paralysis.
5. Lethality and lifespan.
6. Sensory perception, learning, and memory.

4.5.4 Amyotrophic Lateral Sclerosis (ALS) Models

The pathological hallmarks for ALS are mainly reflected by protein aggregation, motor neuron degeneration, and locomotor dysfunction. In *C. elegans*, there are several models raised for assessing the ALS disease. The first group is the TDP-43 models, which were generated by expressing human wild-type or mutant (e.g., A315T, G298S) TDP-43 in nervous system or motor neurons. The phenotypes of these models include: (1) age-dependent protein aggregation (TDP-43 cytoplasmic inclusions), (2) progressive locomotor deficits (such as altered thrashing and crawling speed), (3) motor neuron degeneration, and (4) axon guidance defects. Using these strains, lycorine, methylene blue, and Maple syrup were identified as the candidate compounds and drugs for possible ALS intervention [145–147].

The second group is the FUS models, which were generated by expressing human wild-type or mutant (e.g., R521C, P525L) FUS in nervous system or motor neurons. The phenotypes of FUS models are similar to those of TDP-43 models. Using these strains, for the aim of ALS intervention, methylene blue was identified [148].

The third group is the Dipeptide Repeat (DPR) models, which were generated by expressing toxic DPRs (most commonly poly-GR or poly-PR) from the hexanucleotide repeat expansion [149]. The phenotypes of the corresponding strains include the severe, rapid-onset locomotor paralysis and synaptic dysfunction.

The fourth group is the SOD1 models, which were generated by expressing human mutant SOD1 (e.g., G85R, G93A) in neurons. The phenotypes of these strains include protein aggregation, motor deficits, fecundity defects, and endoplasmic reticulum (ER) stress [150–152].

4.5.5 Machado-Joseph Disease (MJD/SCA3) Models and Friedreich Ataxia (FRDA) Models

In *C. elegans*, the MJD/SCA3 models can be generated by expressing human mutant ataxin-3 with an expanded polyglutamine (polyQ) tract in the MJD1/ATXN3 gene.

The polyglutamine (polyQ) toxicity and selective neuronal vulnerability could be detected in these strains [153–155]. The suggested functional endpoints for the assessment can include:

1. PolyQ aggregation.
2. Motility deficits.
3. Neurodegeneration and neuronal dysfunction.
4. Proteostasis disruption.
5. Lifespan reduction.
6. Chemosensory deficits.

Using these strains, some small compounds (alfacalcidol, chenodiol, cyclophosphamide, fenbufen, and sulfaphenazole), aripiprazole, epigallocatechin-3-gallate, tetracycline, Hyptis spp. leaf extracts, and *Lavado cocoa* flavanols were identified to be helpful for inhibiting the ATXN3 toxicity [156–160].

FRDA is a neurodegenerative disease associated with GAA repeat expansion in the FXN gene and lack of mitochondrial protein frataxin [161–163]. In *C. elegans*, there are two groups of models for FRDA. The first group is the tissue-specific (such as neuronal- or muscle-specific) RNAi or mutation of *frh-1* encoding the Frataxin homologue. The second group is strains which can be generated by expressing human FXN with pathological GAA repeat expansions or human mutant FXN. The suggested functional endpoints for the assessment can include:

1. Mitochondrial dysfunction.
2. Neuronal degeneration.
3. Oxidative stress sensitivity.
4. Locomotor deficits and altered pharyngeal pumping.
5. Lifespan reduction.
6. Developmental delay.

Using these strains, flavin adenine dinucleotide was shown to have the beneficial effect in rescuing phenotypes caused by frataxin deficiency [164].

4.5.6 Adult Neuronal Ceroid Lipofuscinosis (ANCL) Models

The ANCL is characterized by lysosomal-autophagic dysfunction and abnormal lipofuscin-like storage material accumulation. The mutants with mutation of genes encoding homologues of ANCL-causing genes (most commonly DNAJC5/CSPα and CLN3) were employed as the ANCL models [165]. The first is the mutant of *dnj-14* (the ortholog of DNAJC5). The second is the mutant of *cln-3* (the ortholog of CLN-3) [166]. The suggested functional endpoints for the assessment can include:

1. Intestinal lipofuscin accumulation.
2. Lysosomal dysfunction.

3. Neurodegeneration.
4. Locomotion.
5. Lifespan.
6. Hypersensitivity to proteotoxic stress and oxidative stress.

Using these strains, flunarizine, ethosuximide, and resveratrol were identified as the potential therapeutics [165, 167, 168].

4.6 Channelopathy Models

The channelopathy-related models are relevant to the phenotypes of neuronal hyper-excitability, muscle dysfunction, or cellular homeostasis defects caused by the formed aberrant ion flux. The channelopathy models can be classified into two groups. One group can be generated by expressing human wild-type and mutant ion channel genes. Another group is the RNAi or mutation of genes encoding human ion channel homologues. The following several types of ion channels can be considered to generate corresponding channelopathy models.

1. Voltage-gated potassium channel models:
   For example, mutation of *mps-4*, homologue of human *kcne1*, caused abnormal repolarization to provoke arrhythmia like phenotype [169]. Moreover, using transgenic model expressing modified human ether-a-go-go-related gene (hERG), prostratin and ingenol-3,20-dibenzoate could rescue the egg-laying and locomotive defects [170].
2. Voltage-gated sodium channel models:
   Missense mutation of NALCN was linked to infant CLIFAHDD syndrome [171]. For example, the transgene of NALCN variant was pathogenic to *Caenorhabditis elegans*, and caused the phenotypes (such as hypercontraction and uncoordinated movement) similar to the known gain-of-function allele [171].
3. Voltage-gated calcium channel models:
   For example, treatment with both gabapentin (GBP) and pregabalin (PGB) impeded the nocifensive responses by interacting with voltage-gated calcium channels [172]. In addition, mutations of *unc-2* and *unc-36* also exhibited the damaged heat avoidance [172].
4. TRP channel models:
   First of all, the *C. elegans* TRP channel models are helpful for pharmacological study to screen and design the compounds and the drugs in reducing the pain, since these channels-mediated calcium transients in certain defined nociceptor neurons were shown to be required for avoidance behavior [173, 174]. Mutation of TRPC channel genes could cause the defects in the response to nicotine, which could be rescued by transgene of human TRPC channel [175].

The suggested functional endpoints for the assessment can include:

1. Paralysis, locomotion, and sensory perception.
2. Synaptic activity and function.
3. Feeding behavior and muscle development and function.
4. Neurodegeneration.
5. Electrophysiological assay for activity of neurons and body wall muscle.

4.7 Muscle Dystrophy Models

Besides the elucidation of underlying mechanisms, one important aim of establishment of muscle dystrophy models is to screen compounds and drugs having functions to stabilize muscle cell membrane, improve sarcomere integrity, enhance muscle force, and mitigate degeneration [176]. The muscle dystrophy models include:

1. Dystrophin-DGC complex models:
   The *C. elegans* orthologue of Dystrophin is DYS-1, orthologue of β-dystroglycan is STN-1, and orthologue of Sarcoglycan is SGN-1. Similarly, one group of models can be generated by expressing human wild-type and mutant Dystrophin-DGC complex genes [177]. Another group of models is the RNAi or mutation of genes encoding human Dystrophin-DGC complex genes [178, 179]. Using these strains, prednisone, hydrogen sulfide, febuxostat, and mitochonic Acid 5 were shown to reduces muscle degeneration, sulfur amino acid, and improve movement disorder [180–184].
2. Sarcomeric Z-disc and costamere models:
   The *C. elegans* orthologue of Obscurin-like/Titin Kinase is UNC-89, orthologue of Kindlin-2/Mig-2 is UNC-112, and orthologue of Vinculin/α-actinin is DEB-1. With the aid of *unc-112* mutant, calpains was helpful for maintenance of integrin attachment complex in muscle [185].
3. Nuclear envelope (Emery-Dreifuss) models:
   The *C. elegans* orthologue of Lamin is LMN-1, and its mutants show the nuclear envelope abnormalities and progressive muscle degeneration [186].
4. Duchenne muscular dystrophy (DMD) models:
5. The *C. elegans* orthologue of MyoD is HLH-1, which acts as the master regulator of muscle development and cell fate [187]. The *hlh-1* mutant and even the *dys-1; hlh-1* double mutant were suggested to be employed as the muscle dystrophy models [188, 189].
6. Spaceflight induced model: Both spaceflight and microgravity treatment are potentially cause the phenotypes of muscle dystrophy [190].

The suggested functional endpoints for the assessment can include:

1. Muscle development and differentiation.
2. Muscular strength.
3. Muscle cell death.

4. Locomotion.
5. Mitochondrial network integrity.

4.8 Seizure-Like Models

The established seizure models should exhibit the phenotypes of paroxysmal, synchronous, and excessive neuronal firing leading to stereotypic convulsive motor patterns. There are several groups of seizure–like models raised in *C. elegans*. The first group is the chemoconvulsant-induced seizure models:

1. Pentylenetetrazole (PTZ) exposure: PTZ is GABA receptor antagonist to induce the dose-dependent, reversible paralytic seizure state [191]. Using this model, chlorothymol and 2-[(4-methoxyphenylselenyl)methylthio]-5-phenyl-1,3,4-oxadiazole (MPMT-OX) were identified as the potential anticonvulsant [192, 193].
2. Aldicarb exposure: Aldicarb is the acetylcholinesterase inhibitor. To model the cholinergic-induced seizures, aldicarb is used to induce the hypercontraction paralysis via formation of excess acetylcholine [194]. Using this model, the caffeine was suggested to be useful for controlling the seizure [195].

The second group is the genetic models:

1. *unc-49* mutant: UNC-49 is the GABA receptor, and its mutation causes constitutive spastic paralysis. The ethosuximide was identified to be able to inhibit the convulsions induced by PTZ exposure in *unc-49* mutants [196].
2. *unc-18* mutant: UNC-18 is the orthologue of STXBP1, and its mutation induces a paralyzed strain [197].
3. *dnj-17* mutant: DNJ-17 is the orthologous to human DNAJA5. The double mutation of *dnj-1* and *acr-2* was raised to be considered as the seizure–like model, and double mutation of these two genes caused the more severe convulsion phenotype in conjunction with absence of GABA [198].

The suggested functional endpoints for the assessment can include:

1. Paralysis.
2. Convulsive motor patterns: "Curling" or "C-shaped" postures.
3. Hyperkinesia: Bursts of exaggerated, non-propagative thrashing.
4. Electrophysiological recoding the activity and prolonged depolarizations of motor neurons.
5. Time to seizure onset and time to recovery after removal of the convulsant.

4.9 Addiction Models

In *C. elegans*, drug tolerance and conditioned preference can be used for the study of addiction. The addiction models can be generated by long-term exposure or repeated exposure to addiction drugs. After conditioning, the nematodes can show preference for the addiction drug-paired cue (e.g., specific odor or temperature). The following several aspects of addiction behavior and related behaviors can be performed in *C. elegans* for the further pharmacological study:

1. Opioid addiction model:
   The opioid addiction model has been established [199, 200], and this opioid addiction behavior is induced by acting on opioid-like receptor NPR-17 and under the control of microRNAs [199, 201]. With the aid of this model, it is suggested that compounds can be screened and identified to treat the human addiction [202].
2. Nicotine addition:
   After the treatment with nicotine, the nicotine motivated behavior and nicotine adaptation could be observed [203, 204]. The nicotine addiction behavior is induced by acting on PKC-dependent nicotinic receptor and altering microRNA expression profiles and regulated by small heat shock proteins [204–206].
3. Ethanol addition:
   The ethanol addition model was early established, and considered to be helpful for identifying compounds in treating ethanol addiction [207, 208]. The ethanol addiction behavior is regulated by some molecular signals, such as small heat shock proteins, JNK signal and BK channel [206, 209, 210].
4. Withdrawal performance:
   After addiction drug exposure, withdrawal may lead to measurable phenotypes, such as hyperactivity, slowed locomotion, and changes in egg-laying [211]. The withdrawal performance can be regulated by SLO channel and TRP-family channels [175, 211]. Some small molecule modulators of sigma2R/Tmem97 were identified to inhibit the ethanol withdrawal-induced behaviors [212].

The suggested functional endpoints for the assessment can include:

1. Locomotion behaviors.
2. Egg-laying.
3. Electrophysiological recoding the activity of ion channels.
4. Learned preference for the paired cue.

4.10 Anxiety-Like Models

Anxiety-like behaviors can be reflected by the exaggerated avoidance or risk-assessment behavior in response to potentially threatening. In the anxiety-like models, the nematodes respond to a repulsive odorant by first backing up and then can continue to show the forward movement to the repulsive odorant [213, 214]. This anxiety-like or decision-making behavior is controlled by the AIB interneurons [213]. Treatment with combination of heat-treated *Bifidobacterium longum* and *Lacticaseibacillus rhamnosus* might be helpful to inhibit this anxiety-like behavior [215].

4.11 Autism-Like Models

Similarly, the possible autism-like models can be classified into two groups. One group can be generated by expressing human wild-type and mutant autism associated genes. Another group is the mutants of genes encoding human autism associated gene homologues.

1. Synaptic adhesion molecules: such as NRX-1 (Neurexin), NLG-1 (Neuroligin), and SHN-1 (Shank).

The *nrx-1* mutants have the deficits in synaptic connection, promoting spine outgrowth, exploratory capacity, social feeding behavior, sinusoidal postural movements, retrograde synaptic inhibition, foraging activity, and gentle touch response [216–220]. Some of the defects in *nrx-1* mutants could be rescued by human alpha- and beta-NRXN1 isoforms [216].

The *nlg-1* mutants have deficits in retrograde synaptic inhibition, basal slowing response, enhanced slowing response, sensory behaviors, sensory processing, habituation learning, and sensitive to oxidative stress [218, 220–223]. Some of the defects in *nlg-1* mutants were rescued by human and rat NLGN1 genes [224].

The *shn-1* mutants have deficits in defecation rhythm, pharyngeal pumping, SLO-2 currents, and channel clustering [225–227].

2. Chromatin remodeling signals: such as CHD-7 (CHD family)

The *chd-7* mutants have GABAergic and behavioral defects and impaired habituation learning [223, 228], and some of the defects in *chd-7* mutants could be restored by small molecule ephedrine [228].

3. Synaptic assembly signals: such as RPM-1 (MYCBP2), ANC-1 (Nesprin), GLR-1/2 (glutamate receptors), AVR-15 (chloride channel subunit), RBM-26 (RBM26/27), DAF-18 (PTEN), and EGL-19/UNC-2 (CaV1/CaV2 channels).

The *rpm-1* mutants have the defects of axon termination, synapse formation, and habituation [229, 230]. The *anc-1* mutants have deficits in axon termination and synapse formation [230]. The ANC-1 functioned through mitochondria to polarize axon growth and acted downstream of the RPM-1 [230, 231].

The *glr-1* and *glr-2* mutants have the deficit in social feeding behavior, and GLR-1/2 functioned independently of NRX-1 and NLG-1 to modulate the social feeding behavior [220].

The *avr-5* mutants have the defects in glutamatergic neurotransmission and social feeding behavior [220, 232].

The *rbm-26* mutants have the axonal guidance defects, which was linked to the dysfunction of mitoribosomal assembly factor MALS-1 (MALSU1) [233].

Mutation of *daf-18* impaired the GABAergic neurodevelopment and function, and this defect could be ameliorated by β-hydroxybutyrate, a ketone body [234].

The mutation of *egl-19* encoding a CaV1 channel caused the defect of axon targeting and behavior [235]. In addition, mutation of *unc-2* encoding a CaV2 channel resulted in the deficits in retrograde synaptic inhibition and GABA neuron remodeling [236, 237].

4.12 Tumor-Like Models

In *C. elegans*, several tumor-like models have been raised for both mechanism elucidation of tumor formation and pharmacological study.

1. *gld-1* mutants

The *gld-1* is a tumor suppressor gene. GLD-1 regulates the oocyte development and germ cell differentiation [238, 239]. In the strain of JK1466 (*gld-1 (q485)/dpy-5 (e61)unc-13(e51)*), gonad cells are arrested in the mitosis stage to cause the tumors formation [240]. Using this strain, the synthesized tryptophan-betaxanthin derived compounds showed the obvious antitumoral effect [240].

2. Germline apoptosis and DNA damage-related gene mutants

The *egl-1*, *hus-1*, and *cep-1* mutants lack the germline apoptosis [241–243], which allow them to be used to understand cancer predisposition. Using these strains, broccoli-derived carbon quantum dots were as effective photosensitizers for light irradiation [244].

3. Oncogene signaling gene mutants

Mutants of some oncogene signaling genes (such as Notch and RAS/MAPK signaling genes) can be used as the tumor-like models [245]. GLP-1 is the Notch receptor. Although *glp-1(ar202)* shows wild-type phenotype at ≤20 °C, this mutant forms a germline stem cell at 25 °C [246].

In RAS signaling pathway, LET-60 is ortholog of human HRas proto-oncogene. LET-60 functions as switch for vulval induction, and *let-60* mutations cause the formation of up to four ectopic pseudovulvas [247]. The multi-vulva phenotypes of *let-60* mutants could be reduced by novel Re(I) complexes [248].

4. Human EGFR transgenic strain

The conserved EGFR signaling is linked to multiple cancers, and act as therapeutic target of human cancers [249]. The human EGFR transgenic strain shows multi-vulva phenotypes [250], which is similar to that of *let-60* mutants.

4.13 Wound Healing Models and Regeneration Models

The wound healing models can be generated by materials with sharp structure or needle-based mechanical puncture to result in larger integumentary wound [251, 252]. Besides this, wound healing models can also be generated by laser ablation to cause the precise injury on epidermis, neurons, or muscle. Using these models, pyridine carboxamide oxime and almond gum/gelatin-based nanocomposite hydrogel were shown to have the wound healing ability [253, 254]. The suggested endpoints for assessing the wound healing can be set up surrounding the development and function of damage tissues. Besides these, the suggested endpoints also include survival, oxidative stress, restoration of barrier, closure rate, actin cytoskeleton dynamics, and cell migration.

The regeneration models can also be generated by laser ablation or mechanical puncture to cause the precise injury of certain neurons or nervous system [255, 256]. For the regeneration, its primary focus is the neuronal axon regeneration and functional recovery [257, 258]. Using these models, staurosporine (a PKC kinase inhibitor) and vinaxanthone and it derivatives could modulate the neural regeneration [255, 259]. The suggested endpoints for assessing the neural regeneration can be set up surrounding the development and function of damage neurons or nervous system. Besides these, the suggested endpoints also include survival, oxidative stress, axon development and guidance, axon regrowth length, synaptic development, function, and reconnection, and cytoskeletal remodeling in the neurons.

4.14 mtDNA Replicative Polymerase (POLG)-Related Disease Models

The *C. elegans* is considered as a useful model for evaluating the POLG-related disorders [260, 261]. Using the *polg-1* mutants as the animal model, clofilium tosylate was identified as potential therapeutic compound against the POLG-related diseases by improving the behavior and the mtDNA content [262]. Moreover, mitochondrial ATFS-1 governing mitochondrial UPR could promote binding of POLG to deleterious mitochondrial DNA (ΔmtDNA), and the accumulated ΔmtDNA could further induce mitochondrial diseases [263].

4.15 α1-Antitrypsin Deficiency (ATD) Models

The *C. elegans* can be applied as the disease model for ATD, which is characterized by accumulated serpin oligomers and polymers within endoplasmic reticulum (ER) [264]. The ATD model was generated by expressing wild-type and Z mutant form of human serpin [265]. The transgenic strain expressing mutant human serpin showed phenotypes of reduced lifespan, slow growth, reduced brood size, and proteotoxicity [265, 266]. Some potential drug targets have been identified in ATD model based on RNAi screen [267]. After the drug screen, cantharidin, tamoxifen, fluphenazine, and pimozide were identified as the candidate drugs to enhance the autophagy [268]. Among them, the fluphenazine could reduce the proteotoxicity in ATD model [269]. In addition, analog of glibenclamide enhanced the autophagic degradation of misfolded ATD [270].

4.16 Pathogen Infection Models

4.16.1 Bacterial Infection Models

C. elegans has been widely used in the infection with different bacterial pathogens to examine the interaction between bacterial pathogens and host and the innate immune response of hosts and the underlying mechanisms [271–275]. Thus, *C. elegans* can be employed for modeling host-pathogen interactions and the discovery of novel anti-infective compounds [276–278]. Nevertheless, *Pseudomonas*, *Salmonella* infection is the most commonly used bacterial infection model in *C. elegans*. Considering the infection routes, the bacterial pathogen infection models can be classified into two groups. One group is the intestinal pathogen infection [279–281]. Another group is the epidermal pathogen infection [282, 283]. The suggested functional endpoints for assessing bacterial pathogen infection can include:

1. Survival and lifespan.
2. Locomotion, oxidative stress, brood size, and healthspan.
3. Expression of antimicrobial genes.
4. Body pathogen accumulation reflected by colony-forming unit (CFU) and GFP labeled pathogen accumulation in the body.

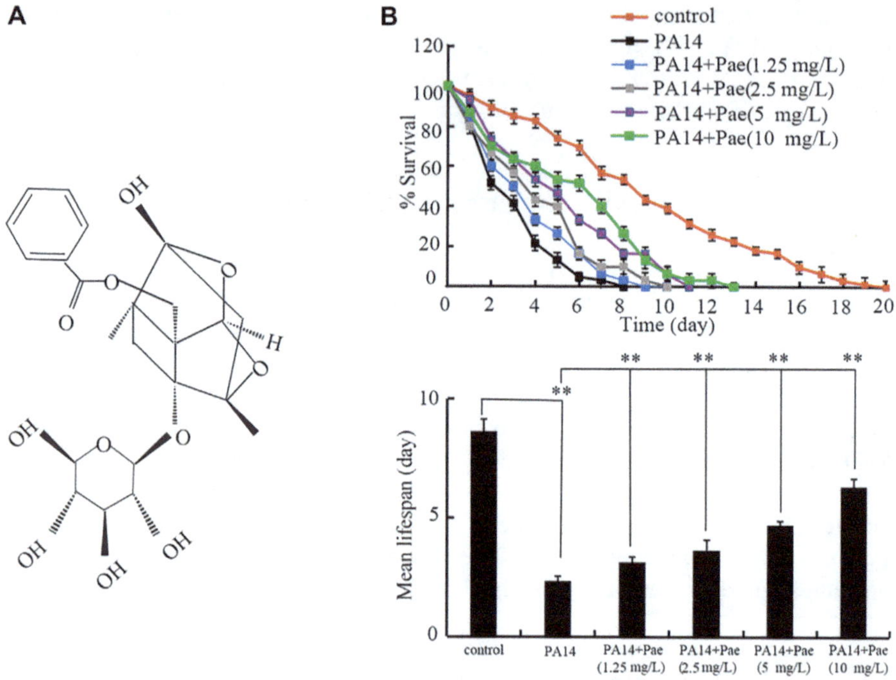

Fig. 4.7 Effect of paeoniflorin treatment in increasing lifespan in *P. aeruginosa* PA14 infected nematodes [289]. (**a**) Chemical structure of paeoniflorin. (**b**) Effect of paeoniflorin treatment on lifespan of nematodes infected with *P. aeruginosa* PA14. Lifespan curve of PA14 showed a significant difference ($p < 0.01$) compared to control. Lifespan curves of PA14 + Pae (1.25 mg/L), PA14 + Pae (2.5 mg/L), PA14 + Pae (5 mg/L), and PA14 + Pae (10 mg/L) showed a significant difference ($p < 0.01$) compared to PA14. Pae paeoniflorin. $^{**}p < 0.01$

Using these bacterial infection models, harmane, 7-ethoxycoumarin, ginkgolide, *Swietenia macrophylla* extract, and *Allium ursinum* leaves aqueous extract were shown to have anti-infective effect in bacterial pathogen-infected nematodes [284–288]. With the paeoniflorin as the example of compounds, its treatment could increase the survival and reduce pathogen body accumulation after infected with *Pseudomonas aeruginosa* (Fig. 4.7) [289].

4.16.2 Fungal Infection Models

Besides acting as the bacterial infection models, *C. elegans* has also been frequently used as the fungal pathogen infection models [290–292]. The most examined fungal pathogen is *Candida albicans*. The examined functional endpoints for assessing fungal pathogen infection were similar to those for bacterial pathogen infection

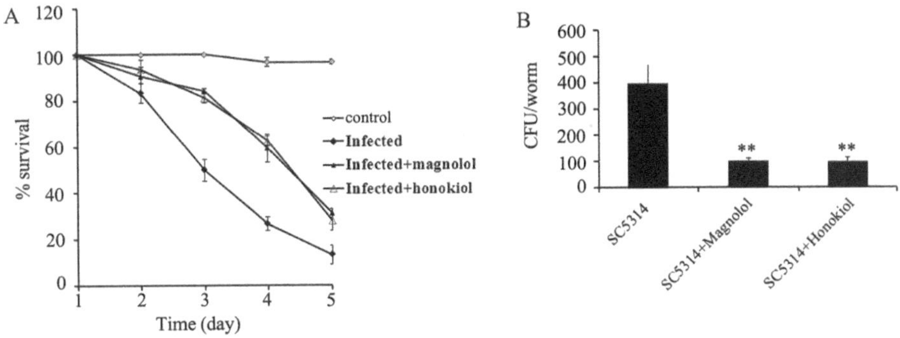

Fig. 4.8 Antifungal activity assay for magnolol and honokiol in nematodes infected with *C. albicans* SC5314 [298]. (**a**) Survival curves of nematodes infected with *C. albicans* SC5314. (**b**) CFU of *C. albicans* SC5314 in nematodes. The concentration for magnolol or honokiol was 16 μg/mL. Control, uninfected. Bars represent means ± S.E.M. $^{**}P < 0.01$

[293, 294]. *C. elegans* is thus considered to be further helpful for screening and discovering antifungal compounds and drugs [295–297]. With the magnolol and honokiol as the examples of compounds, their treatment increased the survival and reduced fungal pathogen accumulation after *Candida albicans* infection (Fig. 4.8) [298].

4.16.3 Virus Infection Models

Recently, *C. elegans* was further found to be helpful for the study of virus–host interactions [299]. The virus infection models can be established, and the widely used virus infection model is the Orsay virus infection model [300, 301].

4.17 Immunosuppression Models

In *C. elegans*, there are at least two types of immunosuppression models to be raised. One immunosuppression model is generated by infection with pathogens, such as bacterial pathogen. The immunosuppression could be induced by *Pseudomonas aeruginosa* infection, and paeoniflorin could increase the survival of *C. elegans* at the immunosuppression stage by activating PMK-1, BAR-1, and EGL-1 signals (Fig. 4.9) [302].

Another immunosuppression model can be generated by chemicals. For example, exposure to some pollutants (such as 6-PPD quinone) could induce the immunosuppression [53, 303]. Moreover, the 6-PPDQ induced immunosuppression could be

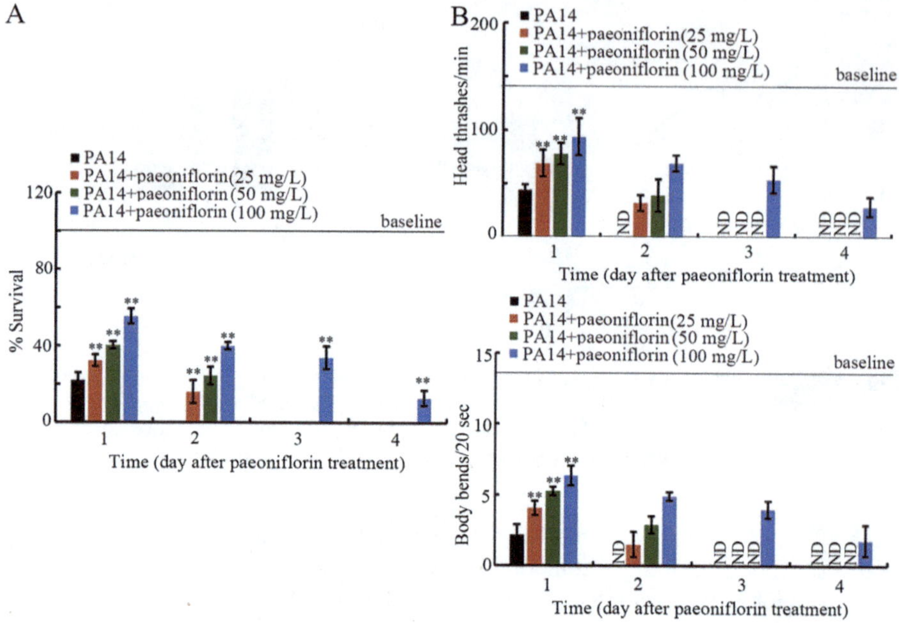

Fig. 4.9 Effect of paeoniflorin treatment on survival (**a**) and locomotion behavior (**b**) in nematodes at the immunosuppression stage (day-5) after *P. aeruginosa* infection [302]. $^{**}P < 0.01$ vs. PA14. ND not done

reduced by following cuminaldehyde treatment (Fig. 4.10) [303]. The suggested functional endpoints for assessing the immunosuppression can include:

1. Expression of antimicrobial genes.
2. Survival.
3. Locomotion and oxidative stress.
4. Body pathogen accumulation reflected by colony-forming unit (CFU) and GFP labeled pathogen accumulation in the body.

4.18 Perspectives

In this chapter, we mainly introduced and discussed the physiological models and the pathological models. Actually, for the aim of pharmacological study, certain phenotypic performance can also be employed as the models for the

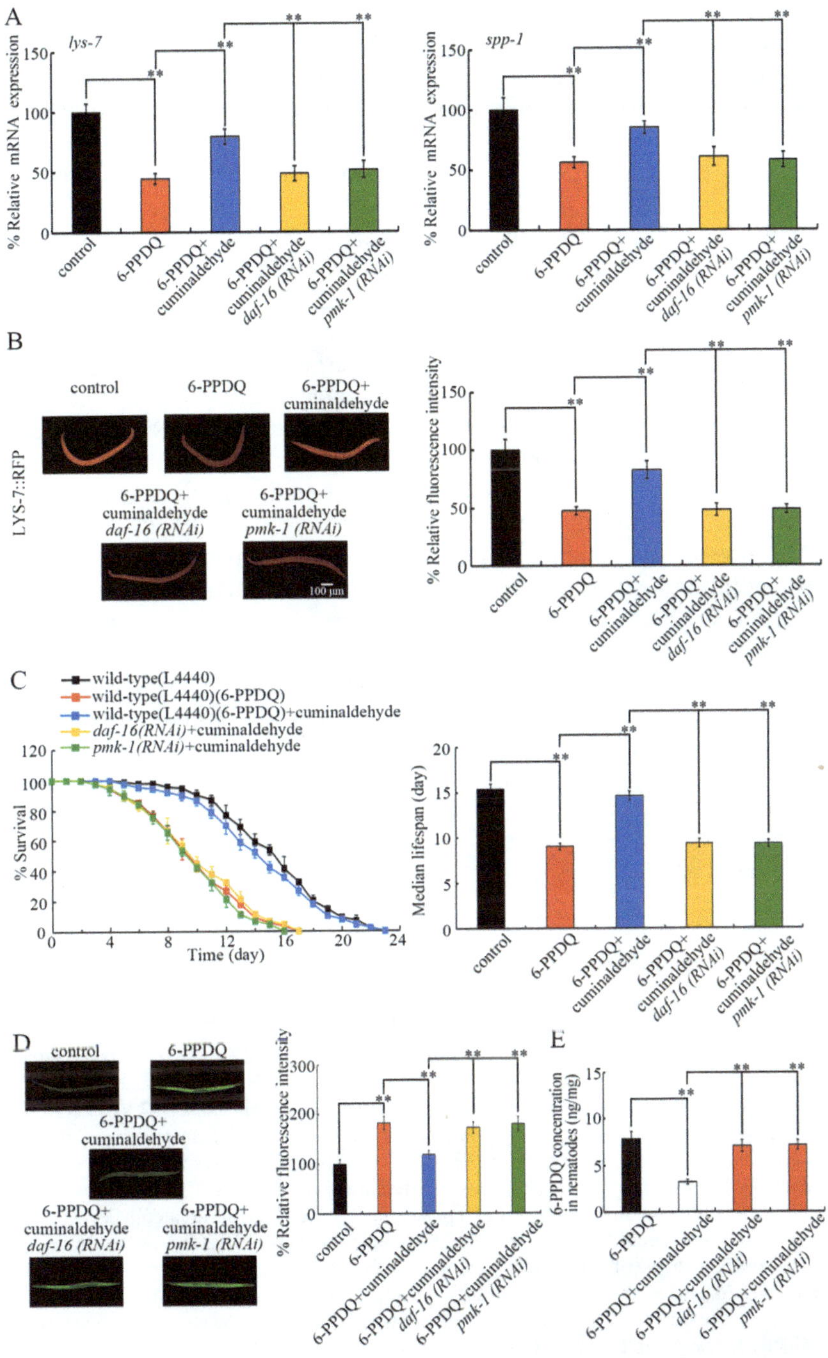

Fig. 4.10 Effect of *daf-16* and *pmk-1* RNAi on beneficial effect of cuminaldehyde on innate immune response, lifespan, ROS production, and 6-PPDQ accumulation in 6-PPDQ exposed nematodes [303]. (**a**) Effect of *daf-16* and *pmk-1* RNAi on beneficial effect of cuminaldehyde

pharmacological screen and identification, which can be defined and considered as the phenotypic models.

Among the introduced physiological models, aging and aging-related changes are the most commonly used physiological models for the pharmacological identification. For the stress models, we only introduced some commonly used stress models, and some other stress models are not covered in this chapter.

For the pathological models, they can be generated at least through three routes. The first is through transgene of human wild-type and mutant gene. The second is to use the *C. elegans* mutant and RNAi strains for genes encoding homologue to human proteins associated with occurrence of certain diseases. The third is induced by some chemicals. Normally, we suggest first use the transgenic strains carrying human wild-type and mutant genes for the pharmacological study. After that, use of *C. elegans* mutant and RNAi strains and chemical induced models can provide the further support and confirmation. Nevertheless, this need to be determined based on certain experimental requirements, experimental design, and actual circumstances. It's worth noting that, for most of the anticipated pathological models, there is a severe lack of corresponding transgenic strains carrying human wild-type and mutant genes. In addition, for some aspects of models, the effort to raise more models are still needed. For example, so far, the raised models for metabolic-related disorders do not cover those on amino acid metabolism, coenzyme and vitamin metabolism, cholesterol metabolism, phospholipid metabolism, and nucleotide metabolism.

Fig. 4.10 (continued) in affecting *lys-7* and *spp-1* expressions in 6-PPDQ exposed nematodes. (**b**) Effect of *daf-16* and *pmk-1* RNAi on beneficial effect of cuminaldehyde in affecting LYS-7::RFP expression in 6-PPDQ exposed nematodes. (**c**) Effect of *daf-16* and *pmk-1* RNAi on beneficial effect of cuminaldehyde on lifespan of 6-PPDQ exposed nematodes. Lifespan curve of 6-PPDQ was significantly ($P < 0.01$) different from that of control. Lifespan curve of 6-PPDQ +cuminaldehyde was significantly ($P < 0.01$) different from that of 6-PPDQ. Lifespan curves of *daf-16(RNAi)*(6-PPDQ+cuminaldehyde) and *pmk-1(RNAi)*(6-PPDQ+cuminaldehyde) were significantly ($P < 0.01$) different from that of 6-PPDQ+cuminaldehyde. (**d**) Effect of *daf-16* and *pmk-1* RNAi on beneficial effect of cuminaldehyde on ROS production in 6-PPDQ exposed nematodes. (**e**) Effect of *daf-16* and *pmk-1* RNAi on beneficial effect of cuminaldehyde in reducing 6-PPDQ accumulation. The 6-PPDQ exposure concentration was 10 μg/L, and the treatment concentration of cuminaldehyde was 7 mg/L. $^{**}P < 0.01$

References

1. DiLoreto R, Murphy CT (2015) The cell biology of aging. Mol Biol Cell 26:4524–4531
2. Panowski SH, Dillin A (2009) Signals of youth: endocrine regulation of aging in *Caenorhabditis elegans*. Trends Endocrinol Metab 20:259–264
3. Murakami S, Tedesco PM, Cypser JR, Johnson TE (2000) Molecular genetic mechanisms of life span manipulation in *Caenorhabditis elegans*. Ann N Y Acad Sci 908:40–49
4. Okoro NO, Odiba AS, Osadebe PO, Omeje EO, Liao G, Fang W, Jin C, Wang B (2021) Bioactive phytochemicals with anti-aging and lifespan extending potentials in *Caenorhabditis elegans*. Molecules 26:7323
5. Ryu D, Mouchiroud L, Andreux PA, Katsyuba E, Moullan N, Nicolet-Dit-Félix AA, Williams EG, Jha P, Lo Sasso G, Huzard D, Aebischer P, Sandi C, Rinsch C, Auwerx J (2016) Urolithin A induces mitophagy and prolongs lifespan in *C. elegans* and increases muscle function in rodents. Nat Med 22:879–888
6. Zhang G, Liu H, Xue T, Kong X, Tian D, Luo L, Yang Y, Xu K, Wei Y, Zhuang Z (2023) Ribavirin extends the lifespan of *Caenorhabditis elegans* through AMPK-TOR signaling. Eur J Pharmacol 946:175548
7. Yu X, Li H, Lin D, Guo W, Xu Z, Wang L, Guan S (2021) Ginsenoside prolongs the lifespan of *C. elegans* via lipid metabolism and activating the stress response signaling pathway. Int J Mol Sci 22:9668
8. Wang J, Liu W, Huang Y, Wang G, Guo X, Shi D, Sun T, Xiao C, Zhang C, Jiang B, Guo Y, Li J (2024) A senomorphlytic three-drug combination discovered in *Salsola collina* for delaying aging phenotypes and extending healthspan. Adv Sci 11:e2401862
9. Carretero M, Gomez-Amaro RL, Petrascheck M (2015) Pharmacological classes that extend lifespan of *Caenorhabditis elegans*. Front Genet 6:77
10. Zhang W, Lv T, Li M, Wu Q, Yang L, Liu H, Sun D, Sun L, Zhuang Z, Wang D (2013) Beneficial effects of wheat gluten hydrolysate to extend lifespan and induce stress resistance in nematode *Caenorhabditis elegans*. PLoS One 8:e74553
11. Petrovic D, Slade L, Paikopoulos Y, D'Andrea D, Savic N, Stancic A, Miljkovic JL, Vignane T, Drekolia MK, Mladenovic D, Sutulovic N, Refeyton A, Kolakovic M, Jovanovic VM, Zivanovic J, Miler M, Vellecco V, Brancaleone V, Bucci M, Casey AM, Yu C, Kasarla SS, Smith KW, Kalfe-Yildiz A, Stenzel M, Miranda-Vizuete A, Hergenröder R, Phapale P, Stanojlovic O, Ivanovic-Burmazovic I, Vlaski-Lafarge M, Bibli SI, Murphy MP, Otasevic V, Filipovic MR (2025) Ergothioneine improves healthspan of aged animals by enhancing cGPDH activity through CSE-dependent persulfidation. Cell Metab 37:542–556
12. Mohammed I, Hollenberg MD, Ding H, Triggle CR (2021) A critical review of the evidence that metformin is a putative anti-aging drug that enhances healthspan and extends lifespan. Front Endocrinol 12:718942
13. Li W, Chen S, Lang J, Luo J, Chen J, Zhang L, Sun Z, Dong D (2024) The clinical antiprotozoal drug nitazoxanide and its metabolite tizoxanide extend *Caenorhabditis elegans* lifespan and healthspan. Acta Pharm Sin B 14:3266–3280
14. Wang L, Li P, Zheng F, Zhu Z, Bai F, Gao R (2024) Collagen peptides from sturgeon swim bladder prolong the lifespan and healthspan in *Caenorhabditis elegans*. J Sci Food Agric 104:5244–5251
15. Hashimoto M, Tanimoto Y, Kage-Nakadai E (2025) *Lactiplantibacillus brownii* strain MH-1 prolongs healthspan and improves stress resistance in *Caenorhabditis elegans*. Biosci Biotechnol Biochem 89:1236–1247
16. Hughes SE, Evason K, Xiong C, Kornfeld K (2007) Genetic and pharmacological factors that influence reproductive aging in nematodes. PLoS Genet 3:e25
17. Parween N, Jabeen A, Prasad B (2022) Eugenol elicits prolongevity by increasing resistance to oxidative stress in *C. elegans*. CNS Neurol Disord Drug Targets 21:841–853

18. Hu Q, Liu Z, Guo Y, Lu S, Du H, Cao Y (2021) Antioxidant capacity of flavonoids from *Folium Artemisiae* Argyi and the molecular mechanism in *Caenorhabditis elegans*. J Ethnopharmacol 279:114398
19. Xiao X, Wu F, Wang B, Cai Z, Wang L, Zhang Y, Yu X, Luo Y (2024) *Clerodendranthus spicatus* (Thunb.) water extracts reduce lipid accumulation and oxidative stress in the *Caenorhabditis elegans*. Int J Mol Sci 25:9655
20. Wang X, Li H, Liu Y, Wu H, Wang H, Jin S, Lu Y, Chang S, Liu R, Peng Y, Guo Z, Wang X (2020) Velvet antler methanol extracts (MEs) protects against oxidative stress in *Caenorhabditis elegans* by SKN-1. Biomed Pharmacother 121:109668
21. Li R, Tao M, Xu T, Huang Y, Zogona D, Pan S, Wu T, Xu X (2022) Artemisia selengensis Turcz. leaf extract promotes longevity and stress resistance in *Caenorhabditis elegans*. J Sci Food Agric 102:4532–4541
22. Paiva FA, Bonomo Lde F, Boasquivis PF, de Paula IT, Guerra JF, Leal WM, Silva ME, Pedrosa ML, Oliveira Rde P (2015) Carqueja (*Baccharis trimera*) protects against oxidative stress and beta-amyloid-induced toxicity in *Caenorhabditis elegans*. Oxidative Med Cell Longev 2015:740162
23. Guan S, Li P, Luo J, Li Y, Huang L, Wang G, Zhu L, Fan H, Li W, Wang L (2010) A deuterohemin peptide extends lifespan and increases stress resistance in *Caenorhabditis elegans*. Free Radic Res 44:813–820
24. Todorova MN, Savova MS, Mihaylova LV, Georgiev MI (2023) Icariin improves stress resistance and extends lifespan in *Caenorhabditis elegans* through *hsf-1* and *daf-2*-driven hormesis. Int J Mol Sci 25:352
25. Zhang Z, Li J, Li F, Wang T, Luo X, Li B, You Y, Wu C, Liu X (2024) *Jujubae Fructus* extract prolongs lifespan and improves stress tolerance in *Caenorhabditis elegans* dependent on DAF-16/SOD-3. Sci Rep 14:13713
26. Osorio-Paz I, Valle-Jiménez X, Brunauer R, Alavez S (2023) Vanillic acid improves stress resistance and substantially extends life span in *Caenorhabditis elegans*. J Gerontol A Biol Sci Med Sci 78:1100–1107
27. Lee EB, Xing MM, Kim DK (2017) Lifespan-extending and stress resistance properties of brazilin from *Caesalpinia sappan* in *Caenorhabditis elegans*. Arch Pharm Res 40:825–835
28. Naß J, Abdelfatah S, Efferth T (2021) Ursolic acid enhances stress resistance, reduces ROS accumulation and prolongs life span in *C. elegans* serotonin-deficient mutants. Food Funct 12: 2242–2256
29. Liu Y, Wu J, Ding Z, Chen L, Liu M, Li B, Li X, Li J, Liu W (2025) Verbascoside from *Callicarpa nudiflora* Hook extends lifespan in *Caenorhabditis elegans* via SKN-1 pathway activation. Biogerontology 27:18
30. Bonomo Lde F, Silva DN, Boasquivis PF, Paiva FA, Guerra JF, Martins TA, de Jesus Torres ÁG, de Paula IT, Caneschi WL, Jacolot P, Grossin N, Tessier FJ, Boulanger E, Silva ME, Pedrosa ML, Oliveira Rde P (2014) Acai (*Euterpe oleracea* Mart.) modulates oxidative stress resistance in *Caenorhabditis elegans* by direct and indirect mechanisms. PLoS One 9:e89933
31. Yang W, Xia W, Zheng B, Li T, Liu RH (2022) DAF-16 is involved in colonic metabolites of ferulic acid-promoted longevity and stress resistance of *Caenorhabditis elegans*. J Sci Food Agric 102:7017–7029
32. Kim SJ, Beak SM, Park SK (2017) Supplementation with triptolide increases resistance to environmental stressors and lifespan in *C. elegans*. J Food Sci 82:1484–1490
33. Jing L, Zhao Y, Jiang L, Song F, An L, Qi E, Fu X, Chen J, Ma J (2025) Unlocking the potential of *Curcumae Rhizoma* aqueous extract in stress resistance and extending lifespan in *Caenorhabditis elegans*. Molecules 30:1668
34. Ye H-Y, Ye B-P, Wang D-Y (2008) Trace administration of vitamin E can retrieve and prevent UV-irradiation- and metal exposure-induced memory deficits in nematode *Caenorhabditis elegans*. Neurobiol Learn Mem 90:10–18

35. Li R, Li H, Wang X, Peng Y (2024) Protective effects of *Velvet Antler* methanol extracts on hypoxia-induced damage in *Caenorhabditis elegans* through HIF-1 and ECH-8 mediated lipid accumulation. Nutrients 16:2257
36. Zhao L, Rui Q, Wang D-Y (2017) Molecular basis for oxidative stress induced by simulated microgravity in nematode *Caenorhabditis elegans*. Sci Total Environ 607–608:1381–1390
37. Liu H-L, Guo D-Q, Kong Y, Rui Q, Wang D-Y (2019) Damage on functional state of intestinal barrier by microgravity stress in nematode *Caenorhabditis elegans*. Ecotoxicol Environ Saf 183:109554
38. Liu P-D, Li D, Li W-J, Wang D-Y (2019) Mitochondrial unfolded protein response to microgravity stress in nematode *Caenorhabditis elegans*. Sci Rep 9:16474
39. Kong Y, Liu H-L, Li W-J, Wang D-Y (2019) Intestine-specific activity of insulin signaling pathway in response to microgravity stress in Caenorhabditis elegans. Biochem Biophys Res Commun 517:278–284
40. Rui Q, Dong S-S, Jiang W-K, Wang D-Y (2019) Response of canonical Wnt/β-catenin signaling pathway in the intestine to microgravity stress in *Caenorhabditis elegans*. Ecotoxicol Environ Saf 186:109782
41. Dhakal R, Yosofvand M, Yavari M, Abdulrahman R, Schurr R, Moustaid-Moussa N, Moussa H (2021) Review of biological effects of acute and chronic radiation exposure on *Caenorhabditis elegans*. Cells 10:1966
42. Liu M, Li N, Shan S, Shi Y, Zhu Y, Lu W (2022) *Acanthopanax senticosus* polysaccharide enhances the pathogen resistance of radiation-damaged *Caenorhabditis elegans* through intestinal p38 MAPK-SKN-1/ATF-7 pathway and stress response. Int J Mol Sci 23:5034
43. Ye K, Ji CB, Lu XW, Ni YH, Gao CL, Chen XH, Zhao YP, Gu GX, Guo XR (2010) Resveratrol attenuates radiation damage in *Caenorhabditis elegans* by preventing oxidative stress. J Radiat Res 51:473–479
44. Yang W, Li Z, Chen X, Wu S, Liu S, Yao L, Zhang J, Liang H, Song J, Ma B (2025) Screening model in *Caenorhabditis elegans* for radioprotective natural products. Int J Radiat Biol 101: 304–313
45. Ullah I, Shakir MZ, Zhou XC, Liu J, Li S, Chen H, Jiang N, Usmani MW, Barkat MQ, Du Q, Zhao Y, Wang N, Liu X (2025) Effect of Hong Huang Tang on memory enhancement and mitigation of microgravity-induced oxidative stress in *C. elegans*. Biogerontology 27:24
46. Hua X, Wang D-Y (2023) Tire-rubber related pollutant 6-PPD quinone: a review of its transformation, environmental distribution, bioavailability, and toxicity. J Hazard Mater 459:132265
47. Hua X, Feng X, Liang G-Y, Chao J, Wang D-Y (2023) Long-term exposure to tire-derived 6-PPD quinone causes intestinal toxicity by affecting functional state of intestinal barrier in *Caenorhabditis elegans*. Sci Total Environ 861:160591
48. Wang Y-X, Liang G-Y, Chao J, Wang D-Y (2024) Comparison of intestinal toxicity in enhancing intestinal permeability and in causing ROS production of six PPD quinones in *Caenorhabditis elegans*. Sci Total Environ 927:172306
49. Hua X, Feng X, Liang G-Y, Chao J, Wang D-Y (2023) Exposure to 6-PPD quinone at environmentally relevant concentrations causes abnormal locomotion behaviors and neurodegeneration in *Caenorhabditis elegans*. Environ Sci Technol 57:4940–4950
50. Hua X, Feng X, Liang G-Y, Chao J, Wang D-Y (2023) Long-term exposure to 6-PPD quinone reduces reproductive capacity by enhancing germline apoptosis associated with activation of both DNA damage and cell corpse engulfment in *Caenorhabditis elegans*. J Hazard Mater 454:131495
51. Liu Z-Y, Bian Q, Wang D-Y (2024) Exposure to 6-PPD quinone causes ferroptosis activation associated with induction of reproductive toxicity in *Caenorhabditis elegans*. J Hazard Mater 471:134356
52. Hua X, Wang D-Y (2023) Exposure to 6-PPD quinone at environmentally relevant concentrations inhibits both lifespan and healthspan in *C. elegans*. Environ Sci Technol 57: 19295–19303

53. Wu J-W, Shen S-H, Wang D-Y (2024) 6-PPD quinone at environmentally relevant concentrations induces immunosenescenece by causing immunosuppression during the aging process. Chemosphere 368:143719
54. Wang Y-X, Hu G-C, Wang D-Y (2025) Increased S-adenosyl methionine strengthens the suppression in mitochondrial unfolded protein response induced by 6-PPD quinone at environmentally relevant concentrations in *Caenorhabditis elegans*. Environ Pollut 386:127231
55. Hu D-Y, Wang Y-X, Hu G-C, Liu R, Wang D-Y (2025) 6-PPD quinone inhibited retinoic acid synthesis mediates toxicity through feedback loop between ALH-3/DHS-19-SEX-1 axis and intestinal signals in *Caenorhabditis elegans*. J Environ Expo Assess 4:40
56. Wang W, Li Y-H, Wang D-Y (2025) Long-term exposure to 6-PPD quinone inhibits glutamate synthesis and glutamate receptor function associated with its toxicity induction in *Caenorhabditis elegans*. Toxics 13:434
57. Wang W, Li Y-H, Wang D-Y (2025) Exposure to 6-PPD quinone disrupts adsorption and catabolism of leucine to cause mitochondrial dysfunction in *Caenorhabditis elegans*. Toxics 13:544
58. Wu J-W, Li L-E, Hu D-Y, Liu R, Bian Q, Wang D-Y (2025) Environmentally relevant concentrations of 6-PPDQ disrupts vitamin D3 adsorption and receptor function in *Caenorhabditis elegans*. Environ Sci Process Impacts 27:2798–2808
59. Wang W, Hu G-C, Li Y-H, Wang D-Y (2026) 6-PPD quinone inhibits ammonia excretion to cause multiple aspects of toxicity in *Caenorhabditis elegans* by activating dual oxidase complex-SKN-1 axis. Environ Pollut 390:127528
60. Liu Z-J, Li Y-H, Wang D-Y (2024) Exposure to 6-PPD quinone disrupts glucose metabolism associated with lifespan reduction by affecting insulin and AMPK signals in *Caenorhabditis elegans*. Chemosphere 363:142975
61. Wang Y-X, Wang D-Y (2024) Exposure to 6-PPD quinone enhances glycogen accumulation in *Caenorhabditis elegans*. Environ Pollut 359:124600
62. Wan X, Liang G-Y, Wang D-Y (2025) 6-PPD quinone at environmentally relevant concentrations disrupts citric acid cycle in *Caenorhabditis elegans*: role of reduction in acetyl CoA and pyruvate contents. Environ Chem Ecotoxicol 7:1119–1129
63. Hua X, Liang G-Y, Chao J, Wang D-Y (2024) Exposure to 6-PPD quinone causes damage on mitochondrial complex I/II associated with lifespan reduction in *Caenorhabditis elegans*. J Hazard Mater 472:134598
64. Hua X, Wang D-Y (2025) 6-PPD quinone causes alteration in ubiquinone-mediated complex III associated with toxicity on mitochondrial function and longevity in *Caenorhabditis elegans*. J Environ Chem Eng 13:116571
65. Hua X, Wang D-Y (2025) 6-PPD quinone at environmentally relevant concentrations activates feedback response of electron transport chain to mediate damage on mitochondrial function and longevity in *Caenorhabditis elegans*. Environ Chem Ecotoxicol 7:2356–2365
66. Hua X, Wang D-Y (2024) 6-PPD quinone at environmentally relevant concentrations induced damage on longevity in *C. elegans*: mechanistic insight from inhibition in mitochondrial UPR response. Sci Total Environ 954:176275
67. Hua X, Wang D-Y (2025) Environmentally relevant concentration of 6-PPD quinone inhibits two types of mitophagy to cause mitochondrial dysfunction and lifespan reduction in *Caenorhabditis elegans*. Environ Sci Process Impacts 27:1928–1940
68. Liu Z-Y, Bian Q, Wang D-Y (2025) 6-PPD quinone induces response of nuclear hormone receptors in the germline associated with formation of reproductive toxicity in *Caenorhabditis elegans*. J Hazard Mater 495:138815
69. Liu Z-J, Li Y-H, Wang D-Y (2024) Transgenerational response of glucose metabolism in *Caenorhabditis elegans* exposed to 6-PPD quinone. Chemosphere 367:143653
70. Wang Y-X, Wu J-W, Wang D-Y (2025) 6-PPD quinone causes lipid accumulation across multiple generations differentially affected by metabolic sensors and components of COMPASS complex in *Caenorhabditis elegans*. Environ Pollut 366:125539

71. Wang Y-X, Wang D-Y (2024) Transgenerational intestinal toxicity of 6-PPD quinone in causing ROS production, enhancement in intestinal permeability and suppression in innate immunity in *C. elegans*. Environ Pollut 363:125208
72. Song M-X, Ruan Q-L, Wang D-Y (2024) Paeoniflorin alleviates toxicity and accumulation of 6-PPD quinone by activating ACS-22 in *Caenorhabditis elegans*. Ecotoxicol Environ Saf 286: 117226
73. Dos Santos FB, Quines CB, Pilissão LEB, Dal Forno AHC, Rodrigues CF, Denardin CC, Farias FM, Ávila DS (2020) Aqueous bark extract of *Ceiba speciosa* (A. St.-Hill) Ravenna protects against glucose toxicity in *Caenorhabditis elegans*. Oxid Med Cell Longev 2020: 1321354
74. Deusing DJ, Beyrer M, Fitzenberger E, Wenzel U (2015) Carnitine protects the nematode *Caenorhabditis elegans* from glucose-induced reduction of survival depending on the nuclear hormone receptor DAF-12. Biochem Biophys Res Commun 460:747–752
75. Shanmugam G, Mohankumar A, Kalaiselvi D, Nivitha S, Murugesh E, Shanmughavel P, Sundararaj P (2017) Diosgenin a phytosterol substitute for cholesterol, prolongs the lifespan and mitigates glucose toxicity via DAF-16/FOXO and GST-4 in *Caenorhabditis elegans*. Biomed Pharmacother 95:1693–1703
76. Deusing DJ, Winter S, Kler A, Kriesl E, Bonnländer B, Wenzel U, Fitzenberger E (2015) A catechin-enriched green tea extract prevents glucose-induced survival reduction in *Caenorhabditis elegans* through *sir-2.1* and *uba-1* dependent hormesis. Fitoterapia 102: 163–170
77. Wang J, Jiang B, Lin X, Zhang J, Xiong L, Feng Y, Cheang WS, Xu B (2026) Genistein ameliorates glucose-induced beta-amyloid toxicity, oxidative stress, and aging in the *C. elegans* model of Alzheimer's disease. Free Radic Biol Med 242:167–187
78. Liu T, Zhuang Z, Wang D (2023) Paeoniflorin mitigates high glucose-induced lifespan reduction by inhibiting insulin signaling in *Caenorhabditis elegans*. Front Pharmacol 14: 1202379
79. Savova MS, Todorova MN, Apostolov AG, Yahubyan GT, Georgiev MI (2022) Betulinic acid counteracts the lipid accumulation in *Caenorhabditis elegans* by modulation of *nhr-49* expression. Biomed Pharmacother 156:113862
80. Qin X, Wang W, Chu W (2021) Antioxidant and reducing lipid accumulation effects of rutin in *Caenorhabditis elegans*. Biofactors 47:686–693
81. Shen P, Yue Y, Kim KH, Park Y (2017) Piceatannol reduces fat accumulation in *Caenorhabditis elegans*. J Med Food 20:887–894
82. Kim AT, Park Y (2025) Esculetin inhibits fat accumulation through insulin/insulin-like growth factor- and AMP-activated protein kinase-dependent pathways in *Caenorhabditis elegans*. Nutrients 17:1565
83. Wang J, He J (2019) Swertiamarin decreases lipid accumulation dependent on 3-ketoacyl-coA thiolase. Biomed Pharmacother 112:108668
84. Zhao Y, Wu C, Bai J, Li J, Cheng K, Zhou X, Dong Y, Xiao X (2020) Fermented barley extracts with lactobacillus plantarum dy-1 decreased fat accumulation of *Caenorhabditis elegans* in a daf-2-dependent mechanism. J Food Biochem 44:e13459
85. Savova MS, Todorova MN, Binev BK, Georgiev MI, Mihaylova LV (2025) Curcumin enhances the anti-obesogenic activity of orlistat through SKN-1/NRF2-dependent regulation of nutrient metabolism in *Caenorhabditis elegans*. Int J Obes 49:516–526
86. Mihaylova LV, Savova MS, Todorova MN, Tonova V, Binev BK, Georgiev MI (2026) Cycloastragenol improves fatty acid metabolism through NHR-49/FAT-7 suppression and potent AAK-2 activation in *Caenorhabditis elegans* obesity model. Int J Mol Sci 27:772
87. Chumphoochai K, Manohong P, Niamnont N, Tamtin M, Sobhon P, Meemon K (2023) Anti-obesity effects of marine macroalgae extract *Caulerpa lentillifera* in a *Caenorhabditis elegans* model. Mar Drugs 21:577
88. Zhang S, Chen L, Wang Y (2025) Tetrahydrocurcumin ameliorates metabolic disorders associated with obesity by regulating gut microbiota homeostasis. J Nutr Biochem 146:110034

89. Hao KX, Zhang YJ, Li YM, Zhong RF, Wang L, Chang X, Jiang JG, Zhu W (2025) Polyphenols from *Mallotus peltatus* (Geiseler) Mull. Arg. ameliorates FFA-induced hepatic steatosis in L02 cells and reduces lipid accumulation in *Caenorhabditis elegans*. Mol Nutr Food Res 69:e202400689
90. Daghar H, Pyman B, Maios C, Doyle J, Perlstein E, Samarut E, Parker JA (2026) CHK1 inhibition rescues abnormal glycogen buildup in a *Caenorhabditis elegans* model for glycogen storage disease III. Commun Biol. https://doi.org/10.1038/s42003-026-09535-9
91. Polyak E, Ostrovsky J, Peng M, Dingley SD, Tsukikawa M, Kwon YJ, McCormack SE, Bennett M, Xiao R, Seiler C, Zhang Z, Falk MJ (2018) N-acetylcysteine and vitamin E rescue animal longevity and cellular oxidative stress in pre-clinical models of mitochondrial complex I disease. Mol Genet Metab 123:449–462
92. Soares ATG, Rodrigues LBLJ, Salgueiro WG, Dal Forno AHC, Rodrigues CF, Sacramento M, Franco J, Alves D, Oliveira RP, Pinton S, Ávila DS (2019) Organoselenotriazoles attenuate oxidative damage induced by mitochondrial dysfunction in *mev-1 Caenorhabditis elegans* mutants. J Trace Elem Med Biol 53:34–40
93. Devagi G, Mohankumar A, Shanmugam G, Nivitha S, Dallemer F, Kalaivani P, Sundararaj P, Prabhakaran R (2018) Organoruthenium(II) complexes ameliorates oxidative stress and impedes the age associated deterioration in *Caenorhabditis elegans* through JNK-1/DAF-16 signalling. Sci Rep 8:7688
94. Lin YT, Hoang H, Hsieh SI, Rangel N, Foster AL, Sampayo JN, Lithgow GJ, Srinivasan C (2006) Manganous ion supplementation accelerates wild type development, enhances stress resistance, and rescues the life span of a short-lived *Caenorhabditis elegans* mutant. Free Radic Biol Med 40:1185–1193
95. Zubovych IO, Straud S, Roth MG (2010) Mitochondrial dysfunction confers resistance to multiple drugs in *Caenorhabditis elegans*. Mol Biol Cell 21:956–969
96. Hihi AK, Beauchamp M, Branicky R, Desjardins A, Casanova I, Guimond M, Carroll M, Ethier M, Kianicka I, McBride K, Hekimi S (2008) Evolutionary conservation of drug action on lipoprotein metabolism-related targets. J Lipid Res 49:74–83
97. Zheng SQ, Huang XB, Xing TK, Ding AJ, Wu GS, Luo HR (2017) Chlorogenic acid extends the lifespan of *Caenorhabditis elegans* via insulin/IGF-1 signaling pathway. J Gerontol A Biol Sci Med Sci 72:464–472
98. Zamberlan DC, Arantes LP, Machado ML, Golombieski R, Soares FA (2014) Diphenyl-diselenide suppresses amyloid-beta peptide in *Caenorhabditis elegans* model of Alzheimer's disease. Neuroscience 278:40–50
99. Huang L, Luo Y, Pu Z, Kong X, Fu X, Xing H, Wei S, Chen W, Tang H (2017) Oxoisoaporphine alkaloid derivative 8-1 reduces Abeta(1-42) secretion and toxicity in human cell and *Caenorhabditis elegans* models of Alzheimer's disease. Neurochem Int 108: 157–168
100. Wang H, Yue Y, Zhao H, Wu H, Jiang K, Li S, Zhao M, Lin F (2021) Neuroprotective effects of 2-substituted 1,3-selenazole amide derivatives on Amyloid-beta-induced toxicity in a transgenic *Caenorhabditis elegans* model of Alzheimer's disease. Neurotox Res 39:841–850
101. Ng LF, Gruber J, Cheah IK, Goo CK, Cheong WF, Shui G, Sit KP, Wenk MR, Halliwell B (2014) The mitochondria-targeted antioxidant MitoQ extends lifespan and improves healthspan of a transgenic *Caenorhabditis elegans* model of Alzheimer disease. Free Radic Biol Med 71:390–401
102. Zhang R, Huang X, Zhou C, Zhang Q, Jia D, Xie X, Zhang J (2024) Network pharmacology-based mechanism analysis of dauricine on the alleviating Abeta-induced neurotoxicity in *Caenorhabditis elegans*. BMC Complement Med Ther 24:321
103. He Q, Huang G, Chen Y, Wang X, Huang Z, Chen Z (2017) The protection of novel 2-arylethenylquinoline derivatives against impairment of associative learning memory induced by neural Abeta in *C. elegans* Alzheimer's disease model. Neurochem Res 42: 3061–3072

104. Shi YC, Pan TM, Liao VH (2016) Monascin from Monascus-fermented products reduces oxidative stress and amyloid-beta toxicity via DAF-16/FOXO in *Caenorhabditis elegans*. J Agric Food Chem 64:7114–7120
105. McCormick AV, Wheeler JM, Guthrie CR, Liachko NF, Kraemer BC (2013) Dopamine D2 receptor antagonism suppresses tau aggregation and neurotoxicity. Biol Psychiatry 73:464–471
106. Miyasaka T, Xie C, Yoshimura S, Shinzaki Y, Yoshina S, Kage-Nakadai E, Mitani S, Ihara Y (2016) Curcumin improves tau-induced neuronal dysfunction of nematodes. Neurobiol Aging 39:69–81
107. Yu YT, Lin HR, Chen XY, Sun ML, Wei CM, Xue MF, Gao YH, Tang WB, Zheng CY, Li S, Wang HB (2024) *Dendrobium officinale* phenolic extract maintains proteostasis by regulating autophagy in a *Caenorhabditis elegans* model of Alzheimer's disease. Fitoterapia 175:105924
108. Sukanya S, Bellver-Sanchis A, Singh Choudhary B, Kumar S, Pérez B, Leandro Martínez Rodríguez A, Brea J, Griñán-Ferré C, Malik R (2024) Design, synthesis, and biological evaluation of tetrahydropyrimidine analogue as GSK-3beta/Abeta aggregation inhibitor and anti-Alzheimer's agent. Bioorg Chem 153:107811
109. Sukanya S, Bellver-Sanchis A, Choudhary BS, Kumar S, Pérez B, Rodríguez ALM, Brea J, Escolano C, Griñán-Ferré C, Malik R (2025) Exploring 2-thioxo 1,2,3,4 tetrahydropyrimidines as dual acting GSK-3beta/Abeta aggregation inhibitors: implications for Alzheimer's disease treatment. Drug Dev Res 86:e70112
110. Fu HJ, Zhou XY, Li YP, Chen X, He YN, Qin DL, Yu L, Yu CL, Wu JM, Wu AG, Zhou XG (2023) The protective effects of *Reineckia carnea* ether fraction against Alzheimer's disease pathology: An exploration in *Caenorhabditis elegans* models. Int J Mol Sci 24:16536
111. Cooper JF, Van Raamsdonk JM (2018) Modeling Parkinson's disease in *C. elegans*. J Parkinsons Dis 8:17–32
112. Muhammad F, Liu Y, Wang N, Zhao L, Zhou Y, Yang H, Li H (2022) Anti-alpha-synuclein toxicity and anti-neurodegenerative role of chrysin in transgenic *Caenorhabditis elegans* models of Parkinson's disease. ACS Chem Neurosci 13:442–453
113. Muhammad F, Liu Y, Wang N, Zhao L, Zhou Y, Yang H, Li H (2023) Rose essential oil diminishes dopaminergic neuron degenerations and reduces alpha-synuclein aggregation in *Caenorhabditis elegans* models of Parkinson's disease. Phytother Res 37:2877–2893
114. Sanguanphun T, Promtang S, Sornkaew N, Niamnont N, Sobhon P, Meemon K (2023) Anti-Parkinson effects of *Holothuria leucospilota*-derived palmitic acid in *Caenorhabditis elegans* model of Parkinson's disease. Mar Drugs 21:141
115. Goya ME, Xue F, Sampedro-Torres-Quevedo C, Arnaouteli S, Riquelme-Dominguez L, Romanowski A, Brydon J, Ball KL, Stanley-Wall NR, Doitsidou M (2020) Probiotic *Bacillus subtilis* protects against alpha-Synuclein aggregation in *C. elegans*. Cell Rep 30:367–380
116. Ma C, Feng Y, Li X, Sun L, He Z, Gan J, He M, Zhang X, Chen X (2023) Potential therapeutic effects of policosanol from insect wax on *Caenorhabditis elegans* models of Parkinson's disease. J Neuroimmune Pharmacol 18:127–144
117. Tsai CW, Tsai RT, Liu SP, Chen CS, Tsai MC, Chien SH, Hung HS, Lin SZ, Shyu WC, Fu RH (2017) Neuroprotective effects of betulin in pharmacological and transgenic *Caenorhabditis elegans* models of Parkinson's disease. Cell Transplant 26:1903–1918
118. He CL, Tang Y, Wu JM, Long T, Yu L, Teng JF, Qiu WQ, Pan R, Yu CL, Qin DL, Wu AG, Zhou XG (2023) Chlorogenic acid delays the progression of Parkinson's disease via autophagy induction in *Caenorhabditis elegans*. Nutr Neurosci 26:11–24
119. Gao W, Wang Y, Wang F, Wu X, Lu F, Liu F (2025) Ergothioneine exerts neuroprotective effects in Parkinson's disease: targeting alpha-synuclein aggregation and oxidative stress. Food Res Int 201:115590
120. Cheon SM, Jang I, Lee MH, Kim DK, Jeon H, Cha DS (2017) *Sorbus alnifolia* protects dopaminergic neurodegeneration in *Caenorhabditis elegans*. Pharm Biol 55:481–486
121. Chalorak P, Sornkaew N, Manohong P, Niamnont N, Malaiwong N, Limboonreung T, Sobhon P, Aschner M, Meemon K (2021) Diterpene glycosides from Holothuria scabra

exert the alpha-synuclein degradation and neuroprotection against alpha-synuclein-mediated neurodegeneration in *C. elegans* model. J Ethnopharmacol 279:114347

122. da Silva LPD, da Cruz Guedes E, Fernandes ICO, Pedroza LAL, da Silva Pereira GJ, Gubert P (2024) Exploring *Caenorhabditis elegans* as Parkinson's disease model: neurotoxins and genetic implications. Neurotox Res 42:11
123. Chen QX, Zhou L, Long T, Qin DL, Wang YL, Ye Y, Zhou XG, Wu JM, Wu AG (2022) Galangin exhibits neuroprotective effects in 6-OHDA-induced models of Parkinson's disease via the Nrf2/Keap1 pathway. Pharmaceuticals 15:1014
124. Ren J, Yuan L, Wang W, Zhang M, Wang Q, Li S, Zhang L, Hu K (2019) Tricetin protects against 6-OHDA-induced neurotoxicity in Parkinson's disease model by activating Nrf2/HO-1 signaling pathway and preventing mitochondria-dependent apoptosis pathway. Toxicol Appl Pharmacol 378:114617
125. Zhu FD, Wang BD, Qin DL, Su XH, Yu L, Wu JM, Law BY, Guo MS, Yu CL, Zhou XG, Wu AG (2024) Carpesii fructus extract exhibits neuroprotective effects in cellular and *Caenorhabditis elegans* models of Parkinson's disease. CNS Neurosci Ther 30:e14515
126. Balkrishna A, Pathak N, Singh R, Gohel V, Varshney Y, Dev R, Varshney A (2025) Neurogrit gold attenuates 6-OHDA-induced dopaminergic neurodegeneration in Parkinson's model of *Caenorhabditis elegans* by reducing alpha-Synuclein accumulation and Pink/Pdr-1 driven mitochondrial dysfunction. CNS Neurosci Ther 31:e70401
127. Chang CH, Wei CC, Ho CT, Liao VH (2021) N-gamma-(L-glutamyl)-L-selenomethionine shows neuroprotective effects against Parkinson's disease associated with SKN-1/Nrf2 and TRXR-1 in *Caenorhabditis elegans*. Phytomedicine 92:153733
128. Tsai RT, Tsai CW, Liu SP, Gao JX, Kuo YH, Chao PM, Hung HS, Shyu WC, Lin SZ, Fu RH (2020) Maackiain ameliorates 6-Hydroxydopamine and SNCA pathologies by modulating the PINK1/Parkin pathway in models of Parkinson's disease in *Caenorhabditis elegans* and the SH-SY5Y cell line. Int J Mol Sci 21:4455
129. Braungart E, Gerlach M, Riederer P, Baumeister R, Hoener MC (2004) *Caenorhabditis elegans* MPP+ model of Parkinson's disease for high-throughput drug screenings. Neurodegener Dis 1:175–183
130. Lee SH, Han YT, Cha DS (2021) Neuroprotective effect of damaurone D in a *C. elegans* model of Parkinson's disease. Neurosci Lett 747:135623
131. Mocko JB, Kern A, Moosmann B, Behl C, Hajieva P (2010) Phenothiazines interfere with dopaminergic neurodegeneration in *Caenorhabditis elegans* models of Parkinson's disease. Neurobiol Dis 40:120–109
132. Parrales V, Arcile G, Laserre L, Normant S, Le Goff G, Da Costa Noble C, Ouazzani J, Callizot N, Haïk S, Rabhi C, Bizat N (2025) Neuroprotective effect of withaferin derivatives toward MPP(+) and 6-OHDA toxicity to dopaminergic neurons. ACS Chem Neurosci 16:802–817
133. Lu XL, Yao XL, Liu Z, Zhang H, Li W, Li Z, Wang GL, Pang J, Lin Y, Xu Z, Chen L, Pei Z, Zeng J (2010) Protective effects of xyloketal B against MPP+-induced neurotoxicity in *Caenorhabditis elegans* and PC12 cells. Brain Res 1332:110–119
134. Kim E, Yang W, Cha DS, Han YT (2024) Synthesis and dopaminergic neuroprotective activity of the proposed structure of bassiamide A, a caffeamide alkaloid derived from Bassia Indica wight. Chem Biodivers 21:e202400836
135. Li X, Hu M, Zhou X, Yu L, Qin D, Wu J, Deng L, Huang L, Ren F, Liao B, Wu A, Fan D (2024) Hederagenin inhibits mitochondrial damage in Parkinson's disease via mitophagy induction. Free Radic Biol Med 224:740–756
136. Hsu YL, Hung HS, Tsai CW, Liu SP, Chiang YT, Kuo YH, Shyu WC, Lin SZ, Fu RH (2021) Peiminine reduces ARTS-mediated degradation of XIAP by modulating the PINK1/Parkin pathway to ameliorate 6-hydroxydopamine toxicity and alpha-Synuclein accumulation in Parkinson's disease models in vivo and in vitro. Int J Mol Sci 22:10240
137. El Din RH, Thabit S (2024) Quinic acid protects against the development of Huntington's disease in *Caenorhabditis elegans* model. BMC Complement Med Ther 24:377

138. Cordeiro LM, Soares MV, da Silva AF, Machado ML, Bicca Obetine Baptista F, da Silveira TL, Arantes LP, Soares FAA (2022) Neuroprotective effects of rutin on ASH neurons in *Caenorhabditis elegans* model of Huntington's disease. Nutr Neurosci 25:2288–2301
139. Boasquívis PF, Silva GMM, Paiva FA, Cavalcanti RM, Nunez CV, de Paula Oliveira R (2018) Guarana (*Paullinia cupana*) extract protects *Caenorhabditis elegans* models for Alzheimer disease and Huntington disease through activation of antioxidant andprotein degradation pathways. Oxidative Med Cell Longev 2018:9241308
140. Fang M, Liu Y, Gao X, Yu J, Tu X, Mo X, Zhu H, Zou Y, Huang C, Fan S (2025) Perillaldehyde alleviates polyQ-induced neurodegeneration through the induction of autophagy and mitochondrial UPR in *Caenorhabditis elegans*. Biofactors 51:e2089
141. Gohil VM, Offner N, Walker JA, Sheth SA, Fossale E, Gusella JF, MacDonald ME, Neri C, Mootha VK (2011) Meclizine is neuroprotective in models of Huntington's disease. Hum Mol Genet 20:294–300
142. Zeng Y, Guo W, Xu G, Wang Q, Feng L, Long S, Liang F, Huang Y, Lu X, Li S, Zhou J, Burgunder JM, Pang J, Pei Z (2016) Xyloketal-derived small molecules show protective effect by decreasing mutant Huntingtin protein aggregates in *Caenorhabditis elegans* model of Huntington's disease. Drug Des Devel Ther 10:1443–1451
143. Cong W, Bai R, Li YF, Wang L, Chen C (2019) Selenium nanoparticles as an efficient nanomedicine for the therapy of Huntington's disease. ACS Appl Mater Interfaces 11:34725–34735
144. Li X, Hernandez I, Koyuncu S, Kis B, Häggblad M, Lidemalm L, Abbas AA, Bendegúz S, Göblös A, Brautigam L, Lucas JJ, Carreras-Puigvert J, Hühn D, Pircs K, Vilchez D, Fernandez-Capetillo O (2024) The anti-leprosy drug clofazimine reduces polyQ toxicity through activation of PPARgamma. EBioMedicine 103:105124
145. Aaron C, Beaudry G, Parker JA, Therrien M (2016) Maple syrup decreases TDP-43 proteotoxicity in a *Caenorhabditis elegans* model of amyotrophic lateral sclerosis (ALS). J Agric Food Chem 64:3338–3344
146. Vaccaro A, Patten SA, Aggad D, Julien C, Maios C, Kabashi E, Drapeau P, Parker JA (2013) Pharmacological reduction of ER stress protects against TDP-43 neuronal toxicity *in vivo*. Neurobiol Dis 55:64–75
147. Wen J, Li Y, Qin Y, Yan L, Zhang K, Li A, Wang Z, Yu F, Lai J, Yang W, Liu YU, Qin D, Su H (2024) Lycorine protects motor neurons against TDP-43 proteinopathy-induced degeneration in cross-species models with amyotrophic lateral sclerosis. Pharmacol Res 210:107518
148. Vaccaro A, Patten SA, Ciura S, Maios C, Therrien M, Drapeau P, Kabashi E, Parker JA (2012) Methylene blue protects against TDP-43 and FUS neuronal toxicity in *C. elegans* and *D. rerio*. PLoS One 7:e42117
149. Snoznik C, Medvedeva V, Mojsilovic-Petrovic J, Rudich P, Oosten J, Kalb RG, Lamitina T (2021) The nuclear ubiquitin ligase adaptor SPOP is a conserved regulator of C9orf72 dipeptide toxicity. Proc Natl Acad Sci USA 118:e2104664118
150. Witan H, Kern A, Koziollek-Drechsler I, Wade R, Behl C, Clement AM (2008) Heterodimer formation of wild-type and amyotrophic lateral sclerosis-causing mutant Cu/Zn-superoxide dismutase induces toxicity independent of protein aggregation. Hum Mol Genet 17:1373–1385
151. Zhai J, Zhang L, Mojsilovic-Petrovic J, Jian X, Thomas J, Homma K, Schmitz A, Famulok M, Ichijo H, Argon Y, Randazzo PA, Kalb RG (2015) Inhibition of cytohesins protects against genetic models of motor neuron disease. J Neurosci 35:9088–9105
152. Lim MA, Selak MA, Xiang Z, Krainc D, Neve RL, Kraemer BC, Watts JL, Kalb RG (2012) Reduced activity of AMP-activated protein kinase protects against genetic models of motor neuron disease. J Neurosci 32:1123–1141
153. Pereira-Sousa J, Ferreira-Lomba B, Bellver-Sanchis A, Vilasboas-Campos D, Fernandes JH, Costa MD, Varney MA, Newman-Tancredi A, Maciel P, Teixeira-Castro A (2021) Identification of the 5-HT(1A) serotonin receptor as a novel therapeutic target in a *C. elegans* model of Machado-Joseph disease. Neurobiol Dis 152:105278

154. Teixeira-Castro A, Jalles A, Esteves S, Kang S, da Silva Santos L, Silva-Fernandes A, Neto MF, Brielmann RM, Bessa C, Duarte-Silva S, Miranda A, Oliveira S, Neves-Carvalho A, Bessa J, Summavielle T, Silverman RB, Oliveira P, Morimoto RI, Maciel P (2015) Serotonergic signalling suppresses ataxin 3 aggregation and neurotoxicity in animal models of Machado-Joseph disease. Brain 138:3221–3237
155. Fardghassemi Y, Tauffenberger A, Gosselin S, Parker JA (2017) Rescue of ATXN3 neuronal toxicity in *Caenorhabditis elegans* by chemical modification of endoplasmic reticulum stress. Dis Model Mech 10:1465–1480
156. Sciandrone B, Palmioli A, Ciaramelli C, Pensotti R, Colombo L, Regonesi ME, Airoldi C (2024) Cell-free and in vivo characterization of the inhibitory activity of *Lavado cocoa* flavanols on the amyloid protein Ataxin-3: toward new approaches against spinocerebellar Ataxia type 3. ACS Chem Neurosci 15:278–289
157. Fardghassemi Y, Maios C, Parker JA (2021) Small molecule rescue of ATXN3 toxicity in *C. elegans* via TFEB/HLH-30. Neurotherapeutics 18:1151–1165
158. Vilasboas-Campos D, Costa MD, Teixeira-Castro A, Rios R, Silva FG, Bessa C, Dias ACP, Maciel P (2021) Neurotherapeutic effect of Hyptis spp. leaf extracts in *Caenorhabditis elegans* models of tauopathy and polyglutamine disease: role of the glutathione redox cycle. Free Radic Biol Med 162:202–215
159. Jalles A, Vieira C, Pereira-Sousa J, Vilasboas-Campos D, Mota AF, Vasconcelos S, Ferreira-Lomba B, Costa MD, Da Silva JD, Maciel P, Teixeira-Castro A (2022) Aripiprazole offsets mutant ATXN3-induced motor dysfunction by targeting dopamine D(2) and serotonin 1A and 2A receptors in *C. elegans*. Biomedicine 10:370
160. Bonanomi M, Natalello A, Visentin C, Pastori V, Penco A, Cornelli G, Colombo G, Malabarba MG, Doglia SM, Relini A, Regonesi ME, Tortora P (2014) Epigallocatechin-3-gallate and tetracycline differently affect ataxin-3 fibrillogenesis and reduce toxicity in spinocerebellar ataxia type 3 model. Hum Mol Genet 23:6542–6552
161. Ventura N, Rea SL, Handerson ST, Condò I, Testi R, Johnson TE (2006) *C. elegans* as a model for Friedreich Ataxia. FASEB J 20:1029–1030
162. Vázquez-Manrique RP, González-Cabo P, Ros S, Aziz H, Baylis HA, Palau F (2006) Reduction of *Caenorhabditis elegans* frataxin increases sensitivity to oxidative stress, reduces lifespan, and causes lethality in a mitochondrial complex II mutant. FASEB J 20:172–174
163. Zarse K, Schulz TJ, Birringer M, Ristow M (2007) Impaired respiration is positively correlated with decreased life span in *Caenorhabditis elegans* models of Friedreich Ataxia. FASEB J 21: 1271–1275
164. Gonzalez-Cabo P, Ros S, Palau F (2010) Flavin adenine dinucleotide rescues the phenotype of frataxin deficiency. PLoS One 5:e8872
165. Barker E, Morgan A, Barclay JW (2023) A *Caenorhabditis elegans* model of autosomal dominant adult-onset neuronal ceroid lipofuscinosis identifies ethosuximide as a potential therapeutic. Hum Mol Genet 32:1772–1785
166. De Voer G, Jansen G, van Ommen GJ, Peters DJ, Taschner PE (2001) *Caenorhabditis elegans* homologues of the CLN3 gene, mutated in juvenile neuronal ceroid lipofuscinosis. Eur J Paediatr Neurol Suppl A 5:115–120
167. Kwon YJ, Falk MJ, Bennett MJ (2017) Flunarizine rescues reduced lifespan in CLN3 triple knock-out *Caenorhabditis elegans* model of batten disease. J Inherit Metab Dis 40:291–296
168. Kashyap SS, Johnson JR, McCue HV, Chen X, Edmonds MJ, Ayala M, Graham ME, Jenn RC, Barclay JW, Burgoyne RD, Morgan A (2014) Caenorhabditis elegans dnj-14, the orthologue of the DNAJC5 gene mutated in adult onset neuronal ceroid lipofuscinosis, provides a new platform for neuroprotective drug screening and identifies a SIR-2.1-independent action of resveratrol. Hum Mol Genet 23:5916–5927
169. Park KH, Sesti F (2007) An arrhythmia susceptibility gene in *Caenorhabditis elegans*. J Biol Chem 282:19799–19807

170. Jiang Q, Li K, Lu WJ, Li S, Chen X, Liu XJ, Yuan J, Ding Q, Lan F, Cai SQ (2018) Identification of small-molecule ion channel modulators in *C. elegans* channelopathy models. Nat Commun 9:3941
171. Bend EG, Si Y, Stevenson DA, Bayrak-Toydemir P, Newcomb TM, Jorgensen EM, Swoboda KJ (2016) NALCN channelopathies: distinguishing gain-of-function and loss-of-function mutations. Neurology 87:1131–1139
172. Sultana J, Castaño JD, Benkhraba K, Del Castillo JRE, Beaudry F (2025) Voltage-gated calcium channels play a central role in *Caenorhabditis elegans* response to noxious heat. Neurochem Res 50:215
173. Ernest-Hoar GF, Simmons ME, Loucks C (2026) *Caenorhabditis elegans* as a model to explore the genetic underpinnings of human pain-related processes: cannabinoid and opioid neuropharmacology as an example. Can J Physiol Pharmacol. https://doi.org/10.1139/cjpp-2025-0303
174. Lindy AS, Parekh PK, Zhu R, Kanju P, Chintapalli SV, Tsvilovskyy V, Patterson RL, Anishkin A, van Rossum DB, Liedtke WB (2014) TRPV channel-mediated calcium transients in nociceptor neurons are dispensable for avoidance behaviour. Nat Commun 5:4734
175. Feng Z, Li W, Ward A, Piggott BJ, Larkspur ER, Sternberg PW, Xu XZ (2006) A *C. elegans* model of nicotine-dependent behavior: regulation by TRP-family channels. Cell 127:621–633
176. Ellwood RA, Piasecki M, Szewczyk NJ (2021) *Caenorhabditis elegans* as a model system for Duchenne muscular dystrophy. Int J Mol Sci 22:4891
177. Shalash R, Levi-Ferber M, Cohen C, Dori A, Brodie C, Henis-Korenblit S (2024) Cross-species modeling of muscular dystrophy in *Caenorhabditis elegans* using patient-derived extracellular vesicles. Dis Model Mech 17:dmm050412
178. Oh KH, Kim H (2013) Reduced IGF signaling prevents muscle cell death in a *Caenorhabditis elegans* model of muscular dystrophy. Proc Natl Acad Sci USA 110:19024–19029
179. Hewitt JE, Pollard AK, Lesanpezeshki L, Deane CS, Gaffney CJ, Etheridge T, Szewczyk NJ, Vanapalli SA (2018) Muscle strength deficiency and mitochondrial dysfunction in a muscular dystrophy model of *Caenorhabditis elegans* and its functional response to drugs. Dis Model Mech 11:dmm036137
180. Gaud A, Simon JM, Witzel T, Carre-Pierrat M, Wermuth CG, Ségalat L (2004) Prednisone reduces muscle degeneration in dystrophin-deficient *Caenorhabditis elegans*. Neuromuscul Disord 14:365–370
181. Yoshina S, Izuhara L, Mashima R, Maejima Y, Kamatani N, Mitani S (2023) Febuxostat ameliorates muscle degeneration and movement disorder of the dystrophin mutant model in *Caenorhabditis elegans*. J Physiol Sci 73:28
182. Wu X, Nagasawa S, Muto K, Ueda M, Suzuki C, Abe T, Higashitani A (2022) Mitochonic acid 5 improves Duchenne muscular dystrophy and Parkinson's disease model of *Caenorhabditis elegans*. Int J Mol Sci 23:9572
183. Ellwood RA, Hewitt JE, Torregrossa R, Philp AM, Hardee JP, Hughes S, van de Klashorst D, Gharahdaghi N, Anupom T, Slade L, Deane CS, Cooke M, Etheridge T, Piasecki M, Antebi A, Lynch GS, Philp A, Vanapalli SA, Whiteman M, Szewczyk NJ (2021) Mitochondrial hydrogen sulfide supplementation improves health in the *C. elegans* Duchenne muscular dystrophy model. Proc Natl Acad Sci USA 118:e2018342118
184. Etheridge T, Oczypok EA, Lehmann S, Fields BD, Shephard F, Jacobson LA, Szewczyk NJ (2012) Calpains mediate integrin attachment complex maintenance of adult muscle in *Caenorhabditis elegans*. PLoS Genet 8:e1002471
185. Ellwood RA, Slade L, Lewis J, Torregrossa R, Sudevan S, Piasecki M, Whiteman M, Etheridge T, Szewczyk NJ (2022) Sulfur amino acid supplementation displays therapeutic potential in a *C. elegans* model of Duchenne muscular dystrophy. Commun Biol 5:1255
186. Mattout A, Pike BL, Towbin BD, Bank EM, Gonzalez-Sandoval A, Stadler MB, Meister P, Gruenbaum Y, Gasser SM (2011) An EDMD mutation in *C. elegans* lamin blocks muscle-specific gene relocation and compromises muscle integrity. Curr Biol 21:1603–1614

187. Bar-Lavan Y, Shemesh N, Dror S, Ofir R, Yeger-Lotem E, Ben-Zvi A (2016) A differentiation transcription factor establishes muscle-specific proteostasis in *Caenorhabditis elegans*. PLoS Genet 12:e1006531
188. Gieseler K, Grisoni K, Ségalat L (2000) Genetic suppression of phenotypes arising from mutations in dystrophin-related genes in *Caenorhabditis elegans*. Curr Biol 10:1092–1097
189. Gieseler K, Grisoni K, Mariol MC, Ségalat L (2002) Overexpression of dystrobrevin delays locomotion defects and muscle degeneration in a dystrophin-deficient *Caenorhabditis elegans*. Neuromuscul Disord 12:371–377
190. Soni P, Edwards H, Anupom T, Rahman M, Lesanpezeshki L, Blawzdziewicz J, Cope H, Gharahdaghi N, Scott D, Toh LS, Williams PM, Etheridge T, Szewczyk N, Willis CRG, Vanapalli SA (2023) Spaceflight induces strength decline in *Caenorhabditis elegans*. Cells 12: 2470
191. Locke CJ, Williams SN, Schwarz EM, Caldwell GA, Caldwell KA (2006) Genetic interactions among cortical malformation genes that influence susceptibility to convulsions in *C. elegans*. Brain Res 1120:23–34
192. Câmara DF, Machado ML, Arantes LP, Silva TC, Silveira TL, Leal JG, Dornelles L, Stefanello ST, Soares FAA (2019) MPMT-OX up-regulates GABAergic transmission and protects against seizure-like behavior in *Caenorhabditis elegans*. Neurotoxicology 74:272–281
193. Jones A, Barker-Haliski M, Ilie AS, Herd MB, Baxendale S, Holdsworth CJ, Ashton JP, Placzek M, Jayasekera BAP, Cowie CJA, Lambert JJ, Trevelyan AJ, White HS, Marson AG, Cunliffe VT, Sills GJ, Morgan A (2020) A multiorganism pipeline for antiseizure drug discovery: identification of chlorothymol as a novel γ-aminobutyric acidergic anticonvulsant. Epilepsia 61:2106–2118
194. Locke CJ, Kautu BB, Berry KP, Lee SK, Caldwell KA, Caldwell GA (2009) Pharmacogenetic analysis reveals a post-developmental role for Rac GTPases in *Caenorhabditis elegans* GABAergic neurotransmission. Genetics 183:1357–1372
195. Di Rocco M, Galosi S, Lanza E, Tosato F, Caprini D, Folli V, Friedman J, Bocchinfuso G, Martire A, Di Schiavi E, Leuzzi V, Martinelli S (2022) *Caenorhabditis elegans* provides an efficient drug screening platform for GNAO1-related disorders and highlights the potential role of caffeine in controlling dyskinesia. Hum Mol Genet 31:929–941
196. Wong SQ, Jones A, Dodd S, Grimes D, Barclay JW, Marson AG, Cunliffe VT, Burgoyne RD, Sills GJ, Morgan A (2018) A *Caenorhabditis elegans* assay of seizure-like activity optimised for identifying antiepileptic drugs and their mechanisms of action. J Neurosci Methods 309: 132–142
197. Zhu B, Mak JCH, Morris AP, Marson AG, Barclay JW, Sills GJ, Morgan A (2020) Functional analysis of epilepsy-associated variants in STXBP1/Munc18-1 using humanized *Caenorhabditis elegans*. Epilepsia 61:810–821
198. Takayanagi-Kiya S, Jin Y (2016) Altered function of the DnaJ family cochaperone DNJ-17 modulates locomotor circuit activity in a *Caenorhabditis elegans* seizure model. G3 6:2165–2171
199. Ide S, Kunitomo H, Iino Y, Ikeda K (2022) *Caenorhabditis elegans* exhibits morphine addiction-like behavior via the opioid-like receptor NPR-17. Front Pharmacol 12:802701
200. Ide S, Ikeda K (2024) *Caenorhabditis elegans* for opioid addiction research. Curr Opin Neurobiol 88:102914
201. He Y, Wang ZJ (2012) *let-7* microRNAs and opioid tolerance. Front Genet 3:110
202. Engleman EA, Steagall KB 2nd, Bredhold KE, Breach M, Kline HL, Bell RL, Katner SN, Neal-Beliveau BS (2018) *Caenorhabditis elegans* show preference for stimulants and potential as a model organism for medications screening. Front Physiol 9:1200
203. Salim C, Batsaikhan E, Kan AK, Chen H, Jee C (2024) Nicotine motivated behavior in *C. elegans*. Int J Mol Sci 25:1634

204. Waggoner LE, Dickinson KA, Poole DS, Tabuse Y, Miwa J, Schafer WR (2000) Long-term nicotine adaptation in *Caenorhabditis elegans* involves PKC-dependent changes in nicotinic receptor abundance. J Neurosci 20:8802–8811
205. Taki FA, Pan X, Zhang B (2014) Chronic nicotine exposure systemically alters microRNA expression profiles during post-embryonic stages in *Caenorhabditis elegans*. J Cell Physiol 229:79–89
206. Johnson JR, Rajamanoharan D, McCue HV, Rankin K, Barclay JW (2016) Small heat shock proteins are novel common determinants of alcohol and nicotine sensitivity in *Caenorhabditis elegans*. Genetics 202:1013–1027
207. Zhu G, Zhang F, Li W (2014) Nematodes feel a craving – using *Caenorhabditis elegans* as a model to study alcohol addiction. Neurosci Bull 30:595–600
208. Scholz H (2025) From natural behavior to drug screening: invertebrates as models to study mechanisms associated with alcohol use disorders. Curr Top Behav Neurosci 71:145–167
209. Jee C, Batsaikhan E (2024) JNK signaling positively regulates acute ethanol tolerance in *C. elegans*. Int J Mol Sci 25:6398
210. Davis SJ, Scott LL, Ordemann G, Philpo A, Cohn J, Pierce-Shimomura JT (2015) Putative calcium-binding domains of the *Caenorhabditis elegans* BK channel are dispensable for intoxication and ethanol activation. Genes Brain Behav 14:454–465
211. Scott LL, Davis SJ, Yen RC, Ordemann GJ, Nordquist SK, Bannai D, Pierce JT (2017) Behavioral deficits following withdrawal from chronic ethanol are influenced by SLO channel function in *Caenorhabditis elegans*. Genetics 206:1445–1458
212. Scott LL, Sahn JJ, Ferragud A, Yen RC, Satarasinghe PN, Wood MD, Hodges TR, Shi T, Prakash BA, Friese KM, Shen A, Sabino V, Pierce JT, Martin SF (2018) Small molecule modulators of sigma2R/Tmem97 reduce alcohol withdrawal-induced behaviors. Neuropsychopharmacology 43:1867–1875
213. Summers PJ, Layne RM, Ortega AC, Harris GP, Bamber BA, Komuniecki RW (2015) Multiple sensory inputs are extensively integrated to modulate nociception in *C. elegans*. J Neurosci 35:10331–10342
214. Chao MY, Komatsu H, Fukuto HS, Dionne HM, Hart AC (2004) Feeding status and serotonin rapidly and reversibly modulate a *Caenorhabditis elegans* chemosensory circuit. Proc Natl Acad Sci USA 101:15512–15517
215. Robles V, Balaguer F, Maicas M, Martínez-Vázquez JM, Martorell P, Tortajada M, Ramón D, Valcarce DG (2024) The effect of the combination of two postbiotics on anxiety-like behavior in animal models. Cells 13:2006
216. Calahorro F, Ruiz-Rubio M (2013) Human alpha- and beta-NRXN1 isoforms rescue behavioral impairments of *Caenorhabditis elegans* neurexin-deficient mutants. Genes Brain Behav 12:453–464
217. Oliver D, Alexander K, Francis MM (2018) Molecular mechanisms directing spine outgrowth and synaptic partner selection in *Caenorhabditis elegans*. J Exp Neurosci 12: 1179069518816088
218. Hu Z, Hom S, Kudze T, Tong XJ, Choi S, Aramuni G, Zhang W, Kaplan JM (2012) Neurexin and neuroligin mediate retrograde synaptic inhibition in *C. elegans*. Science 337:980–984
219. Bastien BL, Cowen MH, Hart MP (2023) Distinct neurexin isoforms cooperate to initiate and maintain foraging activity. Transl Psychiatry 13:367
220. Cowen MH, Haskell D, Zoga K, Reddy KC, Chalasani SH, Hart MP (2024) Conserved autism-associated genes tune social feeding behavior in *C. elegans*. Nat Commun 15:9301
221. Hunter JW, Mullen GP, McManus JR, Heatherly JM, Duke A, Rand JB (2010) Neuroligin-deficient mutants of *C. elegans* have sensory processing deficits and are hypersensitive to oxidative stress and mercury toxicity. Dis Model Mech 3:366–376
222. Izquierdo PG, Calahorro F, Ruiz-Rubio M (2013) Neuroligin modulates the locomotory dopaminergic and serotonergic neuronal pathways of *C. elegans*. Neurogenetics 14:233–242
223. McDiarmid TA, Belmadani M, Liang J, Meili F, Mathews EA, Mullen GP, Hendi A, Wong WR, Rand JB, Mizumoto K, Haas K, Pavlidis P, Rankin CH (2020) Systematic phenomics

analysis of autism-associated genes reveals parallel networks underlying reversible impairments in habituation. Proc Natl Acad Sci USA 117:656–667

224. Calahorro F, Ruiz-Rubio M (2012) Functional phenotypic rescue of *Caenorhabditis elegans* neuroligin-deficient mutants by the human and rat NLGN1 genes. PLoS One 7:e39277
225. Jee C, Lee J, Lee JI, Lee WH, Park BJ, Yu JR, Park E, Kim E, Ahnn J (2004) SHN-1, a Shank homologue in *C. elegans*, affects defecation rhythm via the inositol-1,4,5-trisphosphate receptor. FEBS Lett 561:29–36
226. Oh WC, Song HO, Cho JH, Park BJ (2011) ANK repeat-domain of SHN-1 is indispensable for in vivo SHN-1 function in *C. elegans*. Mol Cells 31:79–84
227. Gao L, Zhao J, Ardiel E, Hall Q, Nurrish S, Kaplan JM (2022) Shank promotes action potential repolarization by recruiting BK channels to calcium microdomains. elife 11:e75140
228. Jamadagni P, Breuer M, Schmeisser K, Cardinal T, Kassa B, Parker A, Pilon N, Samarut E, Patten SA (2021) Chromatin remodeller CHD7 is required for GABAergic neuron development by promoting PAQR3 expression. EMBO Rep 22:e50958
229. AlAbdi L, Desbois M, Rusnac DV, Sulaiman RA, Rosenfeld JA, Lalani S, Murdock DR, Burrage LC, Undiagnosed Diseases Network, Billie Au PY, Towner S, Wilson WG, Wong L, Brunet T, Strobl-Wildemann G, Burton JE, Hoganson G, McWalter K, Begtrup A, Zarate YA, Christensen EL, Opperman KJ, Giles AC, Helaby R, Kania A, Zheng N, Grill B, Alkuraya FS (2023) Loss-of-function variants in MYCBP2 cause neurobehavioural phenotypes and corpus callosum defects. Brain 146:1373–1387
230. Tulgren ED, Turgeon SM, Opperman KJ, Grill B (2014) The Nesprin family member ANC-1 regulates synapse formation and axon termination by functioning in a pathway with RPM-1 and beta-Catenin. PLoS Genet 10:e1004481
231. Fischer NC, Friedman V, Martinez-Reyes MA, Hao H, Chowdhury TA, Starr DA, Quinn CC (2022) The ANC-1 (Nesprin-1/2) organelle-anchoring protein functions through mitochondria to polarize axon growth in response to SLT-1. PLoS Genet 18:e1010521
232. Dent JA, Davis MW, Avery L (1997) *avr-15* encodes a chloride channel subunit that mediates inhibitory glutamatergic neurotransmission and ivermectin sensitivity in *Caenorhabditis elegans*. EMBO J 16:5867–5879
233. Chowdhury TA, Luy DA, Scapellato G, Farache D, Lee ASY, Quinn CC (2024) Ortholog of autism candidate gene RBM27 regulates mitoribosomal assembly factor MALS-1 to protect against mitochondrial dysfunction and axon degeneration during neurodevelopment. PLoS Biol 22:e3002876
234. Giunti S, Blanco MG, De Rosa MJ, Rayes D (2024) The ketone body beta-hydroxybutyrate ameliorates neurodevelopmental deficits in the GABAergic system of daf-18/PTEN *Caenorhabditis elegans* mutants. elife 13:RP94520
235. Buddell T, Friedman V, Drozd CJ, Quinn CC (2019) An autism-causing calcium channel variant functions with selective autophagy to alter axon targeting and behavior. PLoS Genet 15:e1008488
236. Tong XJ, López-Soto EJ, Li L, Liu H, Nedelcu D, Lipscombe D, Hu Z, Kaplan JM (2017) Retrograde synaptic inhibition is mediated by α-Neurexin binding to the α2δ subunits of N-type calcium channels. Neuron 95:326–340
237. Zoga K, Villiere S, Tikiyani V, Edwards-Cintron AF, Thokachichu P, Nicodemus P, Camara PG, Hart MP (2025) Multiple autism genes influence GABA neuron remodeling via distinct developmental trajectories. Genetics 231:iyaf156
238. Francis R, Barton MK, Kimble J, Schedl T (1995) *gld-1*, a tumor suppressor gene required for oocyte development in *Caenorhabditis elegans*. Genetics 139:579–606
239. Ellis RE, Kimble J (1994) Control of germ cell differentiation in *Caenorhabditis elegans*. Ciba Found Symp 182:179–188
240. Henarejos-Escudero P, Méndez-García FF, Hernández-García S, Martínez-Rodríguez P, Gandía-Herrero F (2023) Design, synthesis and gene modulation insights into pigments derived from tryptophan-betaxanthin, which act against tumor development in *Caenorhabditis elegans*. Int J Mol Sci 25:63

241. Galluzzi L, Joza N, Tasdemir E, Maiuri MC, Hengartner M, Abrams JM, Tavernarakis N, Penninger J, Madeo F, Kroemer G (2008) No death without life: vital functions of apoptotic effectors. Cell Death Differ 15:1113–1123
242. Harris J, Lowden M, Clejan I, Tzoneva M, Thomas JH, Hodgkin J, Ahmed S (2006) Mutator phenotype of *Caenorhabditis elegans* DNA damage checkpoint mutants. Genetics 174:601–616
243. Luo J, Shah S, Riabowol K, Mains PE (2009) The *Caenorhabditis elegans ing-3* gene regulates ionizing radiation-induced germ-cell apoptosis in a p53-associated pathway. Genetics 181:473–482
244. Elsherbiny SM, Shao C, Acheampong A, Khalifa MA, Liu C, Huang Q (2022) Green synthesis of broccoli-derived carbon quantum dots as effective photosensitizers for the PDT effect testified in the model of mutant *Caenorhabditis elegans*. Biomater Sci 10:2857–2864
245. Wen J, Yang X, Liu Y, Qiu Y, Jia N, Li J, Zhao C (2025) Anticancer bioactive phytochemicals screening from medicinal-edible homologous materials in *Caenorhabditis elegans*. Phytomedicine 148:157471
246. Deng X, Michaelson D, Tchieu J, Cheng J, Rothenstein D, Feldman R, Lee SG, Fuller J, Haimovitz-Friedman A, Studer L, Powell S, Fuks Z, Hubbard EJ, Kolesnick R (2015) Targeting homologous recombination in Notch-driven *C. elegans* stem cell and human tumors. PLoS One 10:e0127862
247. Beitel GJ, Clark SG, Horvitz HR (1990) *Caenorhabditis elegans* ras gene *let-60* acts as a switch in the pathway of vulval induction. Nature 348:503–509
248. Marco A, Ashoo P, Hernández-García S, Martínez-Rodríguez P, Cutillas N, Vollrath A, Jordan D, Janiak C, Gandía-Herrero F, Ruiz J (2024) Novel Re(I) complexes as potential selective theranostic agents in cancer cells and *in vivo* in *Caenorhabditis elegans* tumoral strains. J Med Chem 67:7891–7910
249. Levantini E, Maroni G, Del Re M, Tenen DG (2022) EGFR signaling pathway as therapeutic target in human cancers. Semin Cancer Biol 85:253–275
250. Akhoon BA, Rathor L, Pandey R (2018) Withanolide A extends the lifespan in human EGFR-driven cancerous *Caenorhabditis elegans*. Exp Gerontol 104:113–117
251. Xu S, Chisholm AD (2014) Methods for skin wounding and assays for wound responses in *C. elegans*. J Vis Exp 94:51959
252. Belougne J, Ozerov I, Caillard C, Bedu F, Ewbank JJ (2020) Fabrication of sharp silicon arrays to wound *Caenorhabditis elegans*. Sci Rep 10:3581
253. Pooranachithra M, Bhaskar JP, Murali D, Das SS, JebaMercy G, Krishnan V, Balamurugan K (2019) Unravelling the wound healing ability and mode of action of pyridine carboxamide oxime using *Caenorhabditis elegans* as potential prescreen wound model. Life Sci 235: 116859
254. Thangavel H, Muthusamy V, Govindhan T, Palanisamy S, Paramasivam P (2025) Almond gum/gelatin-based multifunctional nanocomposite hydrogel: a synergistic hybrid platform for wound healing. ACS Appl Biol Mater 8:10933–10952
255. Samara C, Rohde CB, Gilleland CL, Norton S, Haggarty SJ, Yanik MF (2010) Large-scale in vivo femtosecond laser neurosurgery screen reveals small-molecule enhancer of regeneration. Proc Natl Acad Sci USA 107:18342–18347
256. Behera S, Ghosh Roy A (2026) Exercise-induced differential transcriptional output of AMPK signalling improves axon regeneration and functional recovery. J Neurosci 28:e0124252026
257. Shimizu T, Hisamoto N (2020) Factors regulating axon regeneration via JNK MAP kinase in *Caenorhabditis elegans*. J Biochem 167:433–439
258. El Bejjani R, Hammarlund M (2012) Neural regeneration in *Caenorhabditis elegans*. Annu Rev Genet 46:499–513
259. Chin MR, Zlotkowski K, Han M, Patel S, Eliasen AM, Axelrod A, Siegel D (2015) Expedited access to vinaxanthone and chemically edited derivatives possessing neuronal regenerative effects through ynone coupling reactions. ACS Chem Neurosci 6:542–550

260. Casas RB, Risato G, Zuppardo A, Viscomi C, Argenton F, Doimo M, Facchinello N, Tiso N (2025) Model organisms in POLG-related disorders: insights from yeast to multicellular systems. Cell Death Dis 17:137
261. Meyrick J, Stefanetti RJ, Errington L, McFarland R, Gorman GS, Lax NZ (2025) Model systems informing mechanisms and drug discovery: a review of POLG-related disease models. Wellcome Open Res 8:33
262. Pitayu L, Baruffini E, Rodier C, Rötig A, Lodi T, Delahodde A (2016) Combined use of Saccharomyces cerevisiae, Caenorhabditis elegans and patient fibroblasts leads to the identification of clofilium tosylate as a potential therapeutic chemical against POLG-related diseases. Hum Mol Genet 25:715–727
263. Yang Q, Liu P, Anderson NS, Shpilka T, Du Y, Naresh NU, Li R, Zhu LJ, Luk K, Lavelle J, Zeinert RD, Chien P, Wolfe SA, Haynes CM (2022) LONP-1 and ATFS-1 sustain deleterious heteroplasmy by promoting mtDNA replication in dysfunctional mitochondria. Nat Cell Biol 24:181–193
264. O'Reilly LP, Perlmutter DH, Silverman GA, Pak SC (2014) alpha1-antitrypsin deficiency and the hepatocytes – an elegans solution to drug discovery. Int J Biochem Cell Biol 47:109–112
265. Long OS, Benson JA, Kwak JH, Luke CJ, Gosai SJ, O'Reilly LP, Wang Y, Li J, Vetica AC, Miedel MT, Stolz DB, Watkins SC, Züchner S, Perlmutter DH, Silverman GA, Pak SC (2014) A *C. elegans* model of human alpha1-antitrypsin deficiency links components of the RNAi pathway to misfolded protein turnover. Hum Mol Genet 23:5109–5122
266. Cummings EE, O'Reilly LP, King DE, Silverman RM, Miedel MT, Luke CJ, Perlmutter DH, Silverman GA, Pak SC (2015) Deficient and null variants of SERPINA1 areproteotoxic in a *Caenorhabditis elegans* model of alpha1-antitrypsin deficiency. PLoS One 10:e0141542
267. O'Reilly LP, Long OS, Cobanoglu MC, Benson JA, Luke CJ, Miedel MT, Hale P, Perlmutter DH, Bahar I, Silverman GA, Pak SC (2014) A genome-wide RNAi screen identifies potential drug targets in a *C. elegans* model of alpha1-antitrypsin deficiency. Hum Mol Genet 23:5123–5132
268. Gosai SJ, Kwak JH, Luke CJ, Long OS, King DE, Kovatch KJ, Johnston PA, Shun TY, Lazo JS, Perlmutter DH, Silverman GA, Pak SC (2010) Automated high-content live animal drug screening using *C. elegans* expressing the aggregation prone serpin alpha1-antitrypsin Z. PLoS One 5:e15460
269. Li J, Pak SC, O'Reilly LP, Benson JA, Wang Y, Hidvegi T, Hale P, Dippold C, Ewing M, Silverman GA, Perlmutter DH (2014) Fluphenazine reduces proteotoxicity in *C. elegans* and mammalian models of alpha-1-antitrypsin deficiency. PLoS One 9:e87260
270. Wang Y, Cobanoglu MC, Li J, Hidvegi T, Hale P, Ewing M, Chu AS, Gong Z, Muzumdar R, Pak SC, Silverman GA, Bahar I, Perlmutter DH (2019) An analog of glibenclamide selectively enhances autophagic degradation of misfolded alpha1-antitrypsin Z. PLoS One 14:e0209748
271. Sifri CD, Begun J, Ausubel FM (2005) The worm has turned – microbial virulence modeled in *Caenorhabditis elegans*. Trends Microbiol 13:119–127
272. Alegado RA, Campbell MC, Chen WC, Slutz SS, Tan MW (2003) Characterization of mediators of microbial virulence and innate immunity using the *Caenorhabditis elegans* host-pathogen model. Cell Microbiol 5:435–444
273. Engelmann I, Pujol N (2010) Innate immunity in *C. elegans*. Adv Exp Med Biol 708:105–121
274. Zhi L-T, Yu Y-L, Li X-Y, Wang D-Y (2017) Molecular control of innate immune response to *Pseudomonas aeruginosa* infection by intestinal *let-7* in *Caenorhabditis elegans*. PLoS Pathog 13:e1006152
275. Yu Y-L, Zhi L-T, Wu Q-L, Jing L-N, Wang D-Y (2018) NPR-9 regulates innate immune response in *Caenorhabditis elegans* by antagonizing activity of AIB interneurons. Cell Mol Immunol 15:27–37
276. Kong C, Eng SA, Lim MP, Nathan S (2016) Beyond traditional antimicrobials: a *Caenorhabditis elegans* model for discovery of novel anti-infectives. Front Microbiol 7:1956
277. Hajdú G, Szathmári C, Sőti C (2024) Modeling host-pathogen interactions in *C. elegans*: lessons learned from *Pseudomonas aeruginosa* infection. Int J Mol Sci 25:7034

278. Peterson ND, Pukkila-Worley R (2018) *Caenorhabditis elegans* in high-throughput screens for anti-infective compounds. Curr Opin Immunol 54:59–65
279. Wu C, Karakuzu O, Garsin DA (2021) Tribbles pseudokinase NIPI-3 regulates intestinal immunity in *Caenorhabditis elegans* by controlling SKN-1/Nrf activity. Cell Rep 36:109529
280. Zhi L-T, Yu Y-L, Jiang Z-X, Wang D-Y (2017) *mir-355* functions as an important link between p38 MAPK signaling and insulin signaling in the regulation of innate immunity. Sci Rep 7:14560
281. Liu J, Lei M, Wang M, Chen S, Tu H (2024) Protocol for survival assay of *Caenorhabditis elegans* to *Pseudomonas aeruginosa* PA14 infection. STAR Protoc 5:103070
282. Zou CG, Tu Q, Niu J, Ji XL, Zhang KQ (2013) The DAF-16/FOXO transcription factor functions as a regulator of epidermal innate immunity. PLoS Pathog 9:e1003660
283. Grover M, Gang SS, Troemel ER, Barkoulas M (2024) Proteasome inhibition triggers tissue-specific immune responses against different pathogens in *C. elegans*. PLoS Biol 22:e3002543
284. Jakobsen H, Bojer MS, Marinus MG, Xu T, Struve C, Krogfelt KA, Løbner-Olesen A (2013) The alkaloid compound harmane increases the lifespan of *Caenorhabditis elegans* during bacterial infection, by modulating the nematode' innate immune response. PLoS One 8: e60519
285. Bajire SK, Prabhu A, Bhandary YP, Irfan KM, Shastry RP (2023) 7-Ethoxycoumarin rescued *Caenorhabditis elegans* from infection of COPD derived clinical isolate *Pseudomonas aeruginosa* through virulence and biofilm inhibition via targeting Rhl and Pqs quorum sensing systems. World J Microbiol Biotechnol 39:208
286. Cui Y, Wang R, Li X, Bai G, Xiao Y (2025) Ginkgolide a enhances the resistance to pathogen infection through mitochondrial unfolded protein response. Cell Mol Life Sci 82:349
287. Dharmalingam K, Tan BK, Mahmud MZ, Sedek SA, Majid MI, Kuah MK, Sulaiman SF, Ooi KL, Khan NA, Muhammad TS, Tan MW, Shu-Chien AC (2012) *Swietenia macrophylla* extract promotes the ability of *Caenorhabditis elegans* to survive *Pseudomonas aeruginosa* infection. J Ethnopharmacol 139:657–663
288. Hangweirer K, Sandner G, König A, Kirchsteiger J, Blank-Landeshammer B, Winkler S, Schurr J, Weghuber J (2025) *Allium ursinum* leaves aqueous extract alleviates *Pseudomonas aeruginosa* infection in *Caenorhabditis elegans*. Phytomedicine 146:157100
289. Wang Y-X, Zhang L, Yuan X-A, Wang D-Y (2023) Treatment with paeoniflorin increases lifespan of *Pseudomonas aeruginosa* infected *Caenorhabditis elegans* by inhibiting bacterial accumulation in intestinal lumen and biofilm formation. Front Pharmacol 14:1114219
290. Elkabti AB, Issi L, Rao RP (2018) *Caenorhabditis elegans* as a model host to monitor the *Candida* infection processes. J Fungi 4:123
291. Madende M, Albertyn J, Sebolai O, Pohl CH (2020) *Caenorhabditis elegans* as a model animal for investigating fungal pathogenesis. Med Microbiol Immunol 209:1–13
292. Pukkila-Worley R, Ausubel FM, Mylonakis E (2011) *Candida albicans* infection of *Caenorhabditis elegans* induces antifungal immune defenses. PLoS Pathog 7:e1002074
293. Muhammed M, Coleman JJ, Mylonakis E (2012) *Caenorhabditis elegans*: a nematode infection model for pathogenic fungi. Methods Mol Biol 845:447–454
294. Sun L-M, Zhi L-T, Shakoor S, Liao K, Wang D-Y (2016) microRNAs involved in the control of innate immunity in *Candida* infected *Caenorhabditis elegans*. Sci Rep 6:36036
295. Anastassopoulou CG, Fuchs BB, Mylonakis E (2011) *Caenorhabditis elegans*-based model systems for antifungal drug discovery. Curr Pharm Des 17:1225–1233
296. Pukkila-Worley R, Holson E, Wagner F, Mylonakis E (2009) Antifungal drug discovery through the study of invertebrate model hosts. Curr Med Chem 16:1588–1595
297. Breger J, Fuchs BB, Aperis G, Moy TI, Ausubel FM, Mylonakis E (2007) Antifungal chemical compounds identified using a *C. elegans* pathogenicity assay. PLoS Pathog 3:e18
298. Sun L-M, Liao K, Wang D-Y (2015) Effects of magnolol and honokiol on adhesion, yeast-hyphal transition, and formation of biofilm by *Candida albicans*. PLoS One 10:e0117695
299. Gammon DB (2017) *Caenorhabditis elegans* as an emerging model for virus-host interactions. J Virol 91:e00509–e00517

300. Casorla-Perez LA, Guennoun R, Cubillas C, Peng B, Kornfeld K, Wang D (2022) Orsay virus infection of *Caenorhabditis elegans* is modulated by zinc and dependent on lipids. J Virol 96: e0121122
301. Wu X, Liu H, Guo YR (2025) Insights into virus-host interactions: lessons from *Caenorhabditis elegans*-Orsay virus model. Curr Med Sci 45:69–184
302. Zhang L, Wang Y-X, Wang D-Y (2023) Paeoniflorin increases the survival of *Pseudomonas aeruginosa* infected *Caenorhabditis elegans* at the immunosuppression stage by activating PMK-1, BAR-1, and EGL-1 signals. Arch Pharm Res 46:616–628
303. Shu C-J, Wang W, Cao S-Q, Hou B-W, Gu S-L, Fu L, Zhang F-L, Wang D-Y (2026) 6-PPD quinone induces lifespan reduction by causing immunosuppression via DAF-16/PMK-1 signaling in Caenorhabditis elegans. Environ Chem Ecotoxicol 8:1176–1184

Pharmacological Safety Assessment 5

Abstract

Pharmacological safety assessment will provide an important support for the clinical pharmacological application. We first introduced several aspects needed to be carefully considered and useful endpoints for pharmacological safety assessment in *Caenorhabditis elegans*. After that, the detailed pharmacological safety assessment of plant compounds, plant extracts, Chinese Medicines, clinical drugs, synthesized compounds, and nanomaterials were discussed.

Keywords

Safety assessment · Compound · Drug · *Caenorhabditis elegans*

5.1 Introduction

Finding and confirmation of pharmacological effects of certain compounds and drugs should be supported by the pharmacological safety assessment. In *Caenorhabditis elegans*, the pharmacological safety assessment can be considered as the systematic evaluation of possible adverse effect of candidate compounds and drugs. The established systematic knowledge of toxicology in *C. elegans* provides important support about this [1–4]. *C. elegans* is suitable for the early-stage systematic toxicity assessment, the prioritization of lead compounds, and the raise of potential molecular pathway signals that is targeted by candidate compounds and drugs.

During the pharmacological assessment, the physiochemical properties of compounds and drugs should be carefully considered, especially the solubility and stability of compounds and drugs. Normally, the anticipated compound and drugs solutions should be fully soluble in the exposure medium (typically M9 buffer or K-medium) without precipitating. During the exposure, it should be also considered

D. Wang, *Pharmacology in Caenorhabditis elegans*,
https://doi.org/10.1007/978-981-92-5166-7_5

for the possibility of degradation and transformation of examined compounds and drugs in aqueous solution or under the lighting.

For the exposure of compounds and drugs to nematodes, the following several aspects are suggested to be carefully considered:

1. Exposure Routes: Liquid culture, agar-based assay, and microfluidic can be considered for the choice. Liquid culture is recommended as the first choice. If the amount of candidate compounds and drugs are very limited or the number of examined compounds and drugs is somewhat large, the microfluidic assay system can be selected.
2. Exposure Stages: The acute exposure to compounds and drugs can be performed at larval stages (such as the L1-larvae), young adult or adult day-1, or during the aging process. Based on the research aims and the future considered human populations, the suitable exposure stages can be selected.
3. Exposure Doses: The exposure doses of compounds and drugs are suggested to be designed from nM to μM or even mM to cover enough exposure doses. One of the reasons is the possible formation of hormesis after exposure to candidate compounds and drugs. Another reason is that the examined endpoints normally include both the lethality assay (LC50) and some sublethal endpoints.
4. Exposure Durations: Selection of acute exposure or long-term exposure is determined by examined compounds and drugs and their clinical application scenarios. If the examined compounds and drugs are lipid-soluble or needs to be repeated administrated, long-term exposure is suggested.
5. Internal Concentrations: After collection of nematodes from exposure solutions, the concentration of compounds in the body of nematodes can be determined and considered as the internal concentrations, which may be largely similar to that in liver of the mammals. Nevertheless, for some drugs (such plant extracts, Chinese Medicines, and nanomaterials), it is difficult to examine their exact internal concentrations.
6. Exposure Controls: Usually, both vehicle control and solvent control are needed. Besides these, some positive and negative controls can also be considered to be included.
7. Strains Selection: Normally, the wild-type is used for the pharmacological safety assessment. Nevertheless, considering the possible clinical application of drugs in sensitive and susceptible populations, some mutants (such as stress-sensitive mutants) are suggested to be covered in the used strains.

5.2 Endpoints for Pharmacological Safety Assessment

In principle, we suggest to use a panel of endpoints for the aim of pharmacological safety assessment. Normally, the following several aspects of endpoints can be used for the pharmacological safety assessment:

5.2.1 Lethality

LC50/LD50 refers to the concentration/dose causing 50% population lethality. To assess the LC50/LD50, the high doses of compounds and drugs should be included in the exposure doses. The lethality can be assessed by directly counting number of dead or survival nematodes or using certain fluorescent probe to indicate the dead nematodes.

5.2.2 Lifespan, Health Span, and Immunosuppression

For lifespan assay, both the maximum lifespan and median lifespan should be examined. The endpoint of lifespan can be used to reflect the long-term effect of compounds and drugs.

Health span indicates the physiological state during the aging, such as the assessment at adult day-8. The locomotion, physiology related endpoints (such as pumping rate and defecation cycle length), and stress response can be employed for the assessment of health span.

Immunosuppression refers to the innate immune response during the aging process, which actually reflects a certain aspect of health span. Expression of antimicrobial genes is used to indicate the innate immune response.

5.2.3 Development

Usually, the assessment of developmental toxicity needs to consider the morphological alteration (such as formation of multi-vulva structure), growth rate reflected by body size, and developmental delay (such as the larval arrest). Besides these, certain compounds and drugs may also disrupt the cell lineage.

5.2.4 Intestinal Toxicity

Intestine is the primary target organ for compounds and drugs. At least from these aspects, the intestinal toxicity of compounds and drugs can be assessed:

1. Intestinal Development: This can be assessed by directly observing the structural change based on transmission electron microscopy analysis and evaluating the expression of genes governing the intestinal development at different aspects.
2. Intestinal Function: The intestinal permeability can be evaluated by certain dye staining and then analyzing the dye distribution in different region of the body of nematodes.
3. Intestinal oxidative stress: This can be assessed by intestinal reactive oxygen species (ROS), some biochemical related endpoints, and expression of genes governing the oxidative stress.

5.2.5 Epidermal Toxicity

Normally, the compounds and drugs would not affect the development of epidermis. However, if the examined compounds are corrosive or have sharp physical structure, this aspect needs also to be considered. The epidermal development can be assessed by assay of transmission electron microscopy and scanning electron microscopy or evaluated using some GFP fused transgenic markers to reflect epidermal development at different aspects. For the epidermal function, the ammonia excretion can be considered.

5.2.6 Reproductive Toxicity

Several aspects of reproduction can be assessed after exposure to compounds and drugs:

1. Embryotoxicity: Such as the induction of embryo mortality.
2. Development of Reproductive Organs: Such as the damage on development of gonad, oogenesis, and induction of cell cycle arrest.
3. Induction of germline apoptosis and germline DNA damage.
4. Reproductive Capacity: This is reflected by brood size and number of eggs in the uterus.
5. Egg laying and vulva development.
6. Generation time to reflect the reproductive efficiency.
7. Mating ability.

5.2.7 Neurotoxicity

The following several aspects are suggested to assess the possible neurotoxicity induced by compounds and drugs:

1. Development of Nervous System: The GABAergic and dopaminergic nervous systems are more frequently used to assess the possibility of compounds and drugs in affecting neuronal development and in inducing the neurodegeneration.
2. Locomotion Behaviors and Development of Motor Neurons: Paralysis, head thrash, body bend, forward movement, backward movement, and movement speed can be assessed. The development of motor neurons provides the underlying developmental basis for the altered at least some locomotion behaviors.
3. Development of Sensory Neurons and Sensory Perception: Some of the sensory perceptions, such as chemotaxis, thermotaxis, and avoidance behavior, can be assessed. The development of sensory neurons provides the underlying developmental basis for the altered sensory perceptions.

4. Learning, Memory, and Choice Behaviors and Development of Interneurons: The development of interneurons provides important development basis for learning, memory, and choice behaviors.
5. Neurotransmission: Electrophysiological recording and some drugs experiments (such as aldicarb resistance) can be used to reflect the neurotransmission.

5.2.8 Metabolic Toxicity

After exposure to compounds and drugs, some biochemical metabolisms may be altered or disrupted. For the assessment of metabolic toxicity, normally the metabolic profiling analysis is suggested to be performed. Besides this, some of the following aspects are suggested to be further examined:

1. Lipid metabolism.
2. Glucose metabolism.
3. Citric acid cycle.
4. Electron transport chain reflected by mitochondrial function and enzyme activities and expression of genes for related components in mitochondrial complexes I–V.
5. Pharyngeal pumping rate and defecation cycle length.

5.2.9 Genetoxicity

The possible genetoxicity of compounds and drugs can be assessed by the following endpoints:

1. DNA damage.
2. DNA repair.
3. RNA processing errors.
4. Chromosome abnormality.

5.2.10 Transgenerational Toxicity

The toxicity of some compounds and drugs is sometimes not limited in one generation, and will last for multiple generations. The assessed endpoints to evaluate the transgenerational toxicity of compounds and drugs can be selected from the above introduced endpoints. Meanwhile, the transgenerational epigenetic changes (such as histone and DNA methylation and histone acetylation) are also suggested to be examined.

5.3 Pharmacological Safety Assessment of Plant Compounds

For pharmacological safety assessment of plant compounds, the following aspects need to be specially considered:

1. Solubility: The plant compounds are often hydrophobic (such as flavonoids and terpenoids). DMSO is usually essential as the solvent. Thus, it is required to confirm the examined plant compounds are fully dissolved and do not precipitate in the aqueous exposure medium. Or, the solutions of plant compounds should be refreshed during the exposure process.
2. Light Sensitivity: Some plant compounds are light-sensitive. For these plant compounds, exposure is suggested to be performed in the darkness.
3. Effect of Pigmentation and Autofluorescence: Some plant compounds are colored or have auto-fluorescence, which may interfere with assays of microscopy and fluorescent reporters. Use non-fluorescent phenotypic endpoints are suggested to corroborate this aspect of effects.

In *C. elegans*, accompanied with the identification of beneficial effects of 5-aryl-2-aminoimidazole, justicidin B, procatechuic acid, protocatechuic aldehyde, and N-gamma-(L-glutamyl)-L-selenomethionine, their pharmacological safety was also examined [5–8]. With paeoniflorin as the example of plant compounds, accompanied with the identification of its beneficial role in reducing glucose toxicity, paeoniflorin treatment at the examined doses did not affect lifespan, locomotion behavior, pumping rate, and brood size (Fig. 5.1) [9].

5.4 Pharmacological Safety Assessment of Plant Extracts

For the pharmacological assessment of plant extracts (crude, fractionated, or standardized), the following aspects need to be specially considered:

1. Batch-To-Batch Variability: Different harvests, seasons, and extraction methods can yield different compositions of plant extracts. Thus, it is suggested to always use the same characterized batch for a full study series.
2. Chemical Fingerprint: For the prepared plant extracts, the chemical fingerprint should be provided based on methods of HPLC, LC-MS, or GC-MS. In the prepared plant extracts, the major compound classes need to be documented, and at least 1–2 marker compounds need to be further quantified.
3. Choice of Suitable Solvents: Crude plant extracts may contain both hydrophilic and lipophilic components, which sometimes require the co-solvent system (e.g., DMSO + small percentage of Tween-80 or ethanol) to ensure a clear, homogeneous solution without precipitation after dilution in assay buffer.
4. Concentration Expression: The concentrations of prepared plant extracts can be expressed in two ways: extract weight/volume (e.g., 100 μg/mL certain leaf

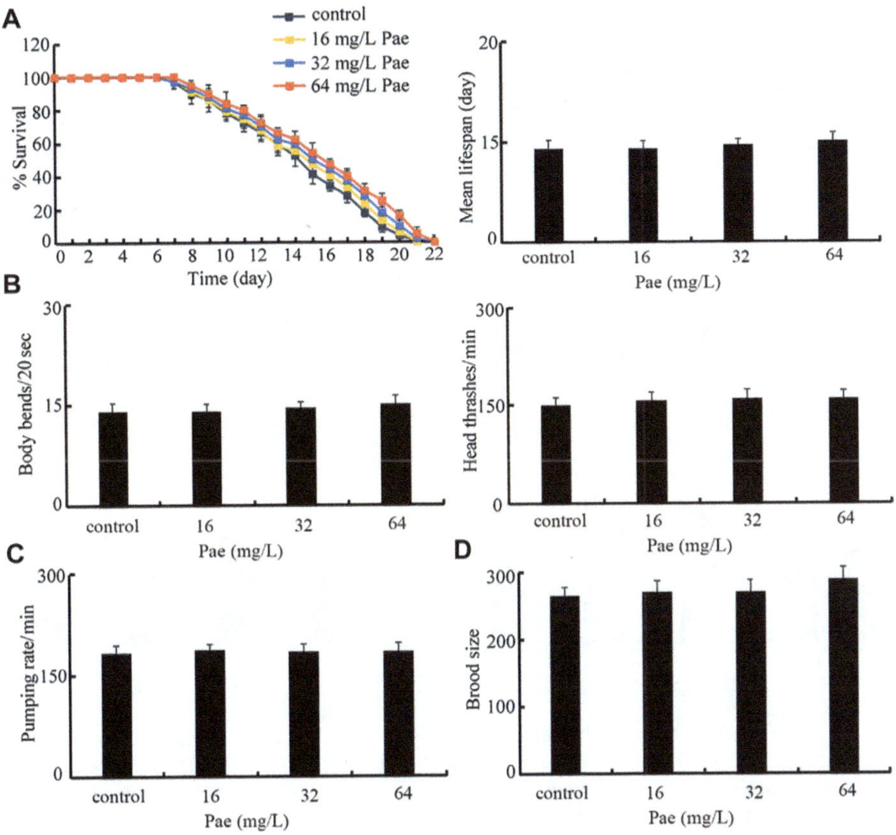

Fig. 5.1 Safety assessment of 16–64 mg/L paeoniflorin on lifespan (**a**), locomotion behavior (**b**), pumping rate (**c**), and brood size (**d**) [9]. Pae paeoniflorin

extract) or marker compound equivalent (e.g., "…containing 100 μg/mL certain compound").

5. Control Groups: Besides the vehicle control and solvent control, the spiked control is suggested by adding a known amount of pure compound to vehicle control solution to exclude the possible effect of solvent to polyphenol-class molecules. For the plant extracts with color of autofluorescence, background in imaging assays is suggested to be measured.
6. The Effect of Changed pH: Some plant crude extracts may alter the pH of the medium, causing the stress to the examined nematodes. Thus, after adding the plant extract, the pH values are suggested to be measured and adjusted to neutral (~6.0–7.0) if necessary.

In *C. elegans*, accompanied with the identification of beneficial effects for extracts of *Allium ursinum* leaves, *Vismia guianensis*, *Tristerix corymbosus*, and *Ceiba speciosa* bark, their exposure was confirmed to be safe [10–13]. With

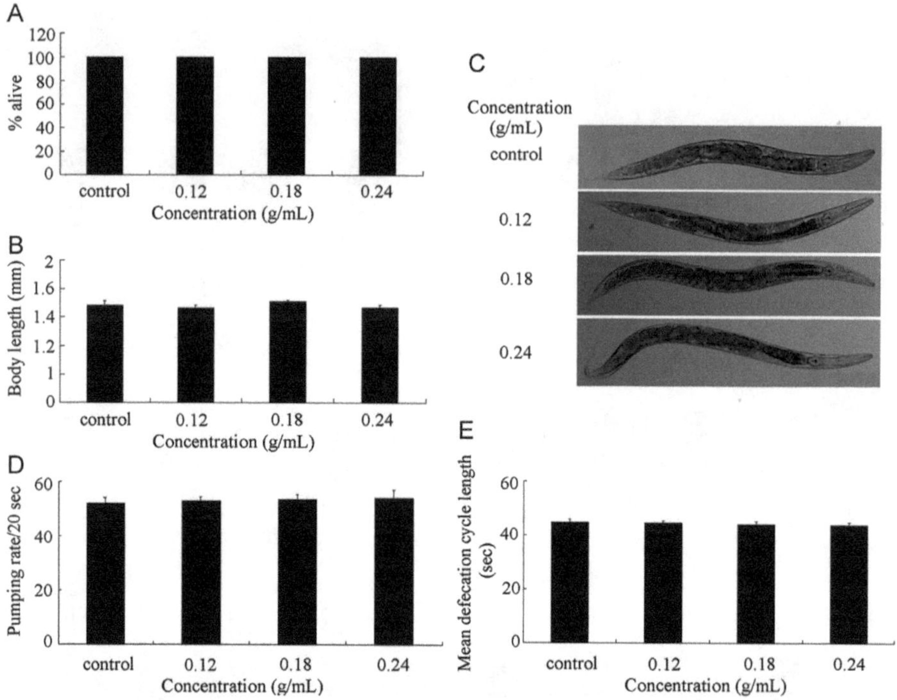

Fig. 5.2 Prolonged exposure to GR did not induce lethality or affect development and metabolism in nematodes [14]. (**a**) effects of GR exposure on lethality. (**b**) effects of GR exposure on body length. (**c**) effects of GR on morphology of nematodes. (**d**) effects of GR exposure on pumping rate. (**e**) effects of GR exposure on mean defecation cycle length. GR, *Glycyrrhizae radix*. *C. elegans* L1-larvae to young adults were exposed to GR. Bars represent means ± S.E.M

Glycyrrhizae radix extract as the example of plant extracts, accompanied with the identification of its function in reducing oxidative damage and extending lifespan, exposure to *Glycyrrhizae radix* extract at examined doses could not affect the growth and the lipid metabolism and induce the lethality (Fig. 5.2) [14].

5.5 Pharmacological Safety Assessment of Chinese Medicines

For the pharmacological assessment of Chinese Medicines, especially the Chinese Medicines formula, the following aspects need to be specially considered:

1. Characterizations: On the one hand, chemical fingerprint should be provided using HPLC or LC-MS methods. On the other hand, 2–4 characteristic markers representing different chemical classes and different herbs in a formula are suggested to be quantified, and marker compound concentrations are suggested to be provided.

2. Sourcing and Authentication: The used plant species are suggested confirmed using morphological or genetic methods. Besides this, the geographical origin and harvest time should be recorded and provided to prove the "Dao-di" (authentic) quality and chemical profile.
3. Processing (Paozhi) Methods: Considering that the processing methods will drastically affect the chemical composition, the Paozhi methods should be carefully documented. The extraction methods (water decoction, ethanol extraction, or other solvent extraction) are suggested to simulate the clinical use.
4. Complex Formula Preparation: The possible chemical interactions induced by co-decoction should be carefully considered. For the dosage ratio, it is suggested to maintain the classic herb ratios from the original formula.
5. Formula Compatibility Testing: For the formula, it is suggested to compare the exposure to full formula and exposure to individual herbs to confirm whether the full formula show lower toxicity than its presumed "toxic" ingredient.
6. Multi-omics Integration: For Chinese Medicines, especially the formula, multi-omics analysis including transcriptome, proteome, and metabolome are strongly suggested to be applied together with the endpoint analysis.

With the Bushenkangshuai Tang (BT), a Chinese medicine formula used to treat premature ovarian failure in the clinical, as the example, accompanied with the identified function in inhibiting UV and oxidative stresses, treatment with BT alone could increase lifespan and slow aging-related decline (Fig. 5.3) [15].

5.6 Pharmacological Safety Assessment of Clinical Drugs

For the pharmacological assessment of clinical drugs in *C. elegans*, the following aspects need to be specially considered:

1. The Further Elucidation of Toxicological Mechanism: The well described genetic backgrounds of *C. elegans* makes it helpful for elucidating the molecular signaling pathways affected by a known clinical drug with the toxicity.
2. Comparative Safety Profiling: The small experimental scale in *C. elegans* makes it suitable to rank the toxicity among the analogues or similar-class drugs.
3. Repurposing Safety Screen: To identify unexpected toxicity for an approved drug being tested for a new indication or for a prodrug before the use in the clinical.
4. Combination Therapy: On the one hand, it can be used to test if a drug can exacerbate or ameliorate toxicity in a disease-model context. On the other hand, it can be used to test the possible drug-drug interactions in relevant pathological models.

Using *C. elegans* as the animal model, genome-wide mutational signature was screened and identified after exposure to mitomycin C, a chemotherapeutic agent [16]. Exposure to some the first and the second-generation antipsychotic drugs could

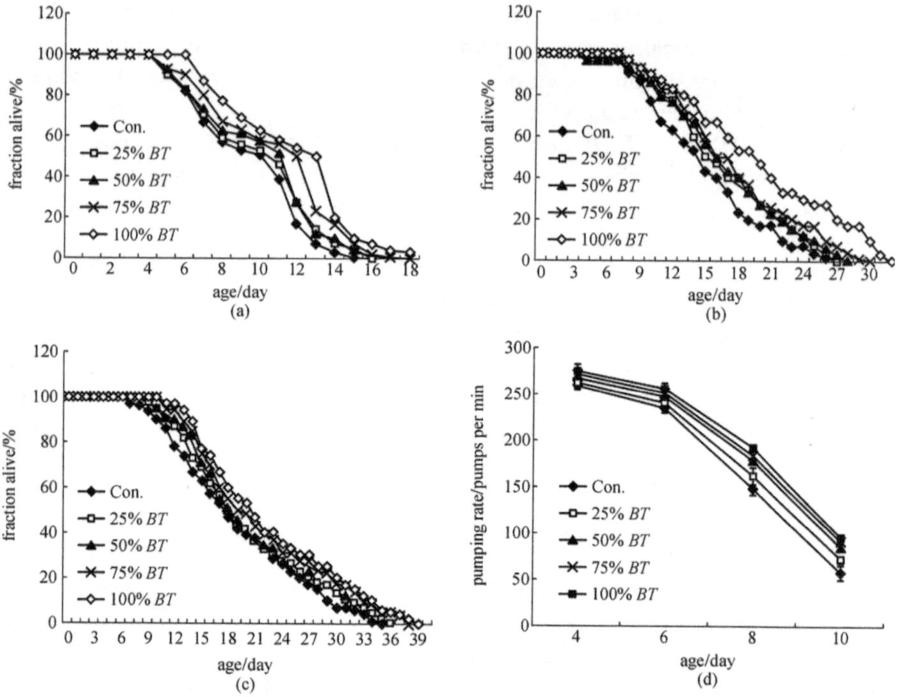

Fig. 5.3 Treatment with BT prolonged lifespan and slowed aging in nematodes [15]. (**a**) Treatment with BT at different concentrations diluted with water prolonged lifespan in exposed animals grown at 25 °C. (**b**) Treatment with BT at different concentrations prolonged lifespan in exposed animals grown at 20 °C. (**c**) Treatment with BT at different concentrations prolonged lifespan in exposed animals grown at 15 °C. (**d**) Treatment with BT at different concentrations slowed the decline in pharynx pumping during aging. Average pumping rates (pumps per min) in 15 animals scored in two trials. Con control, BT Bushenkangshuai Tang. Exposures were performed for 2 h on L4-larva stage animals with 25%, 50%, 75%, and 100% of examined BT. Data were expressed as $x \pm$ sd. *t*-test, treated vs. untreated; day 8, $P < 0.01$ (50% BT, 75% BT, and 100% BT); day 10, $P < 0.01$ (50% BT, 75% BT, and 100% BT). A P value of less than 0.05 was considered to be statistically significant

cause the growth deficits in nematodes [17]. Exposure to certain antidepressants could reduce the lifespan and induce the lipid accumulation in *C. elegans* [18].

5.7 Pharmacological Safety Assessment of Synthesized Compounds

For the pharmacological assessment of synthesized compounds in *C. elegans*, the following aspects need to be specially considered:

1. Structure-Activity Relationship: It is suggested to test structural analogues together or test the synthesized compounds with their parent structure compound

together. These comparisons are helpful for further design of more suitable chemical modifications.
2. Effect of Impurities or Byproducts: For the new synthesized compounds, the possible influence of impurities or byproducts are suggested to be carefully considered. That is, the observed toxicity might be from the impurities or byproducts, and not directly due to the parent compound.
3. Pre-analysis of Chemical Groups: Before the toxicity assessment, the identification for the existence of certain chemically reactive groups showing toxic effects is suggested.

In *C. elegans*, the safety assessment was performed for some synthesized compounds, such as the synthesized 5-(4-methoxyphenyl)-oxazole derivatives and tetrahydroindolone-dihydropyrimidinone hybrids [19, 20].

5.8 Pharmacological Safety Assessment of Nanomaterials

The nanomaterials act as the drug delivery system or the carrier of drugs. For the pharmacological assessment of nanomaterials in *C. elegans*, the following aspects need to be specially considered:

1. Physiochemical Characterizations: For the examined nanomaterials, the detailed physiochemical characterizations must be provided. The physiochemical characterizations include the following aspects:
   - Size and Distribution: The hydrodynamic diameter analyzed by dynamic light scattering (DLS) and aggregation size in suspension solutions should be first provided. Transmission electron microscopy (TEM) can also be used to assess the primary particle size and shape of nanomaterials.
   - Surface Charge: The surface charge of nanomaterials can be reflected by the zeta potential in exposure medium. Normally the nanomaterials with positive charge may have the toxicity.
   - Chemical Compositions and Modification: For nanocomposites, the chemical compositions should bc confirmed. For chemical modified nanomaterials, the existence of chemical modifications should also be confirmed.
2. Stability in Biological Media: Over the exposure period, at least the zeta potential and aggregation size of nanomaterials are suggested to be monitored. For certain nanomaterials, the possible morphological changes should also be observed.
3. Formation of Protein Corona: Once the protein corona is formed on the surface, the biological behavior of nanomaterials may be altered.
4. Dose Expression: The exposure doses of nanomaterials can be expressed by mass/volume, particle number concentration (particles/mL), surface area concentration (μm^2/mL), and molar concentration. Suggest to select the suitable dose expression.
5. Exposure: The gentle, continuous rocking/shaking is suggested for nanomaterials exposure.

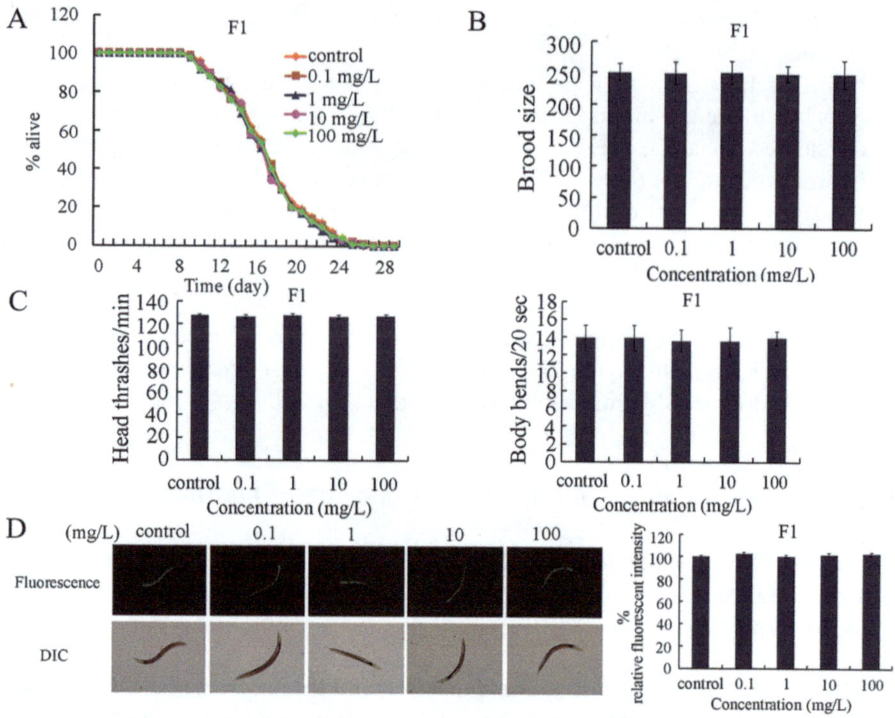

Fig. 5.4 Phenotypic analysis in progeny of nematodes exposed to N-GQDs [21]. (**a**) Lifespan in progeny of nematodes exposed to N-GQDs. (**b**) Brood size in progeny of nematodes exposed to N-GQDs. (**c**) Locomotion behavior in progeny of nematodes exposed to N-GQDs. (**d**) Intestinal ROS production in progeny of nematodes exposed to N-GQDs. Prolonged exposure was performed from L1-larvae to adult day-1. Bars represent means ± standard error of the mean (S.E.M.)

Usually, two possibilities exist for the safety of prepared nanomaterials. One possibility is the nanomaterials themselves are safe. For example, in *C. elegans*, the safety of prepared nitrogen-doped graphene quantum dots was not only confirmed at parental generation, exposure to nitrogen-doped graphene quantum dots also showed the transgenerational safety (Fig. 5.4) [21]. Another possibility is that the toxicity of some nanomaterials can be reduced by certain chemical modifications, such as PEG modification [22]. For example, the toxicity of multi-walled carbon nanotubes (MWCNTs) in inducing intestinal oxidative damage, reducing brood size, and inhibiting locomotion could be inhibited by PEG modification (Fig. 5.5) [23].

5.9 Perspectives

In this chapter, we first introduced several aspects needed to be carefully considered for pharmacological safety assessment of commonly examined compounds and drugs. Besides these, for plant extracts and Chinese Medicines, the transcriptome,

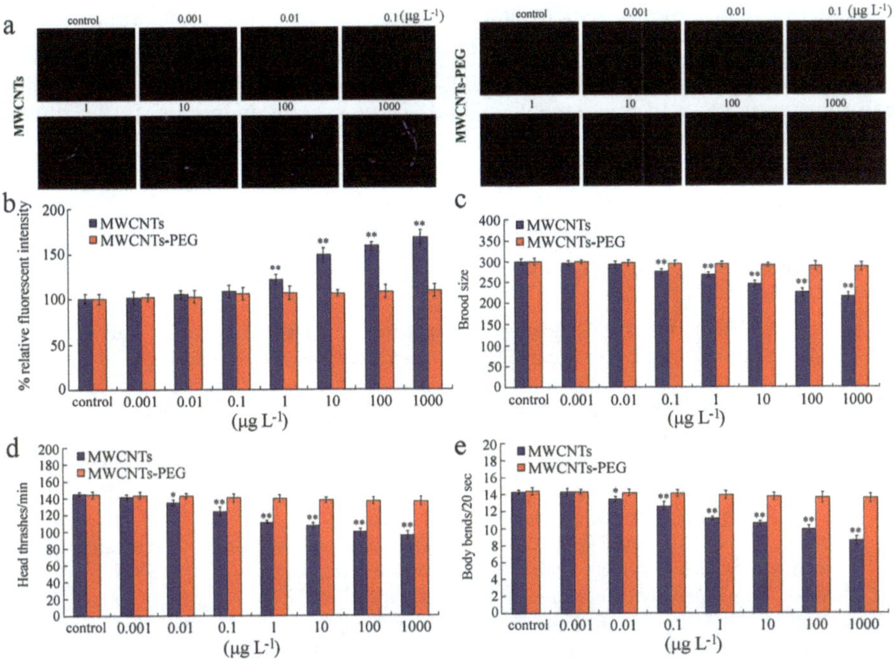

Fig. 5.5 Effects of MWCNTs and MWCNTs-PEG exposure from L1-larvae to adult at predicted environmental relevant concentrations on functions of the primary and secondary targeted organs in nematodes [23]. (**a**) Pictures showing intestinal autofluorescences of MWCNTs and MWCNTs-PEG exposed nematodes. (**b**) Comparison of intestinal autofluorescences in MWCNTs-PEG exposed nematodes from those in MWCNTs exposed nematodes. (**c**) Comparison of brood sizes in MWCNTs-PEG exposed nematodes from those in MWCNTs exposed nematodes. (**d**) Comparison of head thrashes in MWCNTs-PEG exposed nematodes from those in MWCNTs exposed nematodes. (**e**) Comparison of body bends in MWCNTs-PEG exposed nematodes from those in MWCNTs exposed nematodes. Bars represent mean ± S.E.M. $^{*}P < 0.05$, $^{**}P < 0.01$

proteome, and metabolome are suggested to be further determined to comprehensively understand the possible adverse effects induced by these complex samples.

Moreover, the lethal and sub-lethal endpoints were introduced for pharmacological safety assessment in *C. elegans*. However, considering the simple developmental structure of *C. elegans*, the further validation is really needed by performing the comparison with mammalian data or the toxicological analysis in mammals.

We further discussed the pharmacological safety assessment of certain groups of compounds and drugs, including plant compounds, plant extracts, Chinese Medicines, clinical drugs, synthesized compounds, and nanomaterials were discussed. Among them, one of the important hallmarks of phytochemicals is the hormetic effect. That is, low doses of these chemicals often show beneficial effects, whereas their exposure at high doses are toxic. For this kind of compounds or drugs, a very wide concentration range is suggested to be covered to capture this possible biphasic response.

Actually, it is more urgent to achieve the standardization of pharmacological safety assessment. The standardization of pharmacological safety assessment should cover the source and preparation of compounds and drugs, *C. elegans* preparation, exposure methods, and assessment methods using certain endpoints. In the future, the standardization of pharmacological safety assessment is suggested to be further coupled with machine learning or AI-based tools.

References

1. Wang D-Y (2019) Molecular toxicology in *Caenorhabditis elegans*. Springer, Singapore
2. Wang D-Y (2019) Target organ toxicology in *Caenorhabditis elegans*. Springer, Singapore
3. Wang D-Y (2020) Exposure toxicology in *Caenorhabditis elegans*. Springer, Singapore
4. Wang D-Y (2022) Toxicology at environmentally relevant concentrations in *Caenorhabditis elegans*. Springer, Singapore
5. Sciandrone B, Kentsop RAD, Pensotti R, Ottolina G, Mascheretti I, Mattana M, Regonesi ME (2024) Toxicological analysis of the *Arylnaphthalene lignan* justicidin B using a *Caenorhabditis elegans* model. Molecules 29:5516
6. Chang CH, Wei CC, Ho CT, Liao VH (2021) N-gamma-(L-glutamyl)-L-selenomethionine shows neuroprotective effects against Parkinson's disease associated with SKN-1/Nrf2 and TRXR-1 in *Caenorhabditis elegans*. Phytomedicine 92:153733
7. Steenackers H, Dubey A, Robijns S, Ermolat'ev D, Delattin N, Dovgan B, Girandon L, Fröhlich M, De Brucker K, Cammue BP, Thevissen K, Balzarini J, Van der Eycken EV, Vanderleyden J (2014) Evaluation of the toxicity of 5-aryl-2-aminoimidazole-based biofilm inhibitors against eukaryotic cell lines, bone cells and the nematode *Caenorhabditis elegans*. Molecules 19:16707–16723
8. Yuan C-Y, Wang Y-X, Zhang L, Wang D-Y (2024) Procatechuic acid and protocatechuic aldehyde increase survival of *Caenorhabditis elegans* after fungal infection and inhibit fungal virulence. Front Pharmacol 15:1396733
9. Liu T-W, Zhuang Z-H, Wang D-Y (2023) Paeoniflorin mitigates high glucose-induced lifespan reduction by inhibiting insulin signaling in *Caenorhabditis elegans*. Front Pharmacol 14: 1202379
10. Hangweirer K, Sandner G, König A, Kirchsteiger J, Blank-Landeshammer B, Winkler S, Schurr J, Weghuber J (2025) *Allium ursinum* leaves aqueous extract alleviates *Pseudomonas aeruginosa* infection in *Caenorhabditis elegans*. Phytomedicine 146:157100
11. Fernández Moreno K, Maturana G, Blanco-Haros S, Norambuena-Jopia U, Valenzuela-Barra G, Zúñiga-López MC, Bravo Garrido J (2025) Determination of bioactive compounds, antioxidant capacity, safety assessment, and antimicrobial effect of *Tristerix corymbosus* extracts. Molecules 30:4610
12. de Andrades IX, Santos PA, Mendes TL, Marinho LC, Pereira P, von Poser GL, Diel KAP (2025) Safety evaluation of the hexane leaf extract of *Vismia guianensis* (Hypericaceae), a Brazilian medicinal plant, using the *Caenorhabditis elegans* model. Chem Biodivers 22:e01946
13. Dos Santos FB, Quines CB, Pilissão LEB, Dal Forno AHC, Rodrigues CF, Denardin CC, Farias FM, Ávila DS (2020) Aqueous bark extract of *Ceiba speciosa* (A. St.-Hill) Ravenna protects against glucose toxicity in *Caenorhabditis elegans*. Oxid Med Cell Longev 2020:1321354
14. Ruan Q, Qiao Y, Zhao Y, Xu Y, Wang M, Duan J, Wang D (2016) Beneficial effects of *Glycyrrhizae radix* extract in preventing oxidative damage and extending the lifespan of *Caenorhabditis elegans*. J Ethnopharmacol 177:101–110
15. Rui Q, Lu Q, Wang D-Y (2009) Administration of Bushenkangshuai Tang alleviates the UV irradiation- and oxidative stress-induced lifespan defects in nematode *Caenorhabditis elegans*. Front Med China 3:76–90

16. Tam AS, Chu JSC, Rose AM (2016) Genome-wide mutational signature of the chemotherapeutic agent mitomycin C in *Caenorhabditis elegans*. G3 (Bethesda) 6(1):133–140
17. Donohoe DR, Aamodt EJ, Osborn E, Dwyer DS (2006) Antipsychotic drugs disrupt normal development in *Caenorhabditis elegans* via additional mechanisms besides dopamine and serotonin receptors. Pharmacol Res 54:361–372
18. Zarse K, Ristow M (2008) Antidepressants of the serotonin-antagonist type increase body fat and decrease lifespan of adult *Caenorhabditis elegans*. PLoS One 3:e4062
19. Yamamuro D, Uchida R, Ohtawa M, Arima S, Futamura Y, Katane M, Homma H, Nagamitsu T, Osada H, Tomoda H (2015) Synthesis and biological activity of 5-(4-methoxyphenyl)-oxazole derivatives. Bioorg Med Chem Lett 25:313–316
20. Vendrusculo V, Oliveira MPLS, Rodrigues BA, Ramos VS, Menezes JM, Zimmer KR, Russowsky D (2025) Tetrahydroindolone-dihydropyrimidinone hybrids as promising antibiofilm and antibacterial agents. ChemMedChem 20:e202500716
21. Zhao Y-L, Liu Q, Shakoor S, Gong JR, Wang D-Y (2015) Transgenerational safe property of nitrogen-doped graphene quantum dots and the underlying cellular mechanism in *Caenorhabditis elegans*. Toxicol Res 4:270–280
22. Wang D-Y (2018) Nanotoxicology in *Caenorhabditis elegans*. Springer, Singapore
23. Wu Q-L, Li Y-X, Li Y-P, Zhao Y-L, Ge L, Wang H-F, Wang D-Y (2013) Crucial role of biological barrier at the primary targeted organs in controlling translocation and toxicity of multi-walled carbon nanotubes in nematode *Caenorhabditis elegans*. Nanoscale 5:11166–11178

Pharmacological Bioavailability, Translocation, and Metabolism

6

Abstract

The bioavailability and the further dynamic changes are very important for formation of the final pharmacological effects of compounds and drugs. The bioavailability of pharmacological compounds and drugs and the value of dynamic bioavailability assessment were first introduced. The translocation of pharmacological compounds and drugs was next discussed. The metabolism of pharmacological compounds and drugs was discussed from the aspects of phase I metabolism, phase II metabolism, and phase III metabolism, respectively. Moreover, the bioavailability, translocation, and metabolism of pharmacological vectors were introduced.

Keywords

Bioavailability · Translocation · Metabolism · Drugs · *Caenorhabditis elegans*

6.1 Introduction

The pharmacokinetic and pharmacological metabolism are important for understanding the pharmacological effects of examined compounds and drugs. In *Caenorhabditis elegans*, the pharmacokinetic can be understood from two aspects, the pharmacological bioavailability and the pharmacological translocation. The pharmacological bioavailability can help us confirm that the observed negative result after administration with compounds and drugs could mean no effect or no uptake. The dynamic pharmacological bioavailability can further help us determine that the observed limited pharmacological effect of compounds and drugs is due to the not effective uptake. The pharmacological translocation can help us understand that the observed pharmacological effect of compounds and drugs is their direct effect in certain tissues or not. Moreover, the pharmacological metabolism can help us

D. Wang, *Pharmacology in Caenorhabditis elegans*,
https://doi.org/10.1007/978-981-92-5166-7_6

understand the possible different existence forms of compounds and drugs after the uptake.

In this chapter, we mainly introduce and discuss the pharmacological bioavailability, pharmacological translocation, and pharmacological metabolism. Besides these, the bioavailability, translocation, and metabolism of pharmacological vectors are further discussed.

6.2 Pharmacological Bioavailability

6.2.1 Bioavailability of Pharmacological Compounds and Drugs

The pharmacological bioavailability refers to proportion or amount of externally administered compound and drugs that can enter the body of nematodes, which suggests that the examined compounds and drugs become available at their site of action by crossing the biological barriers. For example, accompanied with the observed function in enhancing stress resistance and extending lifespan, the quercetin could be taken up by the nematodes [1].

For the biological barriers, the small, lipophilic compounds and drugs may be somewhat diffused through the epidermal barrier. Nevertheless, at the most cases, the compounds and drugs are bioavailable to nematodes across the intestinal barrier after the oral uptake. That is, after the oral uptake, the intestinal absorption is the critical step for pharmacological compounds and drugs so as to cross the intestinal epithelial cells to enter certain organs or the pseudocoelomic fluid.

Some factors may influence the pharmacological bioavailability of compounds and drugs:

1. Physicochemical Properties of Compounds and Drugs: The molecular weight, lipophilicity, and charge of compounds and drugs may all affect their bioavailability in the body of nematodes.
2. Solvents: The difference in bioavailability of compounds and drugs may be related to the use of certain solvents (e.g., DMSO).
3. Presence of Bacteria in the Exposure Solutions or Medium: Sometimes, the bacteria metabolize or absorb the compounds or drugs. To exclude this aspect of effect, the heat killed *E. coil* OP50 is suggested to be used as the control.
4. Biological Barrier Function: Disrupted intestinal barrier may enhance the uptake of compounds and drugs by nematodes. Thus, accompanied with the observation of beneficial effect of examined compounds and drugs, their possible effects on function of biological barriers and expression of genes governing the function of biological barriers are suggested to be determined.
5. Intestinal Transporters: The uptake of some pharmacological compounds and drugs are also affected by the intestinal efflux/influx transporters (e.g., P-glycoprotein homologs).
6. Feeding Behavior: The feeding behavior can be reflected by change of pumping rate. If the pumping rate is severely decreased, the uptake of examined

compounds and drugs may be inhibited. However, if the pumping rate is not or only moderately affected, the uptake of examined compounds and drugs may not be affected by the feeding behavior.

7. Defecation Behavior: The defecation behavior can be reflected by the defecation cycle length. The accelerated defecation behavior or reduced defecation cycle length will enhance the excretion and reduce the accumulation of examined compounds and drugs.

The pharmacological availability can be quantified using liquid chromatography-mass spectrometry (LC-MS/MS) and HPLC-UV/FLD methods. The LC-MS/MS method is the most powerful and common method, and the HPLC-UV/FLD method is less sensitive but accessible. After quantification, the data for measured compound or drug amount is suggested to be normalize to the total worm protein.

6.2.2 Dynamic Bioavailability of Pharmacological Compounds and Drugs

The pharmacological bioavailability of compounds and drugs is not suitable to be analyzed at just one specific point in time. The dynamic assessment of pharmacological bioavailability of compounds and drugs is suggested. If the examined pharmacological compounds and drugs cannot be accumulated persistently during the exposure, the exposure solutions of compounds and drugs are suggested to be refreshed daily and the repeated administration is suggested for clinical pharmacological treatment. With the small selenium species as the examples, although they all could enhance the resistance to oxidative stress, only SeMet among the examined Se compounds could show the persistent accumulation [2].

6.3 Pharmacological Translocation

Pharmacological translocation refers to the dynamic, multi-step process by which the externally administered compound and drugs will be distributed within the targeted tissues. The pharmacological translocation can help us judge that the observed pharmacological effect of compounds and drugs is their direct effect or not. In the body of *C. elegans*, the pharmacological translocation encompasses the integrated aspects of absorption, distribution, and elimination. Once inside, compounds and drugs may be distributed via the pseudocoelomic fluid, and the movement is possibly driven mainly by diffusion.

Some factors may affect the pharmacological translocation of compounds and drugs:

1. Efflux Transporters in Certain Tissues: The efflux transporters in different tissues may further affect the distribution and translocation of compounds and drugs.

2. Storage and Sequestration: For the organelles such as lysosomes, they can provide the function of storage and sequestration for compounds and drugs in the cells.

Due to the very small body size, normally it is difficult to examine the pharmacological translocation in *C. elegans*. Nevertheless, some indirect methods can be employed to determine this, such as the fluorescent labeling and isotope labeling. For the fluorescent labeling, the binding stability of fluorescent probe to compounds and drugs and the possible effect of fluorescent probe on absorption and activity of compounds and drugs are suggested to be examined before the use. For example, using the fluorescent labeling method, the cholesterol was found to be translocated and accumulated in larval pharynx, nerve ring, excretory gland cell, and gut, and further accumulated in oocytes and spermatozoa of adults [3].

6.4 Pharmacological Metabolism

Pharmacological metabolism refers to the biotransformation of compounds and drugs within the body of nematode, which is mediated by specialized enzymatic systems. This process will in turn lead to the bioactive lifetime and ultimate clearance from the body for the examined compounds and drugs. Considering that *C. elegans* lacks the full cytochrome P450 complexity of a mammalian liver, transgenic *C. elegans* strain carrying corresponding human CYP enzyme genes is suggested as the tool to reflect the possible pharmacological metabolism in mammals [4, 5]. In *C. elegans*, intestine is the major site of xenobiotic metabolism, and has the highest levels of Phase I and Phase II enzymes. Besides the intestine, hypodermis, pharynx, nervous system, excretory cell, and canal are also the metabolic tissues or organs. *C. elegans* has the classic Phase I—Phase II—Phase III paradigm.

6.4.1 Phase I Metabolism

In phase I metabolism, the functionalization reactions can introduce the polar functional groups (such as –OH or –COOH) to compounds and drugs.

1. Cytochrome P450 (CYP)-Mediated Oxidations

C. elegans has many CYP genes, which can be particularly inducible by xenobiotics [6–8]. The CYPs function to drive the reactions of aliphatic & aromatic hydroxylation by adding an –OH group to a carbon chain or ring, N- and O-dealkylation by cleaving alkyl groups from nitrogen or oxygen atoms, epoxidation by forming reactive epoxide intermediates on aromatic rings, and S-oxidation and N-oxidation. Induction of CYP-35d1 caused formation of hydroxylated thiabendazole (TBZ) metabolites from the TBZ, which was mediated by activation of nuclear hormone receptor NHR-176 [9].

2. Flavin-Containing Monooxygenase (FMO) Reactions

These reactions contain the N-oxidation of tertiary amines and the S-oxidation of thioethers to sulfoxides and sulfones, which are catalyzed by FMO homologs (such as FMO-2 and FMO-4) [10–16]. The FMOs induction could be employed as the tool for the screen of pro-longevity drugs [17]. Moreover, the induction of FMO-2 led to the rewiring of endogenous metabolism through changes in one carbon metabolisms [18].

3. Reductive and Hydrolytic Reactions

Carboxylesterases function to hydrolyze the ester and the amide bonds, which possibly allows the activation of prodrugs [19–21]. Epoxide hydrolases act to detoxify reactive epoxide intermediates formed by CYPs and convert them to less reactive dihydrodiols [22, 23].

6.4.2 Phase II Metabolism

In phase II metabolism, the conjugation reactions attach the large, polar, endogenous molecule to Phase I metabolites to increase the water solubility for excretion.

1. UDP-Glucuronosyltransferase (UGT) Mediated Glucuronidation

These reactions can transfer the glucuronic acid from UDP-glucuronic acid to nucleophilic O-, N-, S-, or C- atoms of substrates, which are catalyzed by GUTs [24, 25]. The UGTs mediate the major conjugation pathway for phenols, alcohols, carboxylic acids, and amines [26–28]. Some of the UGTs could modulate the drug sensitivity of propionate and benzimidazole [29, 30].

2. Glutathione S-Transferase (GST) Mediated Glutathione Conjugation

These reactions provide the nucleophilic attack on the electrophilic centers driven by the glutathione (GSH), which is normally the primary defense response for nematodes against reactive, electrophilic metabolites [31–33]. The GSTs catalyze this process to form glutathione conjugates (GS-Xs), which can be processed to the cysteine conjugates for the excretion [33, 34]. The GSTs were shown to be able to metabolize the 4-hydroxynon-2-enal and the 4-hydroxynonenal, a lipid peroxidation product [35, 36].

3. Sulfotransferase (SULT)-Mediated Sulfation

These reactions can transfer the sulfate group from 3′-phosphoadenosine-5′--phosphosulfate (PAPS) to the hydroxyl or amine groups, which is catalyzed by the SULTs [37–42]. The formed sulfate conjugates are highly water-soluble. The transgenic *C. elegans* strain carrying human tyrosylprotein SULT was generated for the examine possible function of human tyrosylprotein SULT [43].

4. Acetylation Mediated by N-acetyltransferases (NATs)

These reactions can transfer the acetyl groups from acetyl-CoA to aromatic amines or hydrazine groups, which is catalyzed by the NATs [44]. In *C. elegans*, the N-acetylation of thialysine [S-(2-aminoethyl)-L-cysteine] is catalyzed by GCN5-related NAT [45]. In addition, the N-acetylation of serotonin can be catalyzed by arylalkylamine NAT to generate the melatonin [46].

5. Methylation Mediated by Methyltransferases

These reactions cause the O-methylation and N-methylation using S-adenosyl methionine (SAM) as the methyl donor, which is catalyzed by methyltransferases [47–50]. In *C. elegans*, the synthesis of coenzyme Q involves the O-methylation process [51]. The methylation reaction can act on drugs with catechol groups catalyzed by catechol-O-methyltransferase [52]. After the uptake, the O-methylation of quercetin to generate quercetin 3′-O-methylether and quercetin 4′-O-methylether may be an important conjugation process [53].

6.4.3 Phase III Metabolism

For glutathione conjugates, the GS-X conjugates are enzymatically converted to cysteinylglycine, cysteine, and finally N-acetylcysteine conjugates for the aim of excretion and elimination [54]. Administration with thiols including the N-acetylcysteine was shown to have beneficial pharmacological effect, such as lifespan extension [55, 56].

In addition, some Phase-II conjugates can act as substrates of ATP-binding cassette (ABC) efflux transporters (such as MRP homologs). With the aid of these transporters, hydrophilic metabolites can be pumped out of cells into the intestinal lumen or pseudocoelomic fluid [57–60]. In *C. elegans*, the ABC transporters could act on the vitamin B12 metabolism [59].

6.5 Pharmacological Vectors

Pharmacological vectors refer to the delivery systems used to introduce pharmacological compounds and drugs into the targets organs or tissues. Among the widely delivery systems, the nanomaterials have been systematically examined in *C. elegans* [61].

6.5.1 Nanomaterials

Nanomaterials include metal-based nanomaterials, carbon-based nanomaterials, and quantum dots. In *C. elegans*, after the uptake, their bioavailability can be examined by directly observation and chemical labeling method. For the bioavailability, at least the intestinal accumulation of some nanomaterials (such as nanoparticles and graphene oxide) can be directly observed under the microscopy with the aid of DIC lens. The bioavailability of carbon-based nanomaterials can be analyzed by labeling them with certain fluorescent probe, such as Rho B. The bioavailability of metal nanoparticles can also be examined by assessing corresponding metal accumulation in the body.

For the translocation of nanomaterials, usually the fluorescence labeling method is used to examine this in nematodes. For example, using Rho B labeling methods, the pristine multi-walled carbon nanotubes (MWCNTs) could not only be accumulated in pharynx and intestine, but also be translocated to spermatheca and embryos in the body from the intestine (Fig. 6.1) [62]. In contrast, the translocation of MWCNTs to spermatheca and embryos was obviously inhibited by carboxylic acid functionalization (Fig. 6.1) [62]. The bioavailability and translocation of quantum dots can be directly visualized based on their fluorescence. For example, CdTe quantum dots (CdTe QDs) having red fluorescence could be not only accumulated in pharynx and intestine, but also translocated to gonad and the region surrounding

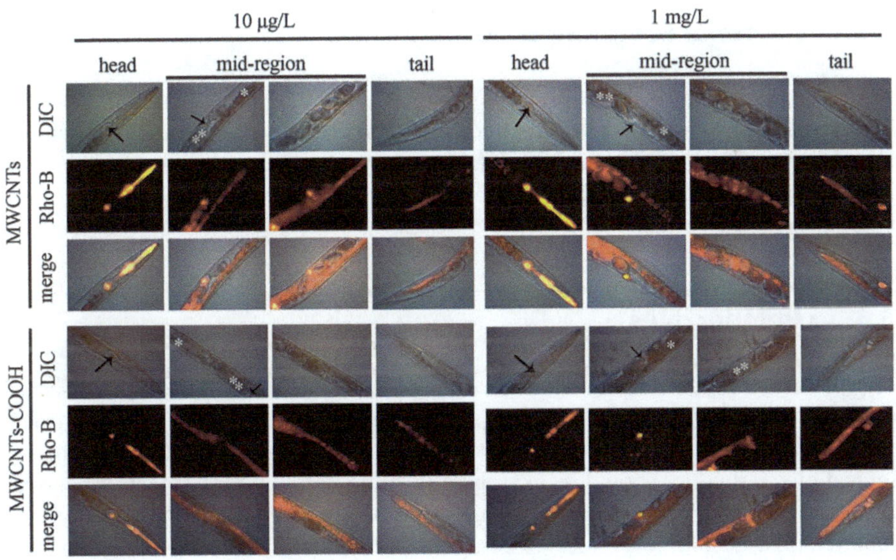

Fig. 6.1 Comparison of translocation and distribution of MWCNTs and MWCNTs–COOH after exposure at different concentrations from L1-larvae to adult [62]. The arrowheads indicate the pharynx and spermatheca, respectively, at the head region or mid-region of nematodes. The intestine (*) and the embryos (**) in the midregion are also indicated. The exposure concentrations for MWCNTs and MWCNTs–COOH were 10 μg L^{-1} and 1 mg L^{-1}, respectively

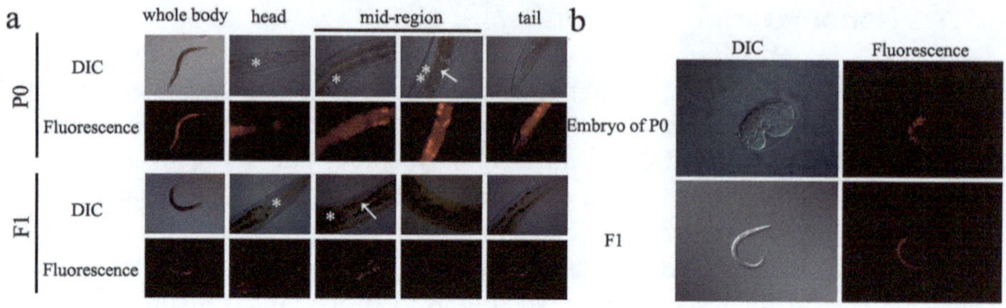

Fig. 6.2 Distribution and translocation of CdTe QDs in exposed wild-type nematodes and their progeny [63]. (**a**) CdTe QDs distribution in exposed wild-type nematodes and their progeny. Arrowheads indicate the intestine. Pharynx (*) in the head, embryos (*) in the mid-region, and gonad (**) in the mid-region were also indicated. (**b**) CdTe QDs distribution in embryos of exposed wild-type nematodes and the animals developed from these embryos. Exposure concentration of CdTe QDs was 20 mg L^{-1}. P0 parents; F1 the filial generation

pharynx (Fig. 6.2) [63]. Moreover, the CdTe QDs could be even transferred to the progeny, such as the embryos of exposed nematodes (Fig. 6.2) [63].

Under certain condition, the metabolism of some nanomaterials (such as QDs) may also occur in the body of nematodes [64]. For example, the metabolism and the degradation of QDs were found in the intestine of nematodes [65].

6.5.2 Bacterial Food-Based Delivery

In *C. elegans*, bacterial food-based delivery is a specific pharmacological vector. That is, the pharmacological compounds and drugs may possibly be mixed with or absorbed by *E. coli* OP50 and further taken up by nematodes after the oral feeding. To a certain degree, the bacterial food-based delivery cannot be considered as a strict pharmacological vector.

6.6 Perspectives

The pharmacological bioavailability can help us understand whether the observed pharmacological effect of compounds and drugs are directly due to their existence. In *C. elegans*, this aspect can be well examined. It is suggested that the pharmacological bioavailability should be examined together with the assessment of pharmacological effects of compounds and drugs. That is, without this aspect of support, the reported pharmacological effects of compounds and drugs have not been guaranteed or confirmed.

The pharmacological translocation can help us understand whether the examined compounds and drugs can directly target to certain organs or tissues. Nevertheless, the small size of *C. elegans* is a huge limit to carefully determine this. Chemical

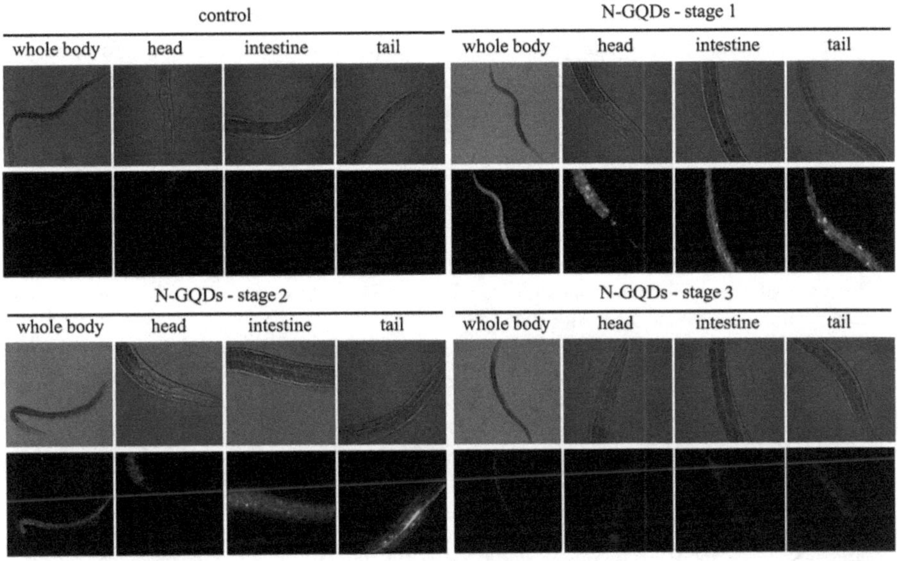

Fig. 6.3 Dynamic metabolism of nitrogen-doped graphene quantum dots (N-GQDs) in nematodes [66]

labeling (such as labeling with fluorescent probe) can solve this question to a certain degree.

The analysis of pharmacological metabolism can help us understand at least three aspects:

1. Whether the compounds and drugs can be easily excreted from the body?
2. Whether the pharmacological effect of compounds and drugs will gradually diminish within the body?
3. Whether the compounds and drugs will exhibit any toxicity within the body?

Nevertheless, at the most cases, it is not or largely unclear how the phase I, phase II, and phase III reactions catalyze the compounds and drugs.

Normally, it is difficult to determine the excretion process in *C. elegans*. Nevertheless, for the chemicals labeled with fluorescence probe or having the fluorescence, their excretion process can be monitored. Nitrogen-doped graphene quantum dots (N-GQDs) have the green fluorescence. Based on the observation of their dynamic green fluorescence signals, excretion process of N-GQDs can be divided into three stages. At the first stage, N-GQDs were strongly accumulated in pharynx, intestine, and tail regions (Fig. 6.3) [66]. At the second stage, N-GQDs were mainly accumulated in intestine and tail region, and meanwhile N-GQDs distribution in the pharynx gradually diminished (Fig. 6.3) [66]. At the third stage, no strong N-GQD green fluorescent signals could be observed in the body (Fig. 6.3) [66].

The bioavailability and the translocation of nanomaterials (pharmacological vector) are similar to those of pharmacological compounds and drugs. However, to a great degree, it is not clear how the different types of nanomaterials are metabolized within the *C. elegans* body.

References

1. Kampkötter A, Timpel C, Zurawski RF, Ruhl S, Chovolou Y, Proksch P, Wätjen W (2008) Increase of stress resistance and lifespan of *Caenorhabditis elegans* by quercetin. Comp Biochem Physiol B Biochem Mol Biol 149:314–323
2. Rohn I, Raschke S, Aschner M, Tuck S, Kuehnelt D, Kipp A, Schwerdtle T, Bornhorst J (2019) Treatment of *Caenorhabditis elegans* with small selenium species enhances antioxidant defense systems. Mol Nutr Food Res 63:e1801304
3. Matyash V, Geier C, Henske A, Mukherjee S, Hirsh D, Thiele C, Grant B, Maxfield FR, Kurzchalia TV (2001) Distribution and transport of cholesterol in *Caenorhabditis elegans*. Mol Biol Cell 12:1725–1736
4. Chen Y, Jiang Y, Sarvanantharajah N, Apirakkan O, Yang M, Milcova A, Topinka J, Abbate V, Arlt VM, Stürzenbaum SR (2024) Genome-modified *Caenorhabditis elegans* expressing the human cytochrome P450 (CYP1A1 and CYP1A2) pathway: an experimental model for environmental carcinogenesis and pharmacological research. Environ Int 194:109187
5. Gonzalez HC, Misare KR, Mendenhall TT, Wolf BJ, Mulholland PJ, Gordon KL, Hartman JH (2024) Transgenic expression of human cytochrome P450 2E1 in *C. elegans* and rat PC-12 cells sensitizes to ethanol-induced locomotor and mitochondrial effects. Biochem Biophys Res Commun 734:150735
6. Lim SYM, Pan Y, Alshagga M, Lim W, Cin K, Alshehade SA, Alshawsh M (2024) CYP14 family in *Caenorhabditis elegans*: mitochondrial function, detoxification, and lifespan. J Appl Toxicol 44:1647–1656
7. Lim SYM, Alshagga M, Kong C, Alshawsh MA, Alshehade SA, Pan Y (2022) CYP35 family in *Caenorhabditis elegans* biological processes: fatty acid synthesis, xenobiotic metabolism, and stress responses. Arch Toxicol 96:3163–3174
8. Menzel R, Bogaert T, Achazi R (2001) A systematic gene expression screen of *Caenorhabditis elegans* cytochrome P450 genes reveals CYP35 as strongly xenobiotic inducible. Arch Biochem Biophys 395:158–168
9. Jones LM, Flemming AJ, Urwin PE (2015) NHR-176 regulates *cyp-35d1* to control hydroxylation-dependent metabolism of thiabendazole in *Caenorhabditis elegans*. Biochem J 466:37–44
10. Jakovljević S, Radojević D, Bajić SS, Brdarić E, Mitrović H, Đokić J, Baumann L, Trifunovic A, Golić N, Dinić M (2025) Bacterial exopolysaccharides mediate activation of flavin-containing monooxygenase 2 to extend lifespan in *Caenorhabditis elegans*. Cell Commun Signal 24:9
11. Bennett CF, Kwon JJ, Chen C, Russell J, Acosta K, Burnaevskiy N, Crane MM, Bitto A, Vander Wende H, Simko M, Pineda V, Rossner R, Wasko BM, Choi H, Chen S, Park S, Jafari G, Sands B, Perez Olsen C, Mendenhall AR, Morgan PG, Kaeberlein M (2017) Transaldolase inhibition impairs mitochondrial respiration and induces a starvation-like longevity response in *Caenorhabditis elegans*. PLoS Genet 13:e1006695
12. Bhat A, Carranza FR, Tuckowski AM, Leiser SF (2024) Flavin-containing monooxygenase (FMO): beyond xenobiotics. BioEssays 46:e2400029
13. Wani KA, Goswamy D, Taubert S, Ratnappan R, Ghazi A, Irazoqui JE (2021) NHR-49/PPAR-alpha and HLH-30/TFEB cooperate for C. elegans host defense via a flavin-containing monooxygenase. elife 10:e62775

14. Kitto ES, Beydoun S, Henry E, Schaller ML, Bhandari M, Easow SA, Tuckowski AM, Howington MB, Bhat A, Sridhar A, Chung E, Evans CR, Leiser SF (2025) Metabolic regulation of behavior by the intestinal enzyme FMO-2. Sci Adv 11:eadx3018
15. Tuckowski AM, Beydoun S, Kitto ES, Bhat A, Howington MB, Sridhar A, Bhandari M, Chambers K, Leiser SF (2025) *fmo-4* promotes longevity and stress resistance via ER to mitochondria calcium regulation in *C. elegans*. elife 13:RP99971
16. Huang S, Howington MB, Dobry CJ, Evans CR, Leiser SF (2021) Flavin-containing monooxygenases are conserved regulators of stress resistance and metabolism. Front Cell Dev Biol 9:630188
17. Huang S, Cox RL, Tuckowski A, Beydoun S, Bhat A, Howington MB, Sarker M, Miller H, Ruwe E, Wang E, Li X, Gardea EA, DeNicola D, Peterson W, Carrier JM, Miller RA, Sutphin GL, Leiser SF (2024) Fmo induction as a tool to screen for pro-longevity drugs. Geroscience 46: 4689–4706
18. Choi HS, Bhat A, Howington MB, Schaller ML, Cox RL, Huang S, Beydoun S, Miller HA, Tuckowski AM, Mecano J, Dean ES, Jensen L, Beard DA, Evans CR, Leiser SF (2023) FMO rewires metabolism to promote longevity through tryptophan and one carbon metabolism in *C. elegans*. Nat Commun 14:562
19. Le HH, Wrobel CJ, Cohen SM, Yu J, Park H, Helf MJ, Curtis BJ, Kruempel JC, Rodrigues PR, Hu PJ, Sternberg PW, Schroeder FC (2020) Modular metabolite assembly in *Caenorhabditis elegans* depends on carboxylesterases and formation of lysosome-related organelles. elife 9: e61886
20. Xu W, Bhar S, Bruner SD, Butcher RA (2026) Expression, purification, and in vitro characterization of the carboxylesterase CEST-9.2 from *Caenorhabditis elegans*. Biosci Rep 46: BSR20253840
21. McGhee JD (1987) Purification and characterization of a carboxylesterase from the intestine of the nematode *Caenorhabditis elegans*. Biochemistry 26:4101–4107
22. Harris TR, Aronov PA, Jones PD, Tanaka H, Arand M, Hammock BD (2008) Identification of two epoxide hydrolases in *Caenorhabditis elegans* that metabolize mammalian lipid signaling molecules. Arch Biochem Biophys 472:139–149
23. Sarparast M, Hinman J, Pourmand E, Vonarx D, Ramirez L, Ma W, Liachko NF, Alan JK, Lee KSS (2023) Cytochrome P450 and epoxide hydrolase metabolites in Abeta and tau-induced neurodegeneration: insights from *Caenorhabditis elegans*. bioRxiv. https://doi.org/10.1101/2023.10.02.560527
24. Hasegawa K, Miwa S, Tsutsumiuchi K, Miwa J (2010) Allyl isothiocyanate that induces GST and UGT expression confers oxidative stress resistance on *C. elegans*, as demonstrated by nematode biosensor. PLoS One 5:e9267
25. Lindblom TH, Dodd AK (2006) Xenobiotic detoxification in the nematode *Caenorhabditis elegans*. J Exp Zool A Comp Exp Biol 305:720–730
26. Schmidt L, Heck NV, Ferreira I, Göethel G, Somacal S, Emanuelli T, Rodrigues E, Garcia SC, Welke JE, Augusti PR (2020) Ochratoxin A presence in *Cabernet Sauvignon* wine changes antioxidant activity in vitro and oxidative stress markers in vivo. Food Addit Contam Part A Chem Anal Control Expo Risk Assess 37:1755–1764
27. Tawfeek N, Sobeh M, Hamdan DI, Farrag N, Roxo M, El-Shazly AM, Wink M (2019) Phenolic compounds from *Populus alba* L. and *Salix subserrata* Willd. (Salicaceae) counteract oxidative stress in *Caenorhabditis elegans*. Molecules 24:1999
28. Wilson MA, Rimando AM, Wolkow CA (2008) Methoxylation enhances stilbene bioactivity in *Caenorhabditis elegans*. BMC Pharmacol 8:15
29. Na H, Zdraljevic S, Tanny RE, Walhout AJM, Andersen EC (2020) Natural variation in a glucuronosyltransferase modulates propionate sensitivity in a *C. elegans* propionic acidemia model. PLoS Genet 16:e1008984
30. Sharma N, Au V, Martin K, Edgley ML, Moerman D, Mains PE, Gilleard JS (2024) Multiple UDP glycosyltransferases modulate benzimidazole drug sensitivity in the nematode *Caenorhabditis elegans* in an additive manner. Int J Parasitol 54:535–549

31. Leiers B, Kampkötter A, Grevelding CG, Link CD, Johnson TE, Henkle-Dührsen K (2003) A stress-responsive glutathione S-transferase confers resistance to oxidative stress in *Caenorhabditis elegans*. Free Radic Biol Med 34:1405–1415
32. Zheng Z, Wu K, Ruan Q, Li D, Liu W, Wang M, Li Y, Xia J, Yang D, Guo J (2022) Suppression of selective voltage-gated calcium channels alleviates neuronal degeneration and dysfunction through glutathione S-transferase-mediated oxidative stress resistance in a *Caenorhabditis elegans* model of Alzheimer's disease. Oxidative Med Cell Longev 2022:8287633
33. Urban N, Tsitsipatis D, Hausig F, Kreuzer K, Erler K, Stein V, Ristow M, Steinbrenner H, Klotz LO (2017) Non-linear impact of glutathione depletion on *C. elegans* life span and stress resistance. Redox Biol 11:502–515
34. Ferguson GD, Bridge WJ (2019) The glutathione system and the related thiol network in *Caenorhabditis elegans*. Redox Biol 24:101171
35. Ayyadevara S, Engle MR, Singh SP, Dandapat A, Lichti CF, Benes H, Shmookler Reis RJ, Liebau E, Zimniak P (2005) Lifespan and stress resistance of *Caenorhabditis elegans* are increased by expression of glutathione transferases capable of metabolizing the lipid peroxidation product 4-hydroxynonenal. Aging Cell 4:257–271
36. Ayyadevara S, Dandapat A, Singh SP, Siegel ER, Shmookler Reis RJ, Zimniak L, Zimniak P (2007) Life span and stress resistance of *Caenorhabditis elegans* are differentially affected by glutathione transferases metabolizing 4-hydroxynon-2-enal. Mech Ageing Dev 128:196–205
37. Kusche-Gullberg M, Kjellén L (2003) Sulfotransferases in glycosaminoglycan biosynthesis. Curr Opin Struct Biol 13:605–611
38. Hattori K, Inoue M, Inoue T, Arai H, Tamura HO (2006) A novel sulfotransferase abundantly expressed in the dauer larvae of *Caenorhabditis elegans*. J Biochem 139:355–362
39. Dierker T, Shao C, Haitina T, Zaia J, Hinas A, Kjellén L (2016) Nematodes join the family of chondroitin sulfate-synthesizing organisms: identification of an active chondroitin sulfotransferase in *Caenorhabditis elegans*. Sci Rep 6:34662
40. Izumikawa T, Dejima K, Watamoto Y, Nomura KH, Kanaki N, Rikitake M, Tou M, Murata D, Yanagita E, Kano A, Mitani S, Nomura K, Kitagawa H (2016) Chondroitin 4-O-sulfotransferase is indispensable for sulfation of chondroitin and plays an important role in maintaining normal life span and oxidative stress responses in nematodes. J Biol Chem 291: 23294–23304
41. Townley RA, Bülow HE (2011) Genetic analysis of the heparan modification network in *Caenorhabditis elegans*. J Biol Chem 286:16824–16831
42. Beisswanger R, Corbeil D, Vannier C, Thiele C, Dohrmann U, Kellner R, Ashman K, Niehrs C, Huttner WB (1998) Existence of distinct tyrosylprotein sulfotransferase genes: molecular characterization of tyrosylprotein sulfotransferase-2. Proc Natl Acad Sci USA 95:11134–11139
43. Ouyang YB, Moore KL (1998) Molecular cloning and expression of human and mouse tyrosylprotein sulfotransferase-2 and a tyrosylprotein sulfotransferase homologue in *Caenorhabditis elegans*. J Biol Chem 273:24770–24774
44. Dorfmueller HC, Fang W, Rao FV, Blair DE, Attrill H, van Aalten DM (2012) Structural and biochemical characterization of a trapped coenzyme A adduct of *Caenorhabditis elegans* glucosamine-6-phosphate N-acetyltransferase 1. Acta Crystallogr D Biol Crystallogr 68: 1019–1029
45. Abo-Dalo B, Ndjonka D, Pinnen F, Liebau E, Lüersen K (2004) A novel member of the GCN5-related N-acetyltransferase superfamily from *Caenorhabditis elegans* preferentially catalyses the N-acetylation of thialysine [S-(2-aminoethyl)-L-cysteine]. Biochem J 384:129–137
46. Migliori ML, Romanowski A, Simonetta SH, Valdez D, Guido M, Golombek DA (2012) Daily variation in melatonin synthesis and arylalkylamine N-acetyltransferase activity in the nematode *Caenorhabditis elegans*. J Pineal Res 53:38–46
47. Pastore B, Hertz HL, Price IF, Tang W (2021) Pre-piRNA trimming and 2'-O-methylation protect piRNAs from 3' tailing and degradation in *C. elegans*. Cell Rep 36:109640

48. Xiao Y, Liu F, Kong Q, Zhu X, Wang H, Li S, Jiang N, Yu C, Yun L (2022) Metformin induces S-adenosylmethionine restriction to extend the *Caenorhabditis elegans* healthspan through H3K4me3 modifiers. Aging Cell 21:e13567
49. Winter AD, Tjahjono E, Beltrán LJ, Johnstone IL, Bulleid NJ, Page AP (2022) Dietary-derived vitamin B12 protects *Caenorhabditis elegans* from thiol-reducing agents. BMC Biol 20:228
50. Xiao Y, Han C, Li X, Zhu X, Li S, Jiang N, Yu C, Liu Y, Liu F (2024) S-Adenosylmethionine (SAM) diet promotes innate immunity via histone H3K4me3 complex. Int Immunopharmacol 131:111837
51. Gomez F, Saiki R, Chin R, Srinivasan C, Clarke CF (2012) Restoring de novo coenzyme Q biosynthesis in *Caenorhabditis elegans coq-3* mutants yields profound rescue compared to exogenous coenzyme Q supplementation. Gene 506:106–116
52. Mia MM, Allaie IM, Zhang X, Li K, Khan SM, Kadotani S, Witola WH (2024) Characterization of a unique as a molecular drug target in parasitic filarial nematodes. PLoS Negl Trop Dis 18: e0012473
53. Surco-Laos F, Cabello J, Gómez-Orte E, González-Manzano S, González-Paramás AM, Santos-Buelga C, Dueñas M (2011) Effects of O-methylated metabolites of quercetin on oxidative stress, thermotolerance, lifespan and bioavailability on *Caenorhabditis elegans*. Food Funct 2:445–456
54. Savion N, Levine A, Kotev-Emeth S, Bening Abu-Shach U, Broday L (2018) S-allylmercapto-N-acetylcysteine protects against oxidative stress and extends lifespan in *Caenorhabditis elegans*. PLoS One 13:e0194780
55. Gusarov I, Shamovsky I, Pani B, Gautier L, Eremina S, Katkova-Zhukotskaya O, Mironov A, Makarov AA, Nudler E (2021) Dietary thiols accelerate aging of *C. elegans*. Nat Commun 12: 4336
56. Guha S, Mathew ND, Konkwo C, Ostrovsky J, Kwon YJ, Polyak E, Seiler C, Bennett M, Xiao R, Zhang Z, Nakamaru-Ogiso E, Falk MJ (2021) Combinatorial glucose, nicotinic acid and N-acetylcysteine therapy has synergistic effect in preclinical *C. elegans* and zebrafish models of mitochondrial complex I disease. Hum Mol Genet 30:536–551
57. Lee A, Asahina K, Okamoto T, Kawaguchi K, Kostsin DG, Kashiwayama Y, Takanashi K, Yazaki K, Imanaka T, Morita M (2014) Role of NH2-terminal hydrophobic motif in the subcellular localization of ATP-binding cassette protein subfamily D: common features in eukaryotic organisms. Biochem Biophys Res Commun 453:612–618
58. Luciani MF, Chimini G (1996) The ATP binding cassette transporter ABC1, is required for the engulfment of corpses generated by apoptotic cell death. EMBO J 15:226–235
59. McDonald MK, Fritz JA, Jia D, Scheuchner D, Snyder FF, Stanislaus A, Curle J, Li L, Stabler SP, Allen RH, Mains PE, Gravel RA (2017) Identification of ABC transporters acting in vitamin B(12) metabolism in *Caenorhabditis elegans*. Mol Genet Metab 122:160–171
60. Tang B, Williams PL, Xue KS, Wang JS, Tang L (2020) Detoxification mechanisms of nickel sulfate in nematode *Caenorhabditis elegans*. Chemosphere 260:127627
61. Wang D-Y (2018) Nanotoxicology in *Caenorhabditis elegans*. Springer, Singapore
62. Wu Q-L, Li Y-X, Li Y-P, Zhao Y-L, Ge L, Wang H-F, Wang D-Y (2013) Crucial role of biological barrier at the primary targeted organs in controlling translocation and toxicity of multi-walled carbon nanotubes in nematode *Caenorhabditis elegans*. Nanoscale 5:11166–11178
63. Liu Z-F, Zhou X-F, Wu Q-L, Zhao Y-L, Wang D-Y (2015) Crucial role of intestinal barrier in the formation of transgenerational toxicity in quantum dots exposed nematodes *Caenorhabditis elegans*. RSC Adv 5:94257–94266
64. Wang D-Y (2016) Biological effects, translocation, and metabolism of quantum dots in the nematode *Caenorhabditis elegans*. Toxicol Res 5:1003–1011

65. Qu Y, Li W, Zhou Y, Liu X, Zhang L, Wang L, Li YF, Iida A, Tang Z, Zhao Y, Chai Z, Chen C (2011) Full assessment of fate and physiological behavior of quantum dots utilizing *Caenorhabditis elegans* as a model organism. Nano Lett 11:3174–3183
66. Zhao Y-L, Liu Q, Shakoor S, Gong JR, Wang D-Y (2015) Transgenerational safe property of nitrogen-doped graphene quantum dots and the underlying cellular mechanism in *Caenorhabditis elegans*. Toxicol Res 4:270–280

Target Organ Pharmacology 7

Abstract

In *Caenorhabditis elegans*, pharmacological compounds and drugs can be translocated not only to the primary targeted organs (such as intestine and pharynx) but also to the secondary targeted organs (such as nervous system, reproductive organs, and muscle). The bioavailability and the pharmacological effects of compounds and drugs on intestine and pharynx were first introduced and discussed. Moreover, we introduced and discussed the bioavailability and the pharmacological effects of compounds and drugs on nervous system, reproductive organs, and muscle.

Keywords

Target organs · Compounds · Drugs · *Caenorhabditis elegans*

7.1 Introduction

The exact pharmacological effects of examined compounds and drugs are mainly determined by their action on targeted organs or tissues. In Chap. 6, we have introduced and discussed the pharmacological translocation of compounds and drugs after the uptake by nematodes. In *C. elegans*, normally pharynx and intestine are the primary targeted organs. Nervous system, gonad, and muscle can be considered as the secondary targeted organs or tissues of compounds and drugs. In this chapter, we thus first introduce and discuss the pharmacological effects in the intestine and the pharynx. Moreover, the pharmacological effects in the nervous system, the reproductive organs, and the muscle were introduced and discussed.

D. Wang, *Pharmacology in Caenorhabditis elegans*,
https://doi.org/10.1007/978-981-92-5166-7_7

7.2 Pharmacological Effects in Intestine and Pharynx

Intestine and pharynx belong to the alimentary system of *C. elegans*. The intestine is the largest organ in *C. elegans* with multiple functions, such as digestion, nutrient absorption, innate immunity, nurturing of germline, and synthesis and storage of macromolecules like fats. Structurally, the intestine contains large and polarized epithelial cells. In *C. elegans* intestine, its apical surface is covered with microvilli (brush border), and provides the surface area for nutrient absorption. The basal surface faces the body cavity (pseudocoelom) and provides the interface for nutrient export to the body. The lateral surfaces function to contact the neighboring cells to organize the intestinal barrier by containing adherens junctions and gap junctions. In intestinal cells, their cytoplasmic contents are rich in mitochondria, rough ER, Golgi, ribosomes, lysosomes, and storage vacuoles, which provides the crucial support for the energy production, protein/lipid synthesis, and intracellular digestion.

The pharynx is a tube-like muscular pump containing epithelial cells, muscle cells, marginal cells, gland cells, valve cells, and its own nervous system. In the pharynx, the bacteria or chemicals can be concentrated, ground, and/or transported to the intestine.

7.2.1 Pharmacological Effects in Intestine

For the water-soluble and lipid-soluble compounds and drugs, they are bioavailable to the intestine after the oral feeding. For the particulate chemicals, if their size is in the range of ng/L, they can be also available to the intestine after the treatment. However, if the size of examined particulate chemicals is in the range of μg/L, their possible bioavailability to the intestine should be carefully examined.

Corresponding to the functions of intestine, the pharmacological effects in the intestine are reflected by at least the following several aspects:

1. Activation of antioxidation system
   The 6-PPD quinone is a tire-rubber related pollutant [1], and can cause multiple toxicities on nematodes, including the intestinal toxicity [2–8]. The 6-PPDQ induced intestinal reactive oxygen species (ROS) and intestinal lipofuscin accumulation could be inhibited by vitamin D3 treatment (Fig. 7.1) [9]. Besides these, the expressions of mitochondrial Mn-SODs (such as SOD-2 and SOD-3) and/or GST-4 could be activated by pharmacological compounds, such as flavonoids and paeoniflorin (Fig. 7.2) [10, 11].
2. Enhancement in mitochondrial function
   The mitochondrial function could be damaged by 6-PPDQ [12]. The 6-PPDQ induced mitochondrial dysfunction reflected by elevated oxygen consumption rate and decreased ATP content could be suppressed by methionine treatment (Fig. 7.3) [13]. Accompanied with these, the suppressed mitochondrial unfolded protein response (mt-UPR) by 6-PPDQ could also be relieved by methionine treatment (Fig. 7.3) [13]. Additionally, the dysregulated expression of genes

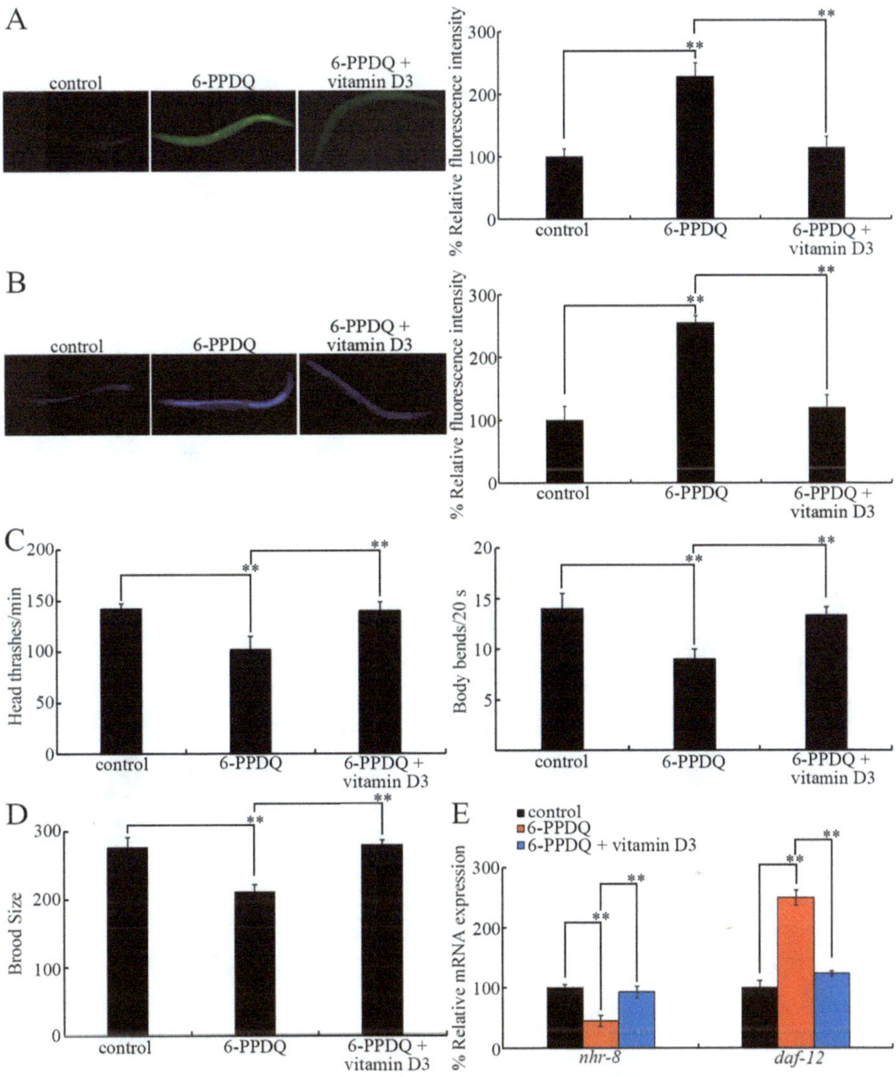

Fig. 7.1 Pharmacological effect of treatment with vitamin D3 on 6-PPDQ toxicity in wild-type nematodes [9]. (**a**) Pharmacological effect of treatment with vitamin D3 on 6-PPDQ toxicity in inducing intestinal ROS generation in wild-type nematodes. (**b**) Pharmacological effect of treatment with vitamin D3 on 6-PPDQ toxicity in inducing intestinal lipofuscin accumulation in wild-type nematodes. (**c**) Pharmacological effect of treatment with vitamin D3 on 6-PPDQ toxicity in causing inhibition of locomotion in wild-type nematodes. (**d**) Pharmacological effect of treatment with vitamin D3 on 6-PPDQ toxicity in causing a reduction in brood size in wild-type nematodes. (**e**) Pharmacological effect of treatment with vitamin D3 on the expressions of *nhr-8* and *daf-12* in 6-PPDQ-exposed wild-type nematodes. After exposure to 10 mg L^{-1} of 6-PPDQ from L1 larvae for 6.5 days, the nematodes were treated with 400 mM of vitamin D3 for 24 h. $^{**}P < 0.01$

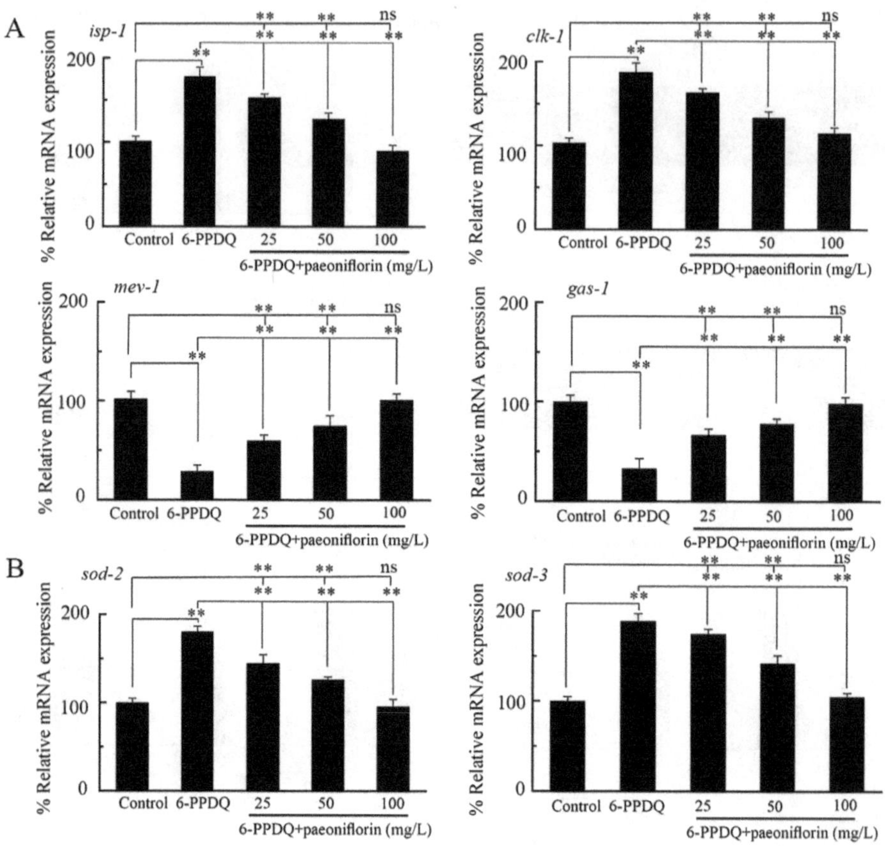

Fig. 7.2 Effect of paeoniflorin treatment on expressions of genes governing oxidative stress (**a**) and genes encoding mitochondrial Mn-SODs (**b**) in 6-PPDQ exposed nematodes [11]. Exposure concentration of 6-PPDQ was 100 μg/L. $^{**}P < 0.01$. ns no significant difference

encoding components of mitochondrial complexes I-III could be rescued by paeoniflorin treatment to a great degree (Fig. 7.2) [11, 14].

3. Enhancement in intestinal barrier function
   Exposure to 6-PPDQ caused damage on intestinal barrier function, which was reflected by the increased intestinal permeability [2]. This disrupted function of intestinal barrier by 6-PPDQ could be relieved by paeoniflorin treatment (Fig. 7.4) [11]. In addition, the expression of *acs-22* governing intestinal barrier function in 6-PPDQ exposed nematodes could be elevated by paeoniflorin treatment (Fig. 7.4) [11].
4. Enhancement in innate immune response
   In *C. elegans*, bacterial pathogen infection could result in immunosuppression [15]. The *Pseudomonas aeruginosa* infection induced immunosuppression reflected by inhibition in *lys-1* and *lys-8* expressions could be suppressed by

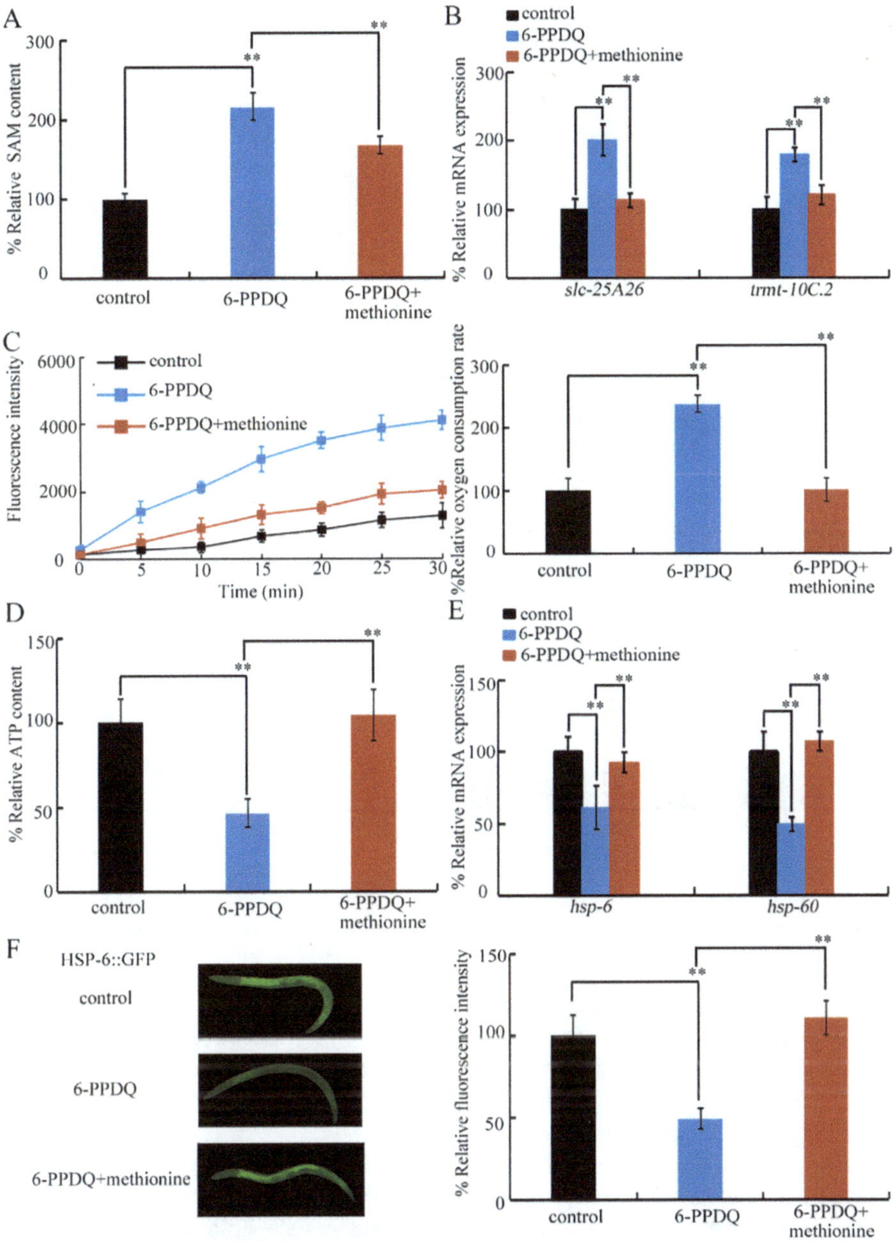

Fig. 7.3 Pharmacological effect of methionine (50 mM) treatment on SAM content and mitochondrial function in 6-PPDQ (10 μg/L) exposed nematodes [13]. (**a**) Effect of methionine treatment on SAM content in 6-PPDQ exposed nematodes. (**b**) Effect of methionine treatment on expressions of *slc-25A26* and *trmt-10C.2* in 6-PPDQ exposed nematodes. (**c**) Effect of methionine treatment on oxygen consumption rate in 6-PPDQ exposed nematodes. (**d**) Effect of methionine treatment on ATP content in 6-PPDQ exposed nematodes. (**e**) Effect of methionine treatment on expressions of *hsp-6* and *hsp-60* in 6-PPDQ exposed nematodes. (**f**) Effect of methionine treatment on expression of HSP-6::GFP in 6-PPDQ exposed nematodes. $^{**}P < 0.01$

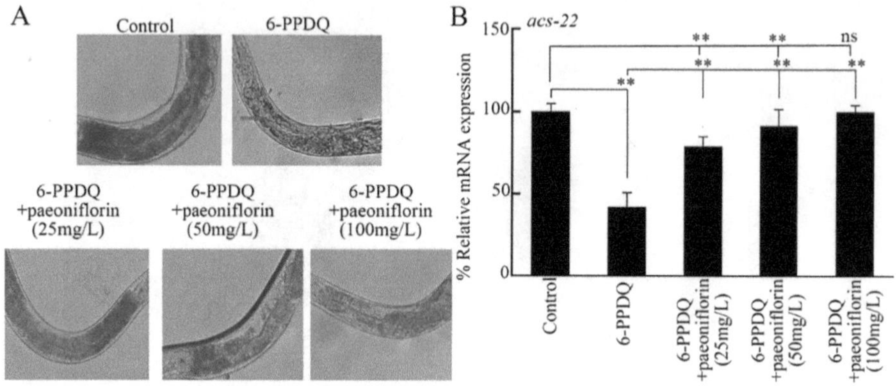

Fig. 7.4 Effect of paeoniflorin treatment on functional state of intestinal barrier in 6-PPDQ exposed nematodes [11]. (**a**) Effect of paeoniflorin treatment on intestinal permeability in 6-PPDQ exposed nematodes. Arrowhead indicates intestinal cell, and asterisk indicates body cavity. (**b**) Effect of paeoniflorin treatment on expression of *acs-22* in 6-PPDQ exposed nematodes. $^{**}P < 0.01$. ns no significant difference. Exposure concentration of 6-PPDQ was 100 μg/L

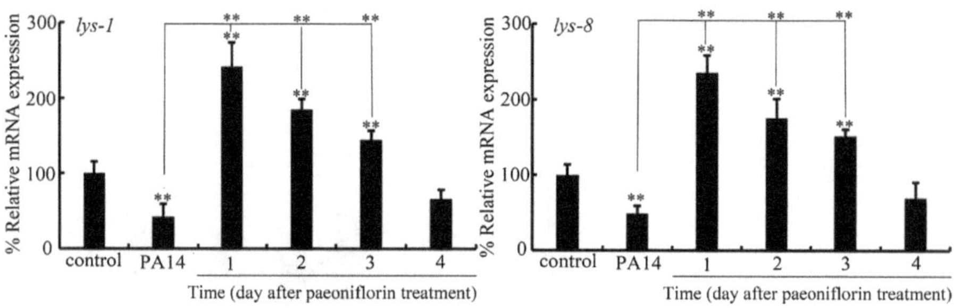

Fig. 7.5 Effect of 100 mg/L paeoniflorin treatment on expression of antimicrobial genes (*lys-1* and *lys-8*) in nematodes at the immunosuppression stage (day-5) after *P. aeruginosa* infection [15]. $^{**}P < 0.01$ vs. control (if not specially indicated)

paeoniflorin treatment (Fig. 7.5) [15]. Meanwhile, the severe *P. aeruginosa* accumulation in body of nematodes could be reduced by paeoniflorin treatment (Fig. 7.6) [15].

5. Reduction in lipid accumulation

   Using the animal model of *C. elegans*, some compounds have been shown to have the function in reducing lipid accumulation. For example, in glucose-induced model, betulinic acid treatment could reduce lipid accumulation and inhibit the expression of genes (*fat-5*, *fat-6*, and *fat-7*) governing lipogenesis and lipid sensor genes (such as *nhr-49*) [16]. Esculetin could reduce the lipid accumulation by modulating insulin and AMPK signals [17].

6. Modulation of defecation behavior

   The organic selenium compounds (such as diphenyl diselenide (DPDS)) showed protective function against the oxidative stress together with the reduced

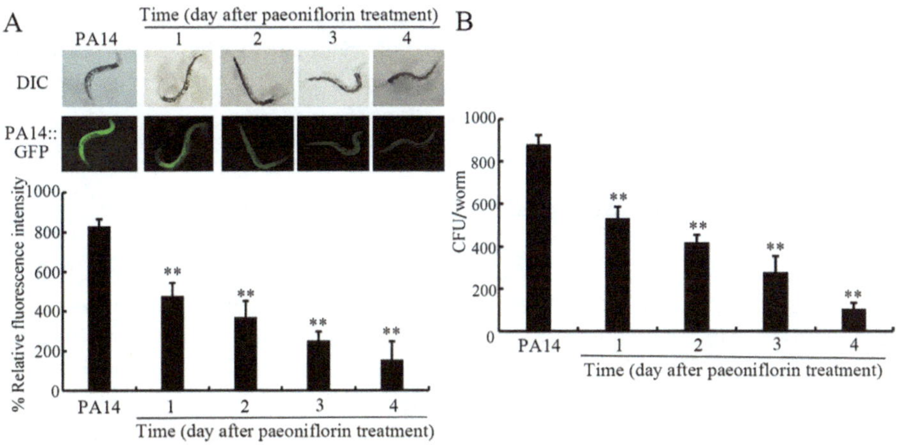

Fig. 7.6 Effect of 100 mg/L paeoniflorin treatment on PA14::GFP accumulation in intestinal lumen A and CFU of PA14 B in nematodes at the immunosuppression stage (day-5) after *P. aeruginosa* infection [15]. $^{**}P < 0.01$ vs. PA14

defecation cycle time [18]. In addition, in graphene oxide exposed nematodes, the normal defecation behavior could be sustained by lactic acid bacteria (LAB) treatment [19].

7.2.2 Pharmacological Effects in Pharynx

The pharmacological effects in pharynx can be reflected by the following several aspects:

1. Reduction in food uptake
   Coffee diterpene kahweol could reduce the food intake, which was related to the reduction in lipid accumulation [20]. In *eat-2* mutants with deficit in pharynx contraction rate, the lipid accumulation was not affected by kahweol [20].
2. Assessment of anesthetic effects
   The pharyngeal mucosa can be employed as the assay system for assessing the local anesthetic effects of some compounds, such as alcohol [21]. The alcohol greater than 100 mM could cause the anesthetic effect and the effect on muscle activity rapidly reached steady-state [22]. Using this assay system, certain compounds and drugs with the anesthetic effect and the underlying mechanisms can be identified and determined, which is helpful to relieve pain (sore throat) or discomfort originating from the (pharyngitis) in the clinical.
3. Evaluation of immunomodulatory effect
   In *C. elegans*, some antimicrobial genes are mainly or strongly expressed in the pharynx, such as ABF-2, SPP-12, and ILYS-3 [23–25]. In the pharynx, the ILYS-3 is required for pharyngeal grinding [25]. In addition, the pharynx HPX-2 further

participated in limiting the intestinal colonization pathogens [26]. These backgrounds suggested that the *C. elegans* pharynx can also be used to evaluate the effect of certain compounds and drugs to modify the immune response within the pharynx.

7.2.3 Pharmacological Effects in Nervous System

The *C. elegans* hermaphrodite has exactly 302 neurons and 56 glial cells, which constitutes nearly 40% of the total somatic cells. The neurons and their support cells are organized into distinct anatomical structures. Neuronal cell bodies are clustered into ganglia located mainly in the head and tail. Head ganglia is the largest collection of neurons surrounds the nerve ring. Tail ganglia is the second largest collection of cell bodies, which are crucial for processing tail sensory information and mating behavior for males. Located in the head and encircling the pharynx region, the nerve ring (the "brain") is the major neuropil to integrate the vast majority of sensory information and to process the commands for behavior. Nerve cords refer to the longitudinal bundles of nerve processes (axons and dendrites) to connect the head and tail ganglia, which contain ventral nerve cord (VNC) with many motor neurons and dorsal nerve cord (DNC) with the function to innervate the body wall muscles for locomotion.

For the water-soluble and lipid-soluble compounds and drugs, they are normally assumed to be bioavailable to the secondary targeted organs, such as nervous system, reproductive organs, and muscle. For the particulate chemicals in the range of ng/L, some indirect evidence has implied their bioavailability to the nervous system [27, 28]. For the particulate chemicals in the range of μg/L, usually they are difficult or not possible to be available to the secondary targeted organs.

The pharmacological effects in the nervous system are reflected by the following several aspects:

1. Inhibition in damage on neuronal development
   The function of compounds and drugs to inhibit the damage on neuronal development can be first examined at the nervous system level. Exposure to Al_2O_3-nanoparticles could induce the neurodegeneration in GABAergic nervous system indicated by neuronal loss and formation of gaps on nerve cords, which could be suppressed by vitamin E treatment [29]. The function of compounds and drugs to inhibit the damage on neuronal development can also be examined at the level of certain type of neurons. Treatment with vitamin E could suppress the damage (such as the formation of abnormal dendrite and cilia) of Al_2O_3-nanoparticles on AFD sensory neurons (Fig. 7.7) [29]. Additionally, the damage of Al_2O_3-nanoparticles on AIY interneurons was further inhibited by vitamin E treatment (Fig. 7.7) [29]. For the neuronal development, the synaptic development and synaptic adhesion are also suggested to be examined after treatment with compounds and drugs.
2. Effect on locomotion, sensory perception, learning, and memory

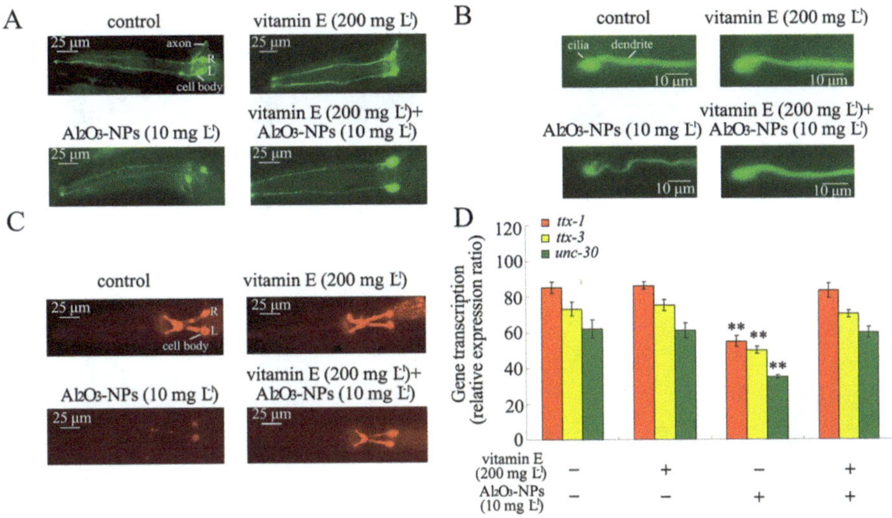

Fig. 7.7 Effects of pretreatment with vitamin E on the development of AFD and AIY neurons of Al_2O_3-NPs exposed nematodes [29]. (**a**) Effects of vitamin E pretreatment on the development of cell body in AFD sensory neurons of Al_2O_3-NPs exposed *adEx1267* transgenic nematodes. (**b**) Effects of vitamin E pretreatment on the development of cilia and dendrites in AFDL sensory neurons of Al_2O_3-NPs exposed *adEx1267* transgenic nematodes. (**c**) Effects of vitamin E pretreatment on the development of AIY interneurons of Al_2O_3-NPs exposed *otIs133* transgenic nematodes. (**d**) Effects of vitamin E pretreatment on the expression pattern of *ttx-1*, *ttx-3*, and *unc-30* genes in Al_2O_3-NPs exposed wild-type N2 nematodes. L, left; R, right. Vitamin E pretreatment was performed at the L2-larvae stage for 24 h. Al_2O_3-NPs exposure was performed from L4-larvae for 24 h. Five replicates were performed. Bars represent means ± SEM. $^{**}P < 0.01$

The 6-PPDQ caused inhibition in locomotion behavior could be suppressed by the following vitamin D3 treatment (Fig. 7.1) [9], and the Al_2O_3-nanoparticles induced inhibition in locomotion behavior could be suppressed by vitamin E treatment (Fig. 7.8) [29]. The Al_2O_3-nanoparticles induced abnormal thermotaxis behavior could be rescued by vitamin E to a great degree (Fig. 7.8) [29]. The Al_2O_3-nanoparticles caused impairment on thermotaxis learning could also be suppressed by vitamin E treatment (Fig. 7.8) [29]. The heavy metal and UV-irradiation caused deficit in memory behavior was also retrieved by vitamin E treatment [30].

3. Effect on neurotransmission and neurotransmitters

Based on analysis of aldicarb and levamisole resistance, exposure to Al_2O_3-nanoparticles might cause deficits in both presynaptic and postsynaptic functions (Fig. 7.9) [29]. Moreover, after vitamin E treatment, the resistance to aldicarb and levamisole induced by Al_2O_3-nanoparticles was obviously inhibited (Fig. 7.9) [29]. The patch clamp-based electrophysiological recording can be further performed to indicate the effect of compounds and drugs on neurotransmission. Besides these, treatment with selegiline, a monoamine oxidase B inhibitor, and tolcapone, a catechol-O-methyltransferase inhibitor, could elevate levels of

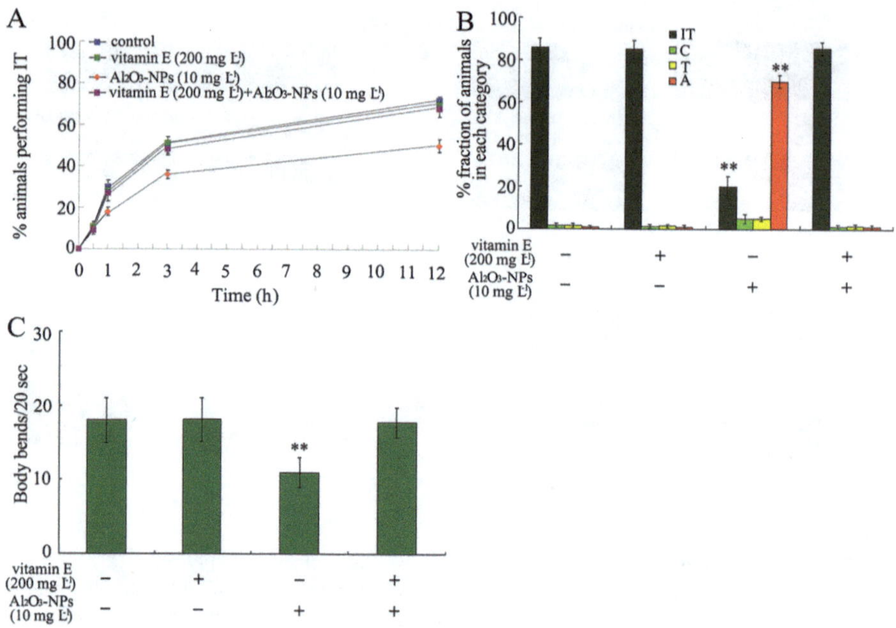

Fig. 7.8 Effects of pretreatment with vitamin E on the behavioral performance in Al_2O_3-NPs exposed wild-type N2 nematodes [29]. (**a**) Effects of vitamin E pretreatment on thermotaxis learning in Al_2O_3-NPs exposed wild-type N2 nematodes. (**b**) Effects of vitamin E pretreatment on thermotaxis perception in Al_2O_3-NPs exposed wild-type N2 nematodes. T thermophilic; C movement to 17 °C; A movement across the thermal gradient (17 °C/25 °C); IT, movement at 20 °C. (**c**) Effects of vitamin E pretreatment on body bend in Al_2O_3-NPs exposed wild-type N2 nematodes. Vitamin E pretreatment was performed at the L2-larvae stage for 24 h. Al_2O_3-NPs exposure was performed from L4-larvae for 24 h. Five replicates were performed. Bars represent means ± SEM. $^{**}P < 0.01$

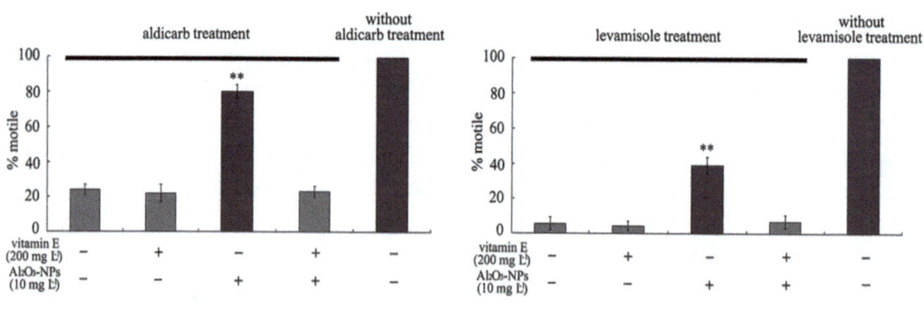

Fig. 7.9 Effects of pretreatment with vitamin E on neurotransmission in Al_2O_3-NPs exposed wild-type N2 nematodes [29]. The presynaptic function was evaluated by aldicarb resistance, and the postsynaptic function was evaluated by levamisole resistance. Vitamin E pretreatment was performed at the L2-larvae stage for 24 h. Al_2O_3-NPs exposure was performed from L4-larvae for 24 h. Five replicates were performed. Bars represent means ± SEM. $^{**}P < 0.01$

dopamine and serotonin [31]. In addition, in radiation-damaged nematodes, long-term memory was enhanced, which was linked to modulation of neuropeptides in sensory neurons [32]. The further effects of compounds and drugs on expression and function of receptors on postsynaptic are suggested to be further examined after the treatment. In addition, the activities of excitability neurons and inhibitory neurons and their balance are also suggested to be carefully considered after treatment with compounds and drugs with the possible neuroprotective effect.

4. Effect on glial cells
   The *C. elegans* glial cells refer to a group of 56 specialized support cells making up about 6% of total somatic cells, and function as active and critical regulators of nervous system development and function. Normally, glial cells form isolated compartments around synapses and sensory endings, and most of them are located around the sensory organs in the head (labial, cephalic, amphid glia) and the tail (phasmid glia) by ensheathing sensory neuron endings. The phenolic compounds were identified to protect the damage of oxidative damage in glial cells by elevating superoxide dismutase expression [33].
5. Effect on neuronal circuit of behaviors
   One of the most significant achievements in *C. elegans* is the complete mapping of the nervous system, known as the connectome, which was reconstructed from serial-section electron micrographs and details every neuron, connection, and synapse with near-perfect invariance between individuals. The hermaphrodite connectome of nervous system includes approximately 6400 chemical synapses, 900 gap junctions, and 1500 neuromuscular junctions (connections to muscles). These backgrounds provide the foundation for identification of detailed neuronal circuits for different behaviors. Accompanied with observation of effect of compounds and drugs on certain behaviors, their possible effect on corresponding neuronal circuit of these behaviors are suggested to be further examined. In nematodes expressing human MSI1, treatment with (−)-gossypol could inhibit MSI1 activity and suppress the rescue of *msi-1(lf)* phenotype in AVA neurons, implying that (−)-gossypol regulated Musashi pathway in the neuronal circuit related to control of memory [34].
6. Neuroprotective effect in the disease models
   In Chap. 4, we have introduced and discussed the neurodegeneration disease models in *C. elegans*, such as Alzheimer' disease (AD) model and Parkinson disease (PD) models. Besides the several aspects described above, in AD models, the pharmacological effects of compounds and drugs on Aβ deposits and Tau aggregation should be also determined. For example, using Aβ expression models, diphenyl-diselenide, oxoisoaporphine alkaloid derivative 8–1, monascin, dauricine, 2-arylethenylquinoline derivatives, and mitochondria-targeted MitoQ showed the possible neuroprotective effect against the AD disease [35–41]. Using Tauopathy models, curcumin, tetrahydropyrimidine analogue, *Dendrobium officinale* phenolic extract, *Reineckia carnea* ether fraction, 2-thioxo-1,2,3,4-tetrahydropyrimidines, and antagonism of dopamine D2 receptor exhibited possible neuroprotective effect against the AD disease [42–47]. In PD models, the pharmacological effects of compounds and drugs on α-Synuclein aggregation

need also to be determined. In α-Synuclein aggregation models, chrysin, policosanol, botulin, palmitic acid, chlorogenic acid, ergothioneine, rose essential oil, and probiotic *Bacillus subtilis* exhibited possible neuroprotective effect against the PD disease [48–55].

7.3 Pharmacological Effects in Reproductive Organs

The hermaphrodite reproductive system contains two U-shaped gonad arms that join at a common uterus. The somatic gonad provides the structural framework for the reproductive system. From the distal tip toward the uterus, the somatic gonad includes:

1. Distal tip cell (DTC): A large cell located at the end of each gonad arm with the function to provide the pool of proliferating germ cells at the distal end by acting as the stem cell niche.
2. Gonadal sheath: A layer of cells enveloping the germline, which are contractile and help the maturing oocytes toward spermatheca.
3. Spermatheca: A structure located between the gonad arm and the uterus with the function to store the sperm and provide the site where oocytes are fertilized.
4. Spermathecal-uterine valve: A structure is located between the spermatheca and the uterus to control the passage of fertilized embryos.
5. Uterus: A specialized compartment where the fertilized embryos initiate development and undergo their early cell divisions before being laid.
6. Germline: It is a large, syncytial structure where many germ cell nuclei are located and share a common cytoplasm. In the germline, germ cell nuclei will develop into mature gametes (sperm and oocytes).

Germ cells develop as they migrate from the distal tip toward the uterus. The mitotic region (distal) is located at the far end, near the DTC, where the germline stem cells actively divide. As germ cells move away from the DTC, they enter the meiosis. In the meiotic region, at the L4-larvae, the first meiotic cells differentiate into large number of sperm. During the oogenesis (proximal), germline switches to producing oocytes.

For the egg-laying apparatus, it contains vulva (egg-laying organ, a slit-like opening in the ventral midbody), uterine and vulval muscles (specialized muscles that contract to push embryos), and certain neurons (HSNs and VC, motor neurons that innervate egg-laying muscles).

The water-soluble and lipid-soluble compounds and drugs can be bioavailable to the reproductive organs. For the particulate chemicals in the range of ng/L, some evidence has been raised for their bioavailability to the reproductive organs, such as gonad and spermatheca [56–61].

The pharmacological effects in the reproductive organs can be reflected by the following several aspects:

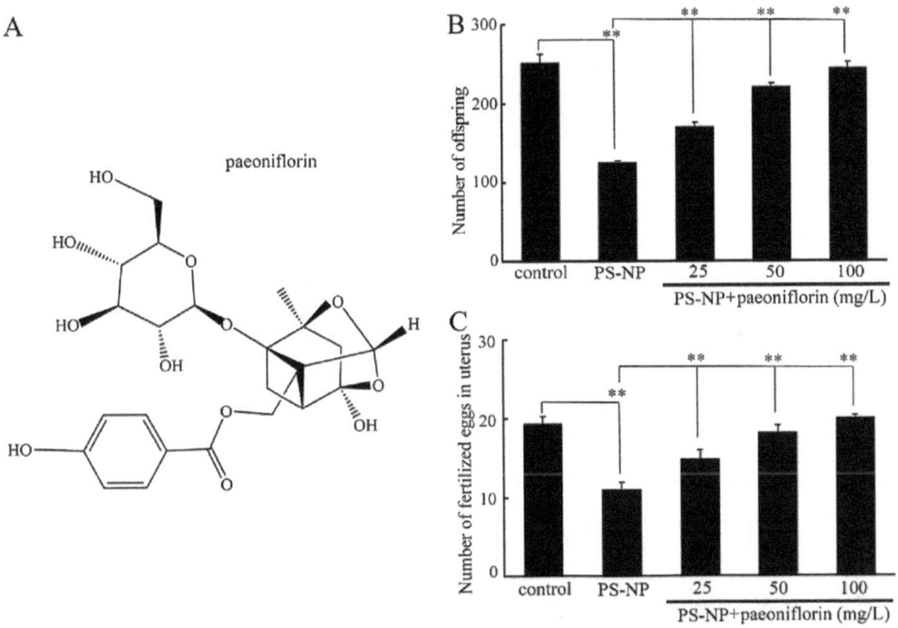

Fig. 7.10 Effect of paeoniflorin treatment against PS-NP toxicity on reproductive capacity [74]. (**a**) Chemical structure of paeoniflorin. (**b**) Effect of paeoniflorin treatment against 100 μg/L PS-NP toxicity on brood size. (**c**) Effect of paeoniflorin treatment against 100 μg/L PS-NP toxicity on number of fertilized eggs in uterus. $^{**}P < 0.01$

1. Effect on reproductive capacity
   Exposure to nanoplastics could cause several aspects of toxicity, such as the neurotoxicity and reproductive toxicity [62–73]. Among the induced reproductive toxicity of nanoplastics, the toxicity in reducing reproductive capacity reflected by decrease in number of offspring and number of fertilized eggs in uterus could be inhibited by paeoniflorin treatment (Fig. 7.10) [74].
2. Effect on gonad development
   The pharmacological effects of compounds and drugs may be at least reflected from two aspects. The first is that the damage on gonad development by toxicants or stresses can be reduced by compounds or drugs treatment. The second is that certain compounds and drugs have the effect against the tumoral gonad development. For example, the synthesized tryptophan methyl ester-betaxanthin and tryptophan benzyl ester-betaxanthin could reduce the tumoral gonad sizes [75].
3. Effect on germline apoptosis
   On the one hand, the toxicants and stresses induced germline apoptosis can be inhibited by some compounds and drugs. For example, the nanoplastics induced germline apoptosis and germline DNA damage could be reduced by the following paeoniflorin treatment (Fig. 7.11) [74]. This beneficial effect of paeoniflorin treatment was largely due to the rescue of dysregulated expression of genes governing germline apoptosis (*celd-3*, *ced-4*, *ced-9*, and *egl-1*) and DNA damage

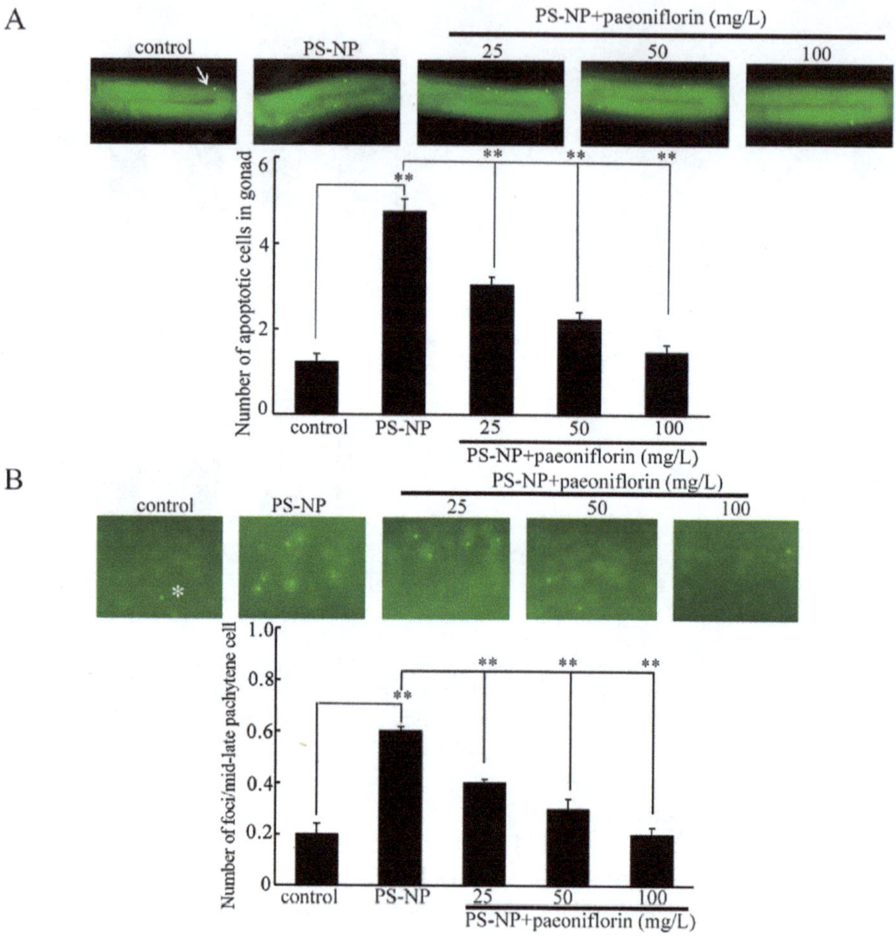

Fig. 7.11 Effect of paeoniflorin treatment on DNA damage induced germline apoptosis in PS-NP exposed nematodes [74]. (**a**) Effect of paeoniflorin treatment against toxicity of 100 μg/L PS-NP in inducing germline apoptosis. The above shows the AO staining images. The below shows the comparison of cell corpse per gonad in nematodes. Arrowhead indicates the germline apoptotic signals. (**b**) Effect of paeoniflorin treatment against toxicity of 100 μg/L PS-NP in inducing germline DNA damage. The images showing the induction of germline DNA damage are given in the above. The below shows the comparison of foci of HUS-1::GFP in the nuclei of germ cells in nematodes. Asterisk indicates the HUS-1::GFP signals. $^{**}P < 0.01$

(*cep-1*, *clk-2*, *hus-1*, and *mrt-2*) (Fig. 7.12) [74], and the possible binding potential of paeoniflorin to DNA damage checkpoints (CLK-2, HUS-1, and MRT-2) [74]. On the other hand, some compounds can act as anti-cancer drugs by enhancing germline apoptosis and DNA damage. For example, aspirin could induce the germline apoptosis to enhance the radio/chemo-therapy sensitivity [76].

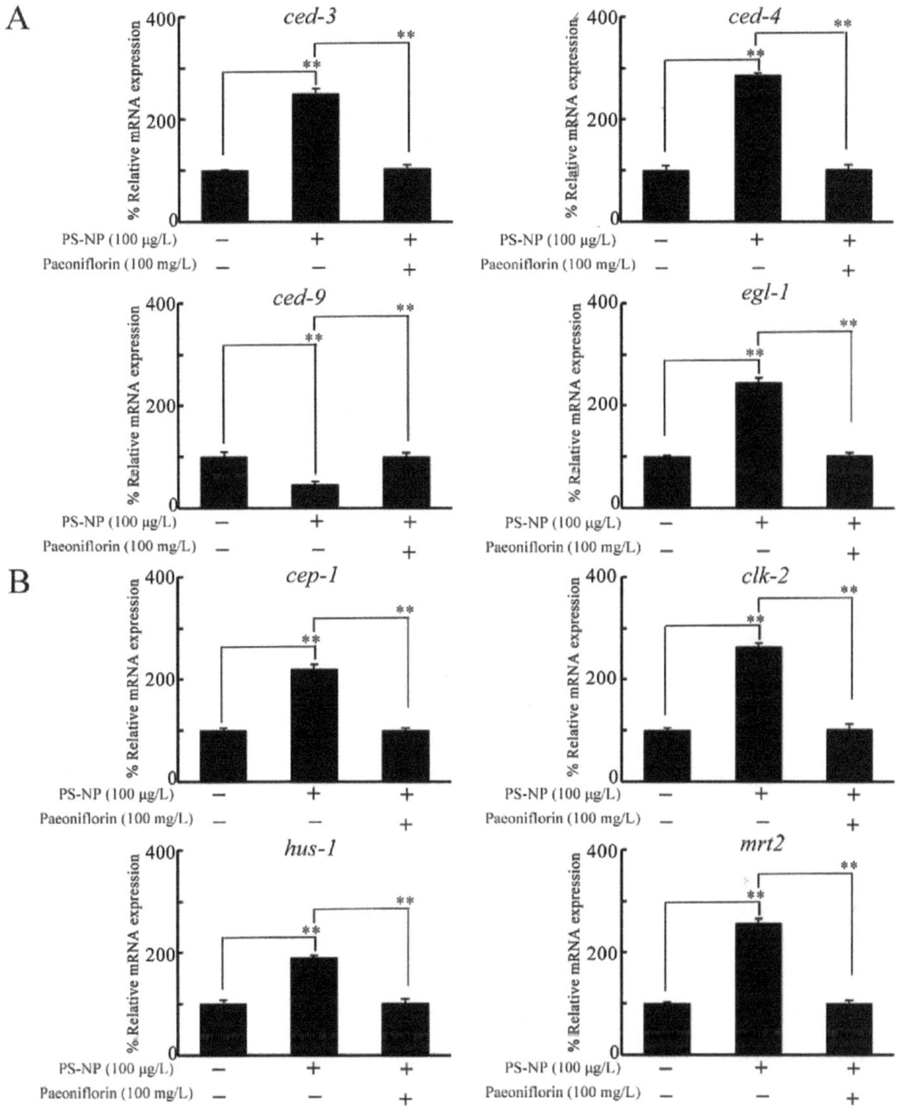

Fig. 7.12 Effect of 100 mg/L paeoniflorin treatment on expressions of genes governing germline apoptosis (**a**) and DNA damage (**b**) in 100 μg/L PS-NP exposed nematodes [74]. $^{**}P < 0.01$

4. Effect on oocyte quality and meiotic maturation
   C. elegans reproductive system can be used as a model to identify the compounds and drugs with the function to modulate and enhance health and development of eggs. Arachidonic acid treatment could extend the reproductive longevity and enhance the oocyte quality [77]. Quercetin treatment could improve the oogenesis and enhance the mitochondrial function [78]. During the aging, pharmacological

inhibition of cathepsin B proteases could suppress the induced deterioration in oocyte quality [79].

5. Effect on egg-laying behavior
   In *C. elegans*, the egg-laying behavior can be modulated by serotonin, fluoxetine, imipramine, and dopamine via targeting distinct serotonin receptors [80]. In addition, the action of fluoxetine and imipramine at the serotonin reuptake transporter (SERT) and at serotonin receptors was confirmed [80].

7.4 Pharmacological Effects in Muscle

C. elegans has two main types of muscle, striated muscles and non-striated muscles. The striated muscles refer to the body wall muscle cells, the most abundant muscle cells that run the length of the nematodes and control the movement. The non-striated muscles refer to the muscles of pharynx, intestine, vulva, and gonad, which are involved in specialized functions. The basic functional unit of body wall muscle is the sarcomere, which is composed of interdigitating thick and thin filaments organized into repeating bands. For the intracellular anchoring, thin filaments of the sarcomere are anchored at the dense bodies, and thick filaments are anchored at the M-lines. On the outside of the muscle cell, the integrin complexes connect to the extracellular matrix (basal lamina) containing perlecan and collagen. The basal lamina is further linked to specialized structures (fibrous organelles) in the hypodermis.

For the muscle cell organization, a single body wall muscle cell has three distinct parts:

1. Contractile filament lattice: This region contains the organized, repeating sarcomeres and is located adjacent to the hypodermis.
2. Muscle belly: This region is bulged into the body cavity (pseudocoelom).
3. Muscle arms: They are projected from the muscle cell toward the nerve cords (dorsal or ventral) to receive the synaptic input from motor neurons at neuromuscular junctions.

We normally assume that the water-soluble and lipid-soluble compounds and drugs can be bioavailable to the muscle. No evidence has been shown the bioavailability of particulate chemicals in the range of ng/L or μg/L to the muscle.

The pharmacological effects in the muscle can be reflected by the following several aspects:

1. Effect on muscle development and activity
   The damage on muscle development and activity by toxicants (such as pollutants) or stresses (such as microgravity) might be able to be inhibited by some compounds and drugs. That is, certain pathological models with damage on muscle development and muscle activity (such as muscle contraction) can be generated by exposure *C. elegans* to some toxicants and stresses, and after that the

compounds and drugs with the function to enhance muscle development and activity can be screened and identified using these models. In *C. elegans*, treatment with sildenafil citrate and Resveramorph-3 (RVM-3) were helpful for rescuing the nematodes from the shock-induced muscle weakness state [81].

2. Beneficial effect in mutants mimicking muscular atrophy
   In *C. elegans*, the *smn-1(cb131)* mutant show the mild form of spinal muscular atrophy, such as slowed swimming [82]. Using this mutant as the model, in a library with 1040 chemical compounds, 4-aminopyridine, gaboxadol hydrochloride, and N-acetylneuraminic acid were shown to rescue certain aspects of deficits in *smn-1* mutant nematodes [82].
3. Beneficial effect in transgenic disease models
   In *C. elegans*, some disease models were generated by transforming human wild-type and mutant genes in the muscle. For example, certain AD disease models were prepared by expressing human $Aß_{1–42}$ in body wall muscle cells, and Alzheimer's therapeutic PBT2 was found to suppress the Aß-induced toxicity on muscle [83].

7.5 Perspectives

In this chapter, we introduced and discussed the pharmacological effects of compounds and drugs on primary targeted organs, such as intestine and pharynx. For the primary targeted organs, the epidermis sometimes also acts as the primary target organ of certain compounds and drugs. We further introduced and discussed the pharmacological effects of compounds and drugs on secondary targeted organs, such as nervous system, reproductive organs, and muscle. Besides these aspects, the possible pharmacological effects of compounds and drugs on excretory system, coelomocyte system, and pericellular structures are suggested to be further carefully considered.

References

1. Hua X, Wang D-Y (2023) Tire-rubber related pollutant 6-PPD quinone: a review of its transformation, environmental distribution, bioavailability, and toxicity. J Hazard Mater 459: 132265
2. Hua X, Feng X, Liang G-Y, Chao J, Wang D-Y (2023) Long-term exposure to tire-derived 6-PPD quinone causes intestinal toxicity by affecting functional state of intestinal barrier in *Caenorhabditis elegans*. Sci Total Environ 861:160591
3. Hua X, Feng X, Liang G-Y, Chao J, Wang D-Y (2023) Exposure to 6-PPD quinone at environmentally relevant concentrations causes abnormal locomotion behaviors and neurodegeneration in *Caenorhabditis elegans*. Environ Sci Technol 57:4940–4950
4. Hua X, Feng X, Liang G-Y, Chao J, Wang D-Y (2023) Long-term exposure to 6-PPD quinone reduces reproductive capacity by enhancing germline apoptosis associated with activation of both DNA damage and cell corpse engulfment in *Caenorhabditis elegans*. J Hazard Mater 454: 131495

5. Hua X, Wang D-Y (2023) Exposure to 6-PPD quinone at environmentally relevant concentrations inhibits both lifespan and healthspan in *C. elegans*. Environ Sci Technol 57: 19295–19303
6. Wang Y-X, Hua X, Wang D-Y (2023) Exposure to 6-PPD quinone enhances lipid accumulation through activating metabolic sensors of SBP-1 and MDT-15 in *Caenorhabditis elegans*. Environ Pollut 333:121937
7. Liu Z-J, Li Y-H, Wang D-Y (2024) Exposure to 6-PPD quinone disrupts glucose metabolism associated with lifespan reduction by affecting insulin and AMPK signals in *Caenorhabditis elegans*. Chemosphere 363:142975
8. Liu Z-Y, Bian Q, Wang D-Y (2024) Exposure to 6-PPD quinone causes ferroptosis activation associated with induction of reproductive toxicity in *Caenorhabditis elegans*. J Hazard Mater 471:134356
9. Wu J-W, Li L-E, Hu D-Y, Liu R, Bian Q, Wang D-Y (2025) Environmentally relevant concentrations of 6-PPDQ disrupts vitamin D3 adsorption and receptor function in *Caenorhabditis elegans*. Environ Sci Process Impacts 27:2798–2808
10. Hu Q, Liu Z, Guo Y, Lu S, Du H, Cao Y (2021) Antioxidant capacity of flavonoids from Folium Artemisiae Argyi and the molecular mechanism in *Caenorhabditis elegans*. J Ethnopharmacol 279:114398
11. Song M-X, Ruan Q-L, Wang D-Y (2024) Paeoniflorin alleviates toxicity and accumulation of 6-PPD quinone by activating ACS-22 in *Caenorhabditis elegans*. Ecotoxicol Environ Saf 286: 117226
12. Hua X, Liang G-Y, Chao J, Wang D-Y (2024) Exposure to 6-PPD quinone causes damage on mitochondrial complex I/II associated with lifespan reduction in *Caenorhabditis elegans*. J Hazard Mater 472:134598
13. Wang Y-X, Hu G-C, Wang D-Y (2025) Increased S-adenosyl methionine strengthens the suppression in mitochondrial unfolded protein response induced by 6-PPD quinone at environmentally relevant concentrations in *Caenorhabditis elegans*. Environ Pollut 386:127231
14. Hua X, Wang D-Y (2025) 6-PPD quinone causes alteration in ubiquinone-mediated complex III associated with toxicity on mitochondrial function and longevity in *Caenorhabditis elegans*. J Environ Chem Eng 13:116571
15. Zhang L, Wang Y-X, Wang D-Y (2023) Paeoniflorin increases the survival of *Pseudomonas aeruginosa* infected *Caenorhabditis elegans* at the immunosuppression stage by activating PMK-1, BAR-1, and EGL-1 signals. Arch Pharm Res 46:616–628
16. Savova MS, Todorova MN, Apostolov AG, Yahubyan GT, Georgiev MI (2022) Betulinic acid counteracts the lipid accumulation in *Caenorhabditis elegans* by modulation of *nhr-49* expression. Biomed Pharmacother 156:113862
17. Kim AT, Park Y (2025) Esculetin inhibits fat accumulation through insulin/insulin-like growth factor- and AMP-activated protein kinase-dependent pathways in *Caenorhabditis elegans*. Nutrients 17:1565
18. Stefanello ST, Gubert P, Puntel B, Mizdal CR, de Campos MM, Salman SM, Dornelles L, Avila DS, Aschner M, Soares FA (2015) Protective effects of novel organic selenium compounds against oxidative stress in the nematode *Caenorhabditis elegans*. Toxicol Rep 2:961–967
19. Zhao Y, Yu X, Jia R, Yang R, Rui Q, Wang D (2015) Lactic acid bacteria protects *Caenorhabditis elegans* from toxicity of graphene oxide by maintaining normal intestinal permeability under different genetic backgrounds. Sci Rep 5:17233
20. Farias-Pereira R, Park CS, Park Y (2020) Kahweol reduces food intake of *Caenorhabditis elegans*. J Agric Food Chem 68:9683–9689
21. Biswas SN, Ebina Y, Okada H, Shingai R (1996) A phenomenological model describing pharyngeal pulsing in the nematode *Caenorhabditis elegans* anesthetized by alcohol. Res Commun Mol Pathol Pharmacol 92:233–244
22. Mitchell PH, Bull K, Glautier S, Hopper NA, Holden-Dye L, O'Connor V (2007) The concentration-dependent effects of ethanol on *Caenorhabditis elegans* behaviour. Pharmacogenomics J 7:411–417

23. Hoeckendorf A, Stanisak M, Leippe M (2012) The saposin-like protein SPP-12 is an antimicrobial polypeptide in the pharyngeal neurons of *Caenorhabditis elegans* and participates in defence against a natural bacterial pathogen. Biochem J 445:205–212
24. Tenor JL, Aballay A (2008) A conserved toll-like receptor is required for *Caenorhabditis elegans* innate immunity. EMBO Rep 9:103–109
25. Gravato-Nobre MJ, Vaz F, Filipe S, Chalmers R, Hodgkin J (2016) The invertebrate lysozyme effector ILYS-3 is systemically activated in response to danger signals and confers antimicrobial protection in *C. Elegans*. PLoS Pathog 12:e1005826
26. Liu Y, Kaval KG, van Hoof A, Garsin DA (2019) Heme peroxidase HPX-2 protects *Caenorhabditis elegans* from pathogens. PLoS Genet 15:e1007944
27. Zhao Y-L, Wang X, Wu Q-L, Li Y-P, Wang D-Y (2015) Translocation and neurotoxicity of CdTe quantum dots in RMEs motor neurons in nematode *Caenorhabditis elegans*. J Hazard Mater 283:480–489
28. Zhao Y-L, Wang X, Wu Q-L, Li Y-P, Tang M, Wang D-Y (2015) Quantum dots exposure alters both development and function of D-type GABAergic motor neurons in nematode *Caenorhabditis elegans*. Toxicol Res 4:399–408
29. Yu X-M, Guang X-M, Wu Q-L, Zhao Y-L, Wang D-Y (2015) Vitamin E ameliorates the neurodegeneration related phenotypes caused by neurotoxicity of Al_2O_3-nanoparticles in *C. elegans*. Toxicol Res 4:1269–1281
30. Ye H-Y, Ye B-P, Wang D-Y (2008) Trace administration of vitamin E can retrieve and prevent UV-irradiation- and metal exposure-induced memory deficits in nematode *Caenorhabditis elegans*. Neurobiol Learn Mem 90:10–18
31. Schumacher F, Chakraborty S, Kleuser B, Gulbins E, Schwerdtle T, Aschner M, Bornhorst J (2015) Highly sensitive isotope-dilution liquid-chromatography-electrospray ionization-tandem-mass spectrometry approach to study the drug-mediated modulation of dopamine and serotonin levels in *Caenorhabditis elegans*. Talanta 144:71–79
32. Liu M, Xiong Y, Shan S, Zhu Y, Zeng D, Shi Y, Zhang Y, Lu W (2020) Eleutheroside E enhances the long-term memory of radiation-damaged *C. elegans* through G-protein-coupled receptor and neuropeptide signaling pathways. J Nat Prod 83:3315–3323
33. Schmidt L, Vargas BK, Sant'Anna Monteiro C, Rodrigues NR, Denardin CC, de Oliveira Mello R, Emanuelli T, ZachiaAyub MA, Moreira JCF, Augusti PR (2025) Effect of bioavailable phenolic compounds from olive pomace against oxidative damage in glial cells and in *Caenorhabditis elegans*. Food Res Int 221:117309
34. Mastrandreas P, Arnold A, Boglari C, de Quervain DJ, Stetak A, Papassotiropoulos A (2023) (−)-Gossypol inhibition of Musashi-mediated forgetting improves memory and age-dependent memory decline in *Caenorhabditis elegans*. Mol Neurobiol 60:820–835
35. Zamberlan DC, Arantes LP, Machado ML, Golombieski R, Soares FA (2014) Diphenyl-diselenide suppresses amyloid-beta peptide in *Caenorhabditis elegans* model of Alzheimer's disease. Neuroscience 278:40–50
36. Huang L, Luo Y, Pu Z, Kong X, Fu X, Xing H, Wei S, Chen W, Tang H (2017) Oxoisoaporphine alkaloid derivative 8-1 reduces Abeta(1-42) secretion and toxicity in human cell and *Caenorhabditis elegans* models of Alzheimer's disease. Neurochem Int 108:157–168
37. Wang H, Yue Y, Zhao H, Wu H, Jiang K, Li S, Zhao M, Lin F (2021) Neuroprotective effects of 2-substituted 1,3-selenazole amide derivatives on Amyloid-beta-induced toxicity in a transgenic *Caenorhabditis elegans* model of Alzheimer's disease. Neurotox Res 39:841–850
38. Ng LF, Gruber J, Cheah IK, Goo CK, Cheong WF, Shui G, Sit KP, Wenk MR, Halliwell B (2014) The mitochondria-targeted antioxidant MitoQ extends lifespan and improves healthspan of a transgenic *Caenorhabditis elegans* model of Alzheimer disease. Free Radic Biol Med 71: 390–401
39. Zhang R, Huang X, Zhou C, Zhang Q, Jia D, Xie X, Zhang J (2024) Network pharmacology-based mechanism analysis of dauricine on the alleviating Abeta-induced neurotoxicity in *Caenorhabditis elegans*. BMC Complement Med Ther 24:321

40. He Q, Huang G, Chen Y, Wang X, Huang Z, Chen Z (2017) The protection of novel 2-arylethenylquinoline derivatives against impairment of associative learning memory induced by neural Abeta in *C. elegans* Alzheimer's disease model. Neurochem Res 42:3061–3072
41. Shi YC, Pan TM, Liao VH (2016) Monascin from monascus-fermented products reduces oxidative stress and amyloid-beta toxicity via DAF-16/FOXO in *Caenorhabditis elegans*. J Agric Food Chem 64:7114–7120
42. McCormick AV, Wheeler JM, Guthrie CR, Liachko NF, Kraemer BC (2013) Dopamine D2 receptor antagonism suppresses tau aggregation and neurotoxicity. Biol Psychiatry 73:464–471
43. Miyasaka T, Xie C, Yoshimura S, Shinzaki Y, Yoshina S, Kage-Nakadai E, Mitani S, Ihara Y (2016) Curcumin improves tau-induced neuronal dysfunction of nematodes. Neurobiol Aging 39:69–81
44. Yu YT, Lin HR, Chen XY, Sun ML, Wei CM, Xue MF, Gao YH, Tang WB, Zheng CY, Li S, Wang HB (2024) *Dendrobium officinale* phenolic extract maintains proteostasis by regulating autophagy in a *Caenorhabditis elegans* model of Alzheimer's disease. Fitoterapia 175:105924
45. Sukanya S, Bellver-Sanchis A, Singh Choudhary B, Kumar S, Pérez B, Rodríguez ALM, Brea J, Griñán-Ferré C, Malik R (2024) Design, synthesis, and biological evaluation of tetrahydropyrimidine analogue as GSK-3beta/Abeta aggregation inhibitor and anti-Alzheimer's agent. Bioorg Chem 153:107811
46. Sukanya S, Bellver-Sanchis A, Choudhary BS, Kumar S, Pérez B, Rodríguez ALM, Brea J, Escolano C, Griñán-Ferré C, Malik R (2025) Exploring 2-thioxo-1,2,3,4-tetrahydropyrimidines as dual acting GSK-3beta/Abeta aggregation inhibitors: implications for Alzheimer's disease treatment. Drug Dev Res 86:e70112
47. Fu HJ, Zhou XY, Li YP, Chen X, He YN, Qin DL, Yu L, Yu CL, Wu JM, Wu AG, Zhou XG (2023) The protective effects of *Reineckia carnea* ether fraction against Alzheimer's disease pathology: an exploration in *Caenorhabditis elegans* models. Int J Mol Sci 24:16536
48. Muhammad F, Liu Y, Wang N, Zhao L, Zhou Y, Yang H, Li H (2022) Anti-alpha-synuclein toxicity and anti-neurodegenerative role of chrysin in transgenic *Caenorhabditis elegans* models of Parkinson's disease. ACS Chem Neurosci 13:442–453
49. Muhammad F, Liu Y, Wang N, Zhao L, Zhou Y, Yang H, Li H (2023) Rose essential oil diminishes dopaminergic neuron degenerations and reduces alpha-synuclein aggregation in *Caenorhabditis elegans* models of Parkinson's disease. Phytother Res 37:2877–2893
50. Sanguanphun T, Promtang S, Sornkaew N, Niamnont N, Sobhon P, Meemon K (2023) Anti-Parkinson effects of *Holothuria leucospilota*-derived palmitic acid in *Caenorhabditis elegans* model of Parkinson's disease. Mar Drugs 21:141
51. Goya ME, Xue F, Sampedro-Torres-Quevedo C, Arnaouteli S, Riquelme-Dominguez L, Romanowski A, Brydon J, Ball KL, Stanley-Wall NR, Doitsidou M (2020) Probiotic *Bacillus subtilis* protects against alpha-Synuclein aggregation in *C. elegans*. Cell Rep 30:367–380
52. Ma C, Feng Y, Li X, Sun L, He Z, Gan J, He M, Zhang X, Chen X (2023) Potential therapeutic effects of policosanol from insect wax on *Caenorhabditis elegans* models of Parkinson's disease. J Neuroimmune Pharmacol 18:127–144
53. Tsai CW, Tsai RT, Liu SP, Chen CS, Tsai MC, Chien SH, Hung HS, Lin SZ, Shyu WC, Fu RH (2017) Neuroprotective effects of betulin in pharmacological and transgenic *Caenorhabditis elegans* models of Parkinson's disease. Cell Transplant 26:1903–1918
54. He CL, Tang Y, Wu JM, Long T, Yu L, Teng JF, Qiu WQ, Pan R, Yu CL, Qin DL, Wu AG, Zhou XG (2023) Chlorogenic acid delays the progression of Parkinson's disease via autophagy induction in *Caenorhabditis elegans*. Nutr Neurosci 26:11–24
55. Gao W, Wang Y, Wang F, Wu X, Lu F, Liu F (2025) Ergothioneine exerts neuroprotective effects in Parkinson's disease: targeting alpha-synuclein aggregation and oxidative stress. Food Res Int 201:115590
56. Wu Q-L, Li Y-X, Li Y-P, Zhao Y-L, Ge L, Wang H-F, Wang D-Y (2013) Crucial role of biological barrier at the primary targeted organs in controlling translocation and toxicity of multi-walled carbon nanotubes in nematode *Caenorhabditis elegans*. Nanoscale 5:11166–11178

57. Nouara A, Wu Q-L, Li Y-X, Tang M, Wang H-F, Zhao Y-L, Wang D-Y (2013) Carboxylic acid functionalization prevents the translocation of multi-walled carbon nanotubes at predicted environmental relevant concentrations into targeted organs of nematode *Caenorhabditis elegans*. Nanoscale 5:6088–6096
58. Wu Q-L, Zhao Y-L, Li Y-P, Wang D-Y (2014) Molecular signals regulating translocation and toxicity of graphene oxide in nematode *Caenorhabditis elegans*. Nanoscale 6:11204–11212
59. Liu Z-F, Zhou X-F, Wu Q-L, Zhao Y-L, Wang D-Y (2015) Crucial role of intestinal barrier in the formation of transgenerational toxicity in quantum dots exposed nematodes *Caenorhabditis elegans*. RSC Adv 5:94257–94266
60. Zhi L-T, Fu W, Wang X, Wang D-Y (2016) ACS-22, a protein homologous to mammalian fatty acid transport protein 4, is essential for the control of toxicity and translocation of multi-walled carbon nanotubes in *Caenorhabditis elegans*. RSC Adv 6:4151–4159
61. Zhao L, Qu M, Wong G, Wang D-Y (2017) Transgenerational toxicity of nanopolystyrene particles in the range of μg/L in nematode *Caenorhabditis elegans*. Environ Sci Nano 4:2356–2366
62. Qu M, Wang D-Y (2020) Toxicity comparison between pristine and sulfonate modified nanopolystyrene particles in affecting locomotion behavior, sensory perception, and neuronal development in *Caenorhabditis elegans*. Sci Total Environ 703:134817
63. Qiu Y-X, Luo L-B, Yang Y-H, Kong Y, Li Y-H, Wang D-Y (2020) Potential toxicity of nanopolystyrene on lifespan and aging process of nematode *Caenorhabditis elegans*. Sci Total Environ 705:135918
64. Qiu Y-X, Liu Y-Q, Li Y-H, Li G-J, Wang D-Y (2020) Effect of chronic exposure to nanopolystyrene on nematode *Caenorhabditis elegans*. Chemosphere 256:127172
65. Qu M, Kong Y, Yuan Y-J, Wang D-Y (2019) Neuronal damage induced by nanopolystyrene particles in nematode *Caenorhabditis elegans*. Environ Sci Nano 6:2591–2601
66. Qu M, Qiu Y-X, Kong Y, Wang D-Y (2019) Amino modification enhances reproductive toxicity of nanopolystyrene on gonad development and reproductive capacity in nematode *Caenorhabditis elegans*. Environ Pollut 254:112978
67. Qu M, Xu K-N, Li Y-H, Wong G, Wang D-Y (2018) Using *acs-22* mutant *Caenorhabditis elegans* to detect the toxicity of nanopolystyrene particles. Sci Total Environ 643:119–126
68. Yang Y-H, Shao H-M, Wu Q-L, Wang D-Y (2020) Lipid metabolic response to polystyrene particles in nematode *Caenorhabditis elegans*. Environ Pollut 256:113439
69. Liu H-L, Tian L-J, Wang S-T, Wang D-Y (2021) Size-dependent transgenerational toxicity induced by nanoplastics in nematode *Caenorhabditis elegans*. Sci Total Environ 790:148217
70. Sun L-M, Liao K, Wang D-Y (2021) Comparison of transgenerational reproductive toxicity induced by pristine and amino modified nanoplastics in *Caenorhabditis elegans*. Sci Total Environ 768:144362
71. Li D, Ji J, Yuan Y-J, Wang D-Y (2020) Toxicity comparison of nanopolystyrene with three metal oxide nanoparticles in nematode *Caenorhabditis elegans*. Chemosphere 245:125625
72. Li D, Deng Y-J, Wang S-T, Du H-H, Xiao G-S, Wang D-Y (2020) Assessment of nanopolystyrene toxicity under fungal infection condition in *Caenorhabditis elegans*. Ecotoxicol Environ Saf 197:110625
73. Zhao Y-Y, Li D, Rui Q, Wang D-Y (2020) Toxicity induction of nanopolystyrene under microgravity stress condition in *Caenorhabditis elegans*. Sci Total Environ 703:135623
74. Hua X, Feng X, Hua Y-S, Wang D-Y (2023) Paeoniflorin attenuates polystyrene nanoparticle-induced reduction in reproductive capacity and increase in germline apoptosis through suppressing DNA damage checkpoints in *Caenorhabditis elegans*. Sci Total Environ 871: 162189
75. Henarejos-Escudero P, Méndez-García FF, Hernández-García S, Martínez-Rodríguez P, Gandía-Herrero F (2023) Design, synthesis and gene modulation insights into pigments derived from tryptophan-betaxanthin, which act against tumor development in *Caenorhabditis elegans*. Int J Mol Sci 25:63

76. Li X, Xu F, Wang R, Shen L, Luo B, Zhou S, Zhang J, Zhang Z, Cao Z, Zhan K, Zhao Y, Zhao G (2024) Aspirin enhances radio/chemo-therapy sensitivity in *C. elegans* by inducing germ cell apoptosis and suppresses RAS overactivated tumorigenesis via mtROS-mediated DNA damage and MAPK pathway. Biochem Biophys Res Commun 735:150828
77. Tang J, Chen Y, Zhou M, Zhao L, Li Y, Li M, Yang F, He Z, Lv L (2025) Arachidonic acid supplementation during reproductive aging prevents female fertility decline. Int J Mol Sci 26: 11332
78. Xiao X, Zheng M, Zhao Y, Huang K, Shi Y, Ren Y, Chen L, Sun J, Ye R, Zheng Z, Ruan Q (2026) Quercetin ameliorates diminished ovarian reserve via modulating intestinal BNIP3/dct-1-mediated mitophagy and subsequently activating the ovarian TGF-β signaling pathway. J Ethnopharmacol 355:120736
79. Templeman NM, Luo S, Kaletsky R, Shi C, Ashraf J, Keyes W, Murphy CT (2018) Insulin signaling regulates oocyte quality maintenance with age via cathepsin B activity. Curr Biol 28: 753–760
80. Dempsey CM, Mackenzie SM, Gargus A, Blanco G, Sze JY (2005) Serotonin (5HT), fluoxetine, imipramine and dopamine target distinct 5HT receptor signaling to modulate *Caenorhabditis elegans* egg-laying behavior. Genetics 169:1425–1436
81. Gonzalez-Lerma PX, Lloyd C, Park SJ, Dawson-Scully K (2026) Anticonvulsant effects of novel and repurposed drugs on docetaxel-induced neuropathy in *C. elegans*. PLoS One 21: e0342236
82. Sleigh JN, Buckingham SD, Esmaeili B, Viswanathan M, Cuppen E, Westlund BM, Sattelle DB (2011) A novel *Caenorhabditis elegans* allele, *smn-1(cb131)*, mimicking a mild form of spinal muscular atrophy, provides a convenient drug screening platform highlighting new and pre-approved compounds. Hum Mol Genet 20:245–260
83. McColl G, Roberts BR, Pukala TL, Kenche VB, Roberts CM, Link CD, Ryan TM, Masters CL, Barnham KJ, Bush AI, Cherny RA (2012) Utility of an improved model of amyloid-beta (Abeta1-42) toxicity in *Caenorhabditis elegans* for drug screening for Alzheimer's disease. Mol Neurodegener 7:57

Target Organelle Pharmacology 8

Abstract

Once the compounds and drugs are bioavailable to organisms and cells, they will further target different organelles and exert the corresponding pharmacological effects. The pharmacological effects of compounds and drugs in nucleus, mitochondria, lysosome, endoplasmic reticulum, peroxisome, proteasome, and Golgi were introduced and discussed, which is helpful for understanding target organelle pharmacology in *Caenorhabditis elegans*.

Keywords

Organelle · Pharmacology · *Caenorhabditis elegans*

8.1 Introduction

In organisms including the *Caenorhabditis elegans*, organelles refer to the highly specialized and membrane-bound subcellular compartments. These subcellular compartments perform distinct biological functions in modulating the cellular homeostasis. The core characteristics of organelles include: (1) compartmentalized function, (2) dynamic structure, (3) developmental regulation, and (4) environmental sensing and signaling. Especially, considering that the organelles can act as primary sensors of environmental stimuli by activating conserved pathways to restore homeostasis or to trigger adaptive responses, the pharmacological effects of compounds and drugs on organelles need to be carefully examined.

For the water-soluble and the lipid-soluble compounds and drugs, it was assumed that they can be bioavailable to different organelles. Nevertheless, the bioavailability and the accumulation of water-soluble and fat-soluble compounds and drugs to or within the organelles are affected by transporters on membranes of organelles and

D. Wang, *Pharmacology in Caenorhabditis elegans*,
https://doi.org/10.1007/978-981-92-5166-7_8

microenvironments in organelles. In addition, some particulate chemicals at the nano size could be accumulated at least in mitochondria and some vesicle structures [1, 2].

Each membrane-bound organelle provides the unique microenvironment and the compartmentalization that optimizes specific reactions and certain biological functions [3, 4]. In this chapter, we successively introduced and discussed the pharmacological effects of compounds and drugs in nucleus, mitochondria, lysosome, endoplasmic reticulum, peroxisome, proteasome, and Golgi.

8.2 Pharmacological Effects in Nucleus

8.2.1 Effects on Development of Nucleus

The nucleus plays the essential functions of storing, replicating, and expressing the genetic information. Structurally, the nucleus contains nuclear pore complexes (NPCs), nuclear lamina, chromatin, and nucleolus. It is assumed that the damage on development of nucleus by environmental pollutants, environmental stresses, aging, or certain pathological processes can be relieved or revered by pharmacological intervention of compounds and drugs. That is, the pharmacological intervention of compounds and drugs may have protective effect to maintain the nuclear structure and integrity. Besides this, the potential beneficial effects of pharmacological intervention of compounds and drugs on nuclear lamins, the structural scaffolding of the nucleus, after exposure to pollutants or stresses or during aging are also suggested to be examined.

8.2.2 Effects on Functions of Transcriptional Factors in Nucleus

The core function of nucleus is the regulation of gene expression via spatial organization of chromatin. During this process, transcription factors control transcriptional rate of genetic information from DNA to messenger RNA and act as "master switches" by binding to and regulating expression of downstream targeted genes after translocation into the nucleus. One of the important mechanisms for compounds and drugs to exert their beneficial functions is to target and regulate the expression and functions of transcriptional factors. Phloretic acid had anti-aging effects by activating transcriptional factors of DAF-16, SKN-1, and HSF-1, which further affected expressions of their downstream target genes [5]. Betulinic acid reduced lipid accumulation by modulating expression and function of NHR-49 [6].

8.2.3 Effects on Functions of Nuclear Hormone Receptors in Nucleus

Nuclear hormone receptors are nuclear proteins that directly regulate gene expression after the response to and the binding by ligands. For fat-soluble compounds and drugs, it is assumed that they exert their functions by binding to the nuclear hormone receptors. Remofuscin extended lifespan by activating a lysosome-to-nucleus signaling pathway, which involved the increased expression of nuclear hormone receptor NHR-234 with the function to upregulate a battery of xenobiotic detoxification genes [7]. Isotschimgine also extended lifespan and improved health span by activating nuclear hormone receptors DAF-12 and NHR-8 [8].

8.2.4 Protection Effects Against Nucleolar Stress

Nucleolar stress is reflected by formation of nucleolar stress body (NoSB) in the nucleoplasm in response to stresses, which is the most prominent and defining feature and visible as a vacuole-like structure [9, 10]. The nucleolar stress is triggered by disturbances in ribosome biogenesis, primarily the inhibition of rRNA transcription or the inactivation of rRNA processing and maturation [11, 12]. In *C. elegans*, the NoSB formation can be controlled by NOSR-1/NUMR-1 signaling axis [13]. The transcription factors NOSR-1 (nucleolar stress response-1) and PHA-4 act as sensors for nucleolar stress as indicated by the increase in expressions upon the stresses [13, 14]. SR-like protein NUMR-1 (nuclear localized metal responsive) is target of NOSR-1 and the core component of the NoSB [13]. The possible beneficial effect of pharmacological compounds and drugs against the nucleolar stress induced by pollutants or stresses or relevant with certain pathological models is suggested to be examined.

8.3 Pharmacological Effects in Mitochondria

8.3.1 Effects on Development of Mitochondria

The mitochondria are double-membraned organelles, which are uniquely designed to maximize energy production for organisms. Structurally, mitochondria contain outer mitochondrial membrane, inner membrane with numerous folds (cristae), and two distinct compartments (the intermembrane space and the matrix). The mitochondrial development can be reflected by transmission electron microscopy analysis and certain transgenic markers fused with fluorescence proteins. The damage on mitochondrial development by pollutants, stresses, aging, or certain pathological processes might be relieved by some compounds and drugs. That is, certain pharmacological interventions can have the preservation function for mitochondrial structure and dynamics. Firstly, some compounds and drugs may have the function to maintain normal shape and organization of mitochondrial network so as to prevent the mitochondrial structural damage induced by pollutants and stresses. Secondly, some compounds and drugs may be helpful to prevent the mitochondrial fragmentation. For example, the mitochonic acid 5 (MA-5) could alleviate some symptoms including the mitochondrial fragmentation in Duchenne muscular dystrophy (DMD) model [15]. The mitochondrial membrane potential reflects the impermeability of inner mitochondrial membrane to allow the cells to pump protons into intermembrane space and to create an electrochemical gradient. Thirdly, some compounds and drugs are useful for stabilizing the mitochondrial membrane potential. Treatment with AP39, a mitochondria-targeted hydrogen sulfide donor, could preserve the mitochondrial membrane potential and increase the movement and the survival [16].

8.3.2 Effects on Functions of Mitochondria

In mitochondria, the outer mitochondrial membrane allows ions and small molecules to pass freely between cytosol and intermembrane, and serves as docking sites for proteins. The intermembrane space acts as the reservoir for protons pumped out of the matrix, and stores the cytochrome c. The main functions of inner mitochondrial membrane are organization of electron transport chain (ETC) complexes (I–IV) by providing site of oxidative phosphorylation (OXPHOS) and generation of ATP at complex V. The contents in matrix contain mitochondrial DNA (mtDNA), mitochondrial ribosomes, and some enzymes responsible for the citric acid cycle. Pharmacological treatment with compounds and drugs may affect the functions of outer mitochondrial membrane, intermembrane space, inner mitochondrial membrane, and/or matrix. Mitochonic acid 5 (MA-5) could elevate ATP level, and reverse the decline in ATP synthesis induced by complex I inhibitor rotenone [15]. The MitoQ, a mitochondria-targeted antioxidant, could protect the complexes I and IV on the electron transport chain [16]. Treatment with RG70, a *Rehmannia glutinosa* polysaccharide, had the anti-aging effect by at least partially influencing the citric acid cycle [17]. In addition, supplementation with oxaloacetate, malate, and fumarate, the citric acid cycle metabolites, could increase the lifespan [18, 19].

8.3.3 Effects on Mitochondrial Unfolded Protein Response and Mitophagy

The mitochondrial unfolded protein response (mt UPR) refers to an adaptive transcriptional program that promotes mitochondrial recovery during stress. In *C. elegans*, the mt UPR is orchestrated by some important transcription factors, such as ATFS-1, UBL-5, and DVE-1 [20]. Under stress conditions, the import efficiency of ATFS-1 into mitochondria is reduced, and ATFS-1 binds to DNA in the nucleus and induces the expression of a specific set of genes aimed at restoring mitochondrial health [21, 22]. These include genes encoding mitochondrial chaperones, such as *hsp-6* (the mitochondrial homolog of HSP70) and *hsp-60*, which are considered as the mt UPR markers [23, 24]. Perillaldehyde alleviated the neurodegeneration induced by polyQ by partially activating mt UPR [25]. Chlorogenic acid enhanced innate immunity, which was linked to activation of mt UPR [26].

The mitophagy refers to the highly selective cellular process in which the cells selectively target damaged, dysfunctional, or surplus mitochondria for degradation by wrapping them in autophagosome and delivering them to lysosome for digestion and recycling. In *C. elegans*, the mitophagy is regulated by two conserved pathways, PINK1/Parkin-dependent mitophagy and receptor mediated mitophagy [27–29]. Urolithin A and catechinic acid extended the lifespan, which was associated with mitophagy activation [30, 31]. The neurodegeneration could be mitigated by 18-norspirostane steroidal saponins via activation of mitophagy [32].

8.3.4 Effects on Mitochondrial Reactive Oxygen Species

Mitochondrial reactive oxygen species (mt ROS) are ROS that are produced as byproduct of energy generation within the mitochondria [33]. In *C. elegans*, generation of mt ROS is genetically at least regulated by insulin/IGF-1 signaling pathway [34, 35]. Lonidamine extended lifespan, which was linked to formation of adaptation response induced by increased mt ROS [36].

8.3.5 Effects on Ion Homeostasis in Mitochondria

In mitochondria, the calcium signaling is intricately linked to mitochondrial function and neurodegeneration [37, 38]. Inhibiting the Na^+/Ca^{2+} exchanger by CGP37157 could cause the reversal of rotenone-induced Parkinson's disease-related phenotypes [39], suggesting that the modulation of calcium signaling is the key pharmacological mechanism for the beneficial effect of CGP37157 by regulating the ion homeostasis.

8.4 Pharmacological Effects in Lysosome

8.4.1 Effects on Activity of Lysosome

Lysosome is a dynamic and multifunctional organelle. The lysosome is a single-membrane enclosed compartment with the function to maintain an acidic internal environment via V-type ATPase proton pumps [40, 41]. In the lumen of lysosome, there are different types of digestive enzymes, including proteases (like cathepsins), lipases, nucleases, and glycosidases [42, 43]. The activity of lysosome is largely reflected by the acidic internal environment and activity of digestive enzymes, such as proteases. The acidity and activity of digestive enzymes are essential for digestive function of lysosome and act as the key indicators of cellular health. In *C. elegans*, lysosome acidity can be measured by cDCFDA staining or assessed by NUC-1::pHTomato to visualize pH changes in lysosomes. The damage on lysosome activity by stresses or induced by pathological processes might be inhibited by certain pharmacological compounds and drugs.

8.4.2 Effects on Development of Lysosome

The lysosome has two primary dynamic morphological states: vesicular (puncta) and tubular networks. The vesicular lysosomes are small and spherical structures, which are the predominant form during most of the worm's life. Under specific conditions, lysosomes can be elongated and fused to form an extensive network of tubules, the tubular lysosomes. The lysosome structure can be observed at least by markers, such as LMP-1::GFP, or by staining with fluorescent probes, such as LysoTracker and DC-Lyso. Similarly, some pharmacological compounds and drugs may be helpful to reduce the damage on lysosome structure caused by stresses or induced by pathological processes.

8.4.3 Effects on Functions of Lysosome

The core function of lysosome is autophagic clearance and cellular homeostasis. That is because the lysosomes contain hydrolytic enzymes; they function to break down macromolecules, damaged organelles, and materials absorbed from the environment, which is essential for basic cellular turnover and nutrient supply [44–46]. In *C. elegans*, lysosomal fusion-competence is conferred by endosomal regulator SNX-3, which is required for the delivery of the small GTPase RAB-7 and the R-SNARE VAMP-8 to the lysosomal membrane [47]. LPLA-2, lysosomal phospholipase A2, is responsible for catalyzing the cleavage of phospholipids in the inner autophagosomal membrane [48]. The HLH-30 acts as the master transcriptional factor to promote lysosome-autophagosome biogenesis [49]. Saikosaponin B2 and hypericin were identified as HLH-30/TFEB activators to enhance protein homeostasis by improving autophagic flux [50]. In addition, Aβ and Tau aggregation could be ameliorated by Ginsenoside composition via modulating autophagy-lysosome pathway [51].

Another important function of lysosome is to act as a hub for metabolic processing, such as the lipid metabolism. During this process, mTORC1 signaling (specifically DAF-15 (RAPTOR homolog) and RSKS-1 (S6 kinase homolog)) is required for the effects of lysosomal inhibition in affecting lipid accumulation [52]. In addition, some transcriptional factors, such as DAF-16, HLH-30, and NHR-49, regulate this aspect of lysosome function [53, 54]. It is assumed that pharmacological interventions can influence lipid accumulation by modulating lysosomal activity and function. Treatment with chloroquine (which raises lysosomal pH and inhibits acid-dependent hydrolases) and leupeptin (a cysteine protease inhibitor) could reduce both basal fat storage and lipid accumulation induced by high glucose or palmitic acid [52].

8.4.4 Effects on Lysosomal Surveillance Response

Lysosomal surveillance response refers to the adaptive transcriptional response typified by a set of transcriptional activations involved in lysosomal function and proteolysis [55]. Among them, when lysosomal function is compromised, ELT-2 activated the lysosomal surveillance response to boost the lysosomal activity and to clear the protein aggregates [55]. This suggests that certain compounds and drugs can enhance lysosomal function by activating the lysosomal surveillance response, which can be employed as strategy for treating lysosome-related diseases.

8.5 Pharmacological Effects in Endoplasmic Reticulum

8.5.1 Effects on Development of Endoplasmic Reticulum

Endoplasmic reticulum is a large, dynamic, and interconnected network of membrane-enclosed tubules and flattened sacs, which is distributed throughout the cytoplasm. There are two types of endoplasmic reticulum: rough endoplasmic reticulum and smooth endoplasmic reticulum. Structurally, endoplasmic reticulum contains single, continuous lipid bilayer membrane and lumen (cisternal space), and the membrane of endoplasmic reticulum is physically continuous with the nuclear envelope. In *C. elegans*, mutation of *rab-5* encoding Rab GTPase phenocopied deficits in endoplasmic reticulum structure, which was linked to depletion of Reticulon proteins (RET-1) and YOP-1 (homologs of Rtn4a and DP1/NogoA) [56]. Certain aspects of damage on endoplasmic reticulum structure or its organization caused by stresses or induced by pathological processes may be reduced by some pharmacological compounds and drugs.

8.5.2 Effects on Functions of Endoplasmic Reticulum

The core function of endoplasmic reticulum is responsible for the synthesis, folding, modification, and transport of proteins. The main function of rough endoplasmic reticulum is the control of protein synthesis and processing. During these processes, Proteins are synthesized by the ribosomes on the endoplasmic reticulum, and further threaded into the endoplasmic reticulum lumen. Inside the lumen, the proteins are folded into their correct three-dimensional shape with the help of chaperones, such as HSP-4 [57], and the newly synthesized proteins will be checked and improperly folded proteins will not be allowed to proceed [58, 59]. The further chemical modifications can be made, such as the addition of sugar molecules (glycosylation) [60]. Some screened compounds with the function to reverse defects of endoplasmic reticulum-to-Golgi trafficking could rescue α-syn-mediated dopaminergic neuron loss in Parkinson's disease models [61]. Endoplasmic reticulum-associated degradation (ERAD) refers to the process by which misfolded proteins are retro-translocated from the endoplasmic reticulum to the cytoplasm for degradation by proteasome

[62]. The accumulation of polyubiquitinated proteins was suppressed by melatonin, suggesting its role as a potential preventive molecule against defective ERAD caused by repeated anesthesia exposure [62].

The second function of endoplasmic reticulum (the smooth endoplasmic reticulum) is the synthesis and the transport of lipids. The smooth endoplasmic reticulum synthesizes phospholipids and cholesterol, which are essential for all cellular membranes. The nuclear envelope localized CNEP-1 played the role of spatial regulation of de novo phospholipid synthesis within the endoplasmic reticulum [63].

The third function of endoplasmic reticulum (the smooth endoplasmic reticulum) is to act as the calcium storage. On endoplasmic reticulum membrane, ITR-1 (IP_3 Receptor) and UNC-68 (ryanodine receptor) control the calcium release [64]. SCA-1 is the calcium uptake pump required for the control of muscle function [65]. CRT-1/ Calreticulin, the calcium-binding chaperone, is located in endoplasmic reticulum lumen [66]. The STIM-1 acts as the calcium sensor on endoplasmic reticulum [67]. It is assumed that the pharmacological modulation of endoplasmic reticulum calcium levels can affect certain biological processes. It was shown that treatment with nimodipine and ethylene glycol tetraacetic acid could reduce the $A\beta_{1-42}$ aggregations and confer resistance to oxidative stress by suppressing the selective voltage-gated calcium channels [68].

8.5.3 Effects on Endoplasmic Reticulum Unfolded Protein Response and ER-Phagy

Endoplasmic reticulum unfolded protein response (ER UPR) is a response activated by endoplasmic reticulum stress, such as accumulation of unfolded proteins. In *C. elegans*, the ER UPR is controlled by three molecular signals:

1. IRE-1/XBP-1 signaling: The endoplasmic reticulum stress triggers the splicing of *xbp-1* mRNA by IRE-1, leading to its functional transcription factor that regulates genes involved in protein folding [69–71].
2. PEK-1 signaling: This signaling reduces the influx of new proteins into the overloaded endoplasmic reticulum by attenuating general protein translation and is required for maintenance of endoplasmic reticulum homeostasis [72, 73].
3. ATF-6 signaling: This signaling is transported to the Golgi apparatus upon endoplasmic reticulum homeostasis, where it is cleaved and further acts as transcription factor to activate specific UPR target genes [74].

The ER UPR can be activated by certain pharmacological compounds and drugs. Treatment with cyclic nucleotide-gated (CNG) channel blocker ZD7288 could activate intestinal ER UPR and extend the lifespan [75]. Asperuloside could enhance innate immunity and xyloketal compound could ameliorate MPP^+-induced neuronal injury by activating IRE-1/XBP-1 mediated ER UPR [76, 77].

ER-phagy refers to the selective degradation of endoplasmic reticulum fragments by autophagy (reticulophagy), which is helpful for the maintenance of endoplasmic reticulum homeostasis. Reticulon long isoform RET-1d acted as the ER-phagy receptor under heat stress conditions by interacting with autophagy protein LGG-1/GABARAP to promote autophagosome biogenesis [78]. The endoplasmic reticulum structure-function could be remodeled by ER-phagy [79].

8.6 Pharmacological Effects in Peroxisome

8.6.1 Effects on Development of Peroxisome

Peroxisome is a multifaceted and dynamic organelle. Structurally, it contains single lipid bilayer, lumen (matrix), and crystalline core (nucleoid). The peroxisome membrane is permeable to small solutes and molecules, and imports larger proteins with the aid of certain membrane proteins (called peroxins, or PEX proteins) [80, 81]. The matrix contains enzymes for metabolic pathways (e.g., fatty acid β-oxidation) [82, 83]. The morphology or structure can be influenced by genetic perturbations or stresses [84, 85]. When aldehyde dehydrogenase gene *alh-4* is lost, toxic fatty aldehydes accumulated [86]. This deficiency in intestinal cells triggered peroxisome proliferation to compensate for aldehyde dehydrogenase deficiency, which was promoted by nuclear hormone receptors NHR-49 and NHR-79 [86].

8.6.2 Effects on Functions of Peroxisome

The peroxisome is a hub for catabolic and anabolic metabolisms. The core function of peroxisome is the control of lipid accumulation by modulating fatty acid β-oxidation to oxidize very-long-chain fatty acids (VLCFAs) and branched-chain fatty acids [87, 88]. This process is catalyzed by related fatty acid β-oxidation enzymes, such as ACOX-3, MAOC-1, DHS-28 and DAF-22, and mutation of these enzyme genes caused rapid fusion of adjacent lipid droplets to form giant lipid droplets [89]. Some transcriptional factors, such as SBP-1 and NHR-49, govern the expression of the enzyme genes regulating peroxisome fatty acid β-oxidation [90–92]. The lipid accumulation was reduced by Borago officinalis seed oil via activation of peroxisome fatty acid β-oxidation [93]. In addition, dihomo-gamma-linolenic acid also suppressed lipid accumulation by activating peroxisome fatty acid β-oxidation in an omega-3-independent manner [94].

The second important function of peroxisome is redox balance & detoxification. The peroxisome catalases (CTL-1 and CTL-2) decompose the hydrogen peroxide (H_2O_2), a toxic byproduct of metabolism, to protect the cells from oxidative damage [95]. Corresponding to this, treatment with catalase mimetics showed protective function against the oxidative damage and to extend the lifespan [96, 97]. Additionally, curcumin acetylsalicylate could extend the lifespan partially by activating peroxisome catalase genes, which was under the control of transcriptional factor DAF-16 [98].

8.6.3 Effects on Pexophagy

The pexophagy refers to selective autophagic process by which peroxisomes are degraded at lysosomes [99]. In *C. elegans*, this can be assessed with the aid of mCherry-GFP-SKL [99]. That is, the tubular lysosomes are degradative site for peroxisomes. The *prx-11* is a pro-pexophagy gene to promote the age-dependent pexophagy [99]. The pexophagy may be activated by certain pharmacological compounds and drugs.

8.7 Pharmacological Effects in Proteasome

8.7.1 Effects on Development of Proteasome

Proteasome is a massive and multi-subunit complex, which is essential for maintaining the protein homeostasis (proteostasis). Structurally, it contains two main sub-complexes, 20S core particle and 19S regulatory particle [100, 101]. The 20S core particle is the central catalytic chamber to make the proteins broken down. The 20S core particle contains α subunits (1–7) serving as the docking site for the regulatory particles and β subunits (1–7) having the proteolytic activity. The 19S regulatory particle is one or two cap-like structures that bind to the ends of the 20S core to regulate its activity. The 19S regulatory particle has two sub-complexes, the Base and the Lid. The Base contains six AAA-ATPases that form a heterohexameric ring, and three non-ATPase subunits that serve as the receptor for ubiquitin chains. The Lid contains at least nine non-ATPase subunits to exert the functions of ubiquitin recognition and deubiquitination. In *C. elegans*, some fluorescent tools have been developed for assessment of change in proteasome structure. The change in proteasome structure can be directly visualized by creating strains with endogenously tagged subunits or measured by functional output of the complex using activity reporters, such as UbG76V-Dendra2. These tools are helpful for the evaluation of pharmacological effects of certain compounds and drugs on proteasome structure under stress and pathological conditions.

8.7.2 Effects on Functions of Proteasome

The core function of proteasome is protein degradation or maintenance of proteostasis. In proteasome, the protein degradation is carried out through two main pathways: the ubiquitin-dependent 26S pathway and the ubiquitin-independent 20S pathway. In the ubiquitin-dependent 26S pathway, the vast majority of short-lived and regulatory proteins are degraded. During this process, the proteins are marked for degradation by a chain of ubiquitin molecules, and this process is ATP-dependent [102]. In the ubiquitin-independent 20S pathway, the degradation of proteins is independent of ubiquitin and ATP, and the preferred substrates are intrinsically disordered proteins and oxidatively damaged proteins, which can directly diffuse into the 20S core [103]. The proteostasis maintenance is essential for normal cellular and organismal health by controlling protein quality. In *C. elegans*, the photoconvertible UbG76V-Dendra2 UPS reporter can be used to measure the protein degradation to reflect the state of ubiquitin-proteasome system [104, 105].

In 26S proteasome (ubiquitin-dependent) of *C. elegans*, the polyubiquitin tag is recognized by receptors, such as RPN-10 [106]. EEL-1, an E3 ubiquitin ligase, contributed to the degradation of unloaded PRG-1, suggesting its role in modulating the protein quality [107]. In addition, when moderate proteasome dysfunction was formed, cells activated backup systems to improve protein homeostasis and organismal health by modulating SKN-1/Nrf2 and ELT-2/GATA transcription factors [108].

In *C. elegans*, the pharmacological effects in the proteasome center on the modulation of ubiquitin-proteasome system (UPS). That is, pharmacological interventions can enhance or inhibit proteasome activity to cause profound consequences for proteostasis. The floxuridine (FUdR), a chemotherapeutic agent, could significantly enhance the UPS activity, and then contribute to the enhancement in survival under the low-temperature stress condition [109]. The 18alpha-glycyrrhetinic acid could act as proteasome activator to reduce aging and to inhibit Alzheimer's disease progression [110]. Based on bis(phenylmethylen)-cycloalkanones and their heterocyclic analogues, some small-molecule compounds were prepared to act as activators of NRF1 and showed the functions in increasing proteasome activity and decreasing size and number of protein aggregates [111]. In addition, the polyQ aggregation could be inhibited by the metabolites from *Peganum harmala* L. polysaccharides via proteasome-mediated protein degradation [112].

8.8 Pharmacological Effects in Golgi

8.8.1 Effects on Development of Golgi

Structurally, Golgi is organized by a series of stacked cisternae (cis to trans) near the nucleus. In *C. elegans*, SQL-1 (homolog of GMAP210) is localized to the Golgi apparatus and contributed to the maintenance of Golgi organization [113]. The wrmScarlet::GOLG-4 is a well-established and bright red marker labeling *C. elegans* Golgi [114], which can help us assess the potential pharmacological effects of some compounds and drugs on Golgi structure and organization under stress and pathological conditions.

8.8.2 Effects on Functions of Golgi

The functions of Golgi include those for modifying, sorting, and packaging proteins and lipids received from the endoplasmic reticulum [115, 116]. Disruption of this process is relevant to different diseases [117, 118]. For the protein modification on Golgi, protein glycosylation is an important form by adding and processing sugar chains [119, 120]. In *C. elegans*, COGC-3 and COGC-1, two components of conserved oligomeric Golgi (COG) complex, were required for glycosylation [121]. The Golgi glycosylation was shown to be required for early embryo development, secretion, and mitosis [122, 123]. The Golgi glycosylation might be affected by certain pharmacological compounds and drugs to influence different biological processes.

Another important function of Golgi is the vesicular trafficking by constantly receiving (retrograde) and sending (anterograde) vesicles to maintain organelle identity and transport cargo [124, 125]. The Golgi trafficking machinery is closely related to the protein glycosylation [126]. In *C. elegans*, besides the COG complex [127], this vesicular trafficking was also regulated by TMEM131 family proteins, small GTPases, and MON2:DOPEY2:ATP9A complex [128–130]. During *C. elegans* post-Golgi exocytic trafficking, RAB-10 cooperated with EHBP-1 to capture the vesicular carriers [131]. In the nervous system, trafficking of vesicular GABA transporter required the function of UNC-46 [132]. Some screened compounds with the function to re-establish endoplasmic reticulum-to-Golgi trafficking could reduce the toxicity of alpha-syn [61].

Considering the involvement of endoplasmic reticulum-to-Golgi trafficking in cellular homeostasis [133, 134], the membrane stress response has received attention. The Golgi can act as a sensor for membrane lipid imbalances [135], and disruption of this aspect of Golgi function activates a specific set of stress responses, such as innate immune response [136]. In *C. elegans*, pharmacological treatment with Brefeldin A (BFA), a protein transport inhibitor, indicated that the Golgi membrane can provide PtdIns4P for generation of PtdIns(4,5)P2 at the sites for epidermal membrane wounding [137].

8.9 Perspectives

In this chapter, the pharmacological effects of compounds and drugs in nucleus, mitochondria, lysosome, endoplasmic reticulum, peroxisome, proteasome, and Golgi were successively introduced and discussed (Fig. 8.1). This is largely due the fact that the organelles development and function are dynamic and plastic, and their morphology and functions can be remodeled in response to stresses and different metabolic states. In the animal model of *C. elegans*, besides the introduced and discussed membrane-bound organelles, some forms of membrane-less organelles have been defined [138, 139], and the possible pharmacological effects on those membrane-less organelles by compounds and drugs are suggested to be paid attention to. Moreover, the inter-organelle communication is widely present in cells to form the physical and functional networks and to coordinate cellular and organismal responses [140–143]. The pharmacological intervention may not only affect development and functions of organelles, but also modulate the inter-organelle communications.

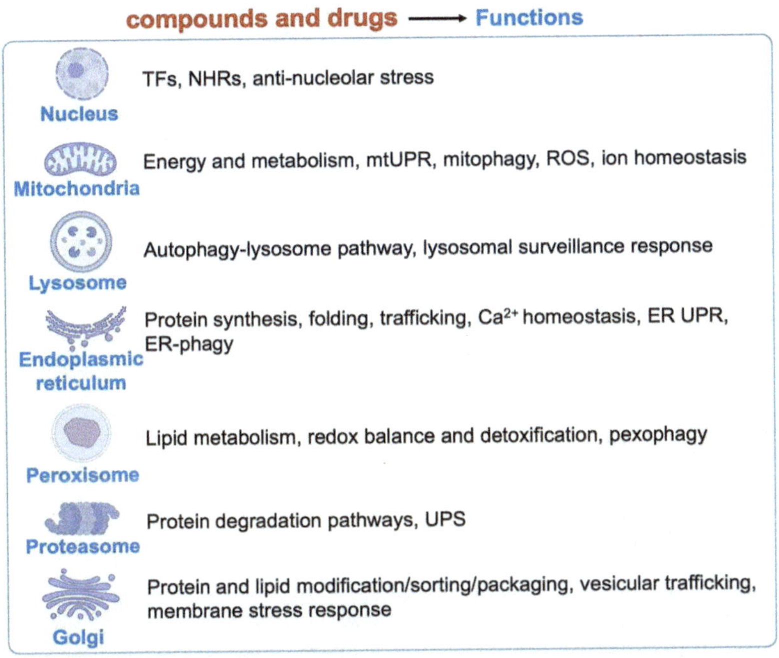

Fig. 8.1 Pharmacological effects of compounds and drugs on functions of organelles in *C. elegans*

References

1. Zhao Y-L, Wu Q-L, Tang M, Wang D-Y (2014) The *in vivo* underlying mechanism for recovery response formation in nano-titanium dioxide exposed *Caenorhabditis elegans* after transfer to the normal condition. Nanomed Nanotechnol Biol Med 10:89–98
2. Yang J-N, Zhao Y-L, Wang Y-W, Wang H-F, Wang D-Y (2015) Toxicity evaluation and translocation of carboxyl functionalized graphene in *Caenorhabditis elegans*. Toxicol Res 4: 1498–1510
3. Prinz WA, Toulmay A, Balla T (2020) The functional universe of membrane contact sites. Nat Rev Mol Cell Biol 21:7–24
4. Bar-Peled L, Kory N (2022) Principles and functions of metabolic compartmentalization. Nat Metab 4:1232–1244
5. Li B, Dong L, Meng W, Xiong S, Wu G, Ma W, Luo H (2024) Phloretic acid requires the insulin/IGF-1 pathway and autophagy to enhance stress resistance and extend the lifespan of *Caenorhabditis elegans*. Front Pharmacol 15:1384227
6. Savova MS, Todorova MN, Apostolov AG, Yahubyan GT, Georgiev MI (2022) Betulinic acid counteracts the lipid accumulation in *Caenorhabditis elegans* by modulation of *nhr-49* expression. Biomed Pharmacother 156:113862
7. Oh M, Yeom J, Schraermeyer U, Julien-Schraermeyer S, Lim Y (2022) Remofuscin induces xenobiotic detoxification via a lysosome-to-nucleus signaling pathway to extend the *Caenorhabditis elegans* lifespan. Sci Rep 12:7161
8. Shi H, Gao X, Yu J, Zhang L, Fan B, Liu Y, Wang X, Fan S, Huang C (2024) Isotschimgine promotes lifespan, healthspan and neuroprotection of *Caenorhabditis elegans* via the activation of nuclear hormone receptors. Biogerontology 26:2

9. Maehama T, Nishio M, Otani J, Mak TW, Suzuki A (2023) Nucleolar stress: molecular mechanisms and related human diseases. Cancer Sci 114:2078–2086
10. Yang K, Yang J, Yi J (2018) Nucleolar stress: hallmarks, sensing mechanism and diseases. Cell Stress 2:125–140
11. Yang Z, Yan X, Zhang J, Liang L, Gao C, Zhang P, Liu Y, Sun J, Ruan B, Duan J, Wang R, Feng X, Che B, Xiao T, Han H (2023) Repression of rRNA gene transcription by endothelial SPEN deficiency normalizes tumor vasculature via nucleolar stress. J Clin Invest 133:e159860
12. Ponti D (2025) The nucleolus: a central hub for ribosome biogenesis and cellular regulatory signals. Int J Mol Sci 26:4174
13. Hong M, Zhou X, Zeng C, Xu D, Xu T, Liao S, Wang K, Zhu C, Shan G, Huang X, Chen X, Feng X, Guang S (2024) Nucleolar stress induces nucleolar stress body formation via the NOSR-1/NUMR-1 axis in *Caenorhabditis elegans*. Nat Commun 15:7256
14. Wu J, Jiang X, Li Y, Zhu T, Zhang J, Zhang Z, Zhang L, Zhang Y, Wang Y, Zou X, Liang B (2018) PHA-4/FoxA senses nucleolar stress to regulate lipid accumulation in *Caenorhabditis elegans*. Nat Commun 9:1195
15. Wu X, Nagasawa S, Muto K, Ueda M, Suzuki C, Abe T, Higashitani A (2022) Mitochonic acid 5 improves Duchenne Muscular Dystrophy and Parkinson's Disease model of *Caenorhabditis elegans*. Int J Mol Sci 23:9572
16. Ng LF, Gruber J, Cheah IK, Goo CK, Cheong WF, Shui G, Sit KP, Wenk MR, Halliwell B (2014) The mitochondria-targeted antioxidant MitoQ extends lifespan and improves healthspan of a transgenic *Caenorhabditis elegans* model of Alzheimer disease. Free Radic Biol Med 71:390–401
17. Liang L, Yue Y, Zhong L, Liang Y, Shi R, Luo R, Zhao M, Cao X, Yang M, Du J, Shen X, Wang Y, Shu Z (2023) Anti-aging activities of *Rehmannia glutinosa* Libosch. crude polysaccharide in *Caenorhabditis elegans* based on gut microbiota and metabonomic analysis. Int J Biol Macromol 253:127647
18. Edwards CB, Copes N, Brito AG, Canfield J, Bradshaw PC (2013) Malate and fumarate extend lifespan in *Caenorhabditis elegans*. PLoS One 8:e58345
19. Williams DS, Cash A, Hamadani L, Diemer T (2009) Oxaloacetate supplementation increases lifespan in *Caenorhabditis elegans* through an AMPK/FOXO-dependent pathway. Aging Cell 8:765–768
20. Pellegrino MW, Nargund AM, Haynes CM (2013) Signaling the mitochondrial unfolded protein response. Biochim Biophys Acta 1833:410–416
21. Haynes CM, Hekimi S (2022) Mitochondrial dysfunction, aging, and the mitochondrial unfolded protein response in *Caenorhabditis elegans*. Genetics 222:iyac160
22. Torres AK, Fleischhart V, Inestrosa NC (2024) Mitochondrial unfolded protein response (UPR(mt)): what we know thus far. Front Cell Dev Biol 12:1405393
23. Anderson NS, Haynes CM (2020) Folding the mitochondrial UPR into the integrated stress response. Trends Cell Biol 30:428–439
24. Haynes CM, Ron D (2010) The mitochondrial UPR – protecting organelle protein homeostasis. J Cell Sci 123:3849–3855
25. Fang M, Liu Y, Gao X, Yu J, Tu X, Mo X, Zhu H, Zou Y, Huang C, Fan S (2025) Perillaldehyde alleviates polyQ-induced neurodegeneration through the induction of autophagy and mitochondrial UPR in *Caenorhabditis elegans*. Biofactors 51:e2089
26. Xiao Y, Li L, Han C, Huang T, Ren S, Wang X, Xing Q, Liu F (2025) Chlorogenic acid inhibits pseudomonas toxin pyocyanin and activates mitochondrial UPR to protect host against pathogen infection. Sci Rep 15:5508
27. Earnshaw R, Zhang YT, Heymann G, Fujisawa K, Hui S, Kapadia M, Kalia LV, Kalia SK (2024) Disease-associated mutations in C-terminus of HSP70 interacting protein (CHIP) impair its ability to negatively regulate mitophagy. Neurobiol Dis 200:106625
28. Hua X, Wang D-Y (2025) An environmentally relevant concentration of 6-PPD quinone inhibits two types of mitophagy to cause mitochondrial dysfunction and lifespan reduction in *Caenorhabditis elegans*. Environ Sci Process Impacts 27:1928–1940

29. Lim Y, Berry B, Viteri S, McCall M, Park EC, Rongo C, Brookes PS, Nehrke K (2021) FNDC-1-mediated mitophagy and ATFS-1 coordinate to protect against hypoxia-reoxygenation. Autophagy 17:3389–3401
30. Ryu D, Mouchiroud L, Andreux PA, Katsyuba E, Moullan N, Nicolet-Dit-Félix AA, Williams EG, Jha P, Lo Sasso G, Huzard D, Aebischer P, Sandi C, Rinsch C, Auwerx J (2016) Urolithin a induces mitophagy and prolongs lifespan in *C. elegans* and increases muscle function in rodents. Nat Med 22:879–888
31. Wu X, Al-Amin M, Zhao C, An F, Wang Y, Huang Q, Teng H, Song H (2020) Catechinic acid, a natural polyphenol compound, extends the lifespan of *Caenorhabditis elegans* via mitophagy pathways. Food Funct 11:5621–5634
32. Wu A, Yong Y, He C, Li Y, Zhou X, Yu L, Chen Q, Lan C, Liu J, Yu C, Qin D, Wu J, Zhou X (2024) Novel 18-norspirostane steroidal saponins: extending lifespan and mitigating neurodegeneration through promotion of mitophagy and mitochondrial biogenesis in *Caenorhabditis elegans*. Mech Ageing Dev 218:111901
33. Sanz A (2016) Mitochondrial reactive oxygen species: Do they extend or shorten animal lifespan? Biochim Biophys Acta 1857:1116–1126
34. Ristow M, Zarse K (2010) How increased oxidative stress promotes longevity and metabolic health: the concept of mitochondrial hormesis (mitohormesis). Exp Gerontol 45:410–418
35. Ristow M, Schmeisser S (2011) Extending life span by increasing oxidative stress. Free Radic Biol Med 51:327–336
36. Schmeisser S, Zarse K, Ristow M (2011) Lonidamine extended lifespan of adult *Caenorhabditis elegans* by increasing the formation of mitochondrial reactive oxygen species. Horm Metab Res 43:687–692
37. Pizzo P (2021) Cell calcium: mitochondria: function and disease. Cell Calcium 96:102370
38. Garbincius JF, Elrod JW (2022) Mitochondrial calcium exchange in physiology and disease. Physiol Rev 102:893–992
39. Romero-Sanz S, Caldero-Escudero E, Álvarez-Illera P, Santo-Domingo J, de la Fuente S, García-Casas P, Fonteriz RI, Montero M, Álvarez J (2025) Rescue of a rotenone model of Parkinson's disease in *C. elegans* by the mitochondrial Na^+/Ca^{2+} exchanger inhibitor CGP37157. Int J Mol Sci 26:3371
40. Hu M, Chen J, Liu S, Xu H (2023) The acid gate in the lysosome. Autophagy 19:1368–1370
41. Mindell JA (2012) Lysosomal acidification mechanisms. Annu Rev Physiol 74:69–86
42. Schröder B, Saftig P (2016) Intramembrane proteolysis within lysosomes. Ageing Res Rev 32: 51–64
43. Turk V, Turk B, Turk D (2001) Lysosomal cysteine proteases: facts and opportunities. EMBO J 20:4629–4633
44. Settembre C, Fraldi A, Medina DL, Ballabio A (2013) Signals from the lysosome: a control centre for cellular clearance and energy metabolism. Nat Rev Mol Cell Biol 14:283–296
45. Xu Y, Wang Q, Wang J, Qian C, Wang Y, Lu S, Song L, He Z, Liu W, Wan W (2025) The cGAS-STING pathway activates transcription factor TFEB to stimulate lysosome biogenesis and pathogen clearance. Immunity 58:309–325
46. Rudinskiy M, Molinari M (2025) Portioning organelles for autophagic clearance. Autophagy 8:1–3
47. Kang Q, Liu Z, Xiang L, Yang S, Yi P, Zhang R (2026) SNX-3 confers lysosomal fusion-competence to sustain basal autophagy. Cell Mol Life Sci 83:85
48. Li Y, Wang X, Li M, Yang C, Wang X (2022) M05B5.4 (lysosomal phospholipase A2) promotes disintegration of autophagic vesicles to maintain *C. elegans* development. Autophagy 18:595–607
49. O'Rourke EJ, Ruvkun G (2013) MXL-3 and HLH-30 transcriptionally link lipolysis and autophagy to nutrient availability. Nat Cell Biol 15:668–676
50. Lin H, Zhang C, Gao Y, Zhou Y, Ma B, Jiang J, Long X, Yimamu N, Zhong K, Li Y, Cui X, Wang H (2024) HLH-30/TFEB modulates autophagy to improve proteostasis in Abeta transgenic *Caenorhabditis elegans*. Front Pharmacol 15:1433030

51. Zhao C, Yue J, Xie Y, Liu B, Xu S, Zhi D, Wang D (2025) A Ginsenoside composition ameliorated Abeta and tau aggregation via autophagy lysosome pathway. Mol Neurobiol 62: 11822–11833
52. Lu R, Chen J, Wang F, Wang L, Liu J, Lin Y (2022) Lysosome inhibition reduces basal and nutrient-induced fat accumulation in *Caenorhabditis elegans*. Mol Cells 45:649–659
53. Ricaurte-Perez C, Gill JP, Wall PK, Dubuisson O, DeLeo KR, Bohnert KA, Johnson AE (2025) DAF-16/FOXO and HLH-30/TFEB comprise a cooperative regulatory axis controlling tubular lysosome induction in *C. elegans*. Nat Commun 16:9878
54. Huang WM, Li ZY, Xu YJ, Wang W, Zhou MG, Zhang P, Liu PS, Xu T, Wu ZX (2014) PKG and NHR-49 signalling co-ordinately regulate short-term fasting-induced lysosomal lipid accumulation in *C. elegans*. Biochem J 461:509–520
55. Li TY, Gao AW, Yang R, Sun Y, Lei Y, Li X, Chen L, Liu YJ, Arey RN, Morales K, Liu RB, Wang W, Zhou A, Zhao TJ, Li W, Lalou A, Wang Q, Lima T, Houtkooper RH, Auwerx J (2025) A lysosomal surveillance response to stress extends healthspan. Nat Cell Biol 27:1083–1097
56. Audhya A, Desai A, Oegema K (2007) A role for Rab5 in structuring the endoplasmic reticulum. J Cell Biol 178:43–56
57. Urano F, Calfon M, Yoneda T, Yun C, Kiraly M, Clark SG, Ron D (2002) A survival pathway for *Caenorhabditis elegans* with a blocked unfolded protein response. J Cell Biol 158:639–646
58. Wiseman RL, Mesgarzadeh JS, Hendershot LM (2022) Reshaping endoplasmic reticulum quality control through the unfolded protein response. Mol Cell 82:477–1491
59. Phillips BP, Gomez-Navarro N, Miller EA (2020) Protein quality control in the endoplasmic reticulum. Curr Opin Cell Biol 65:96–102
60. Vincenz L, Hartl FU (2014) Sugarcoating ER stress. Cell 156:1125–1127
61. Su LJ, Auluck PK, Outeiro TF, Yeger-Lotem E, Kritzer JA, Tardiff DF, Strathearn KE, Liu F, Cao S, Hamamichi S, Hill KJ, Caldwell KA, Bell GW, Fraenkel E, Cooper AA, Caldwell GA, McCaffery JM, Rochet JC, Lindquist S (2010) Compounds from an unbiased chemical screen reverse both ER-to-Golgi trafficking defects and mitochondrial dysfunction in Parkinson's disease models. Dis Model Mech 3:194–208
62. Shin H, Koo B, Yoon J, Kim H, Do S, Na H (2022) Melatonin reduces the endoplasmic reticulum stress and polyubiquitinated protein accumulation induced by repeated anesthesia exposure in *Caenorhabditis elegans*. Sci Rep 12:5783
63. Bahmanyar S (2015) Spatial regulation of phospholipid synthesis within the nuclear envelope domain of the endoplasmic reticulum. Nucleus 6:102–106
64. Xu K, Tavernarakis N, Driscoll M (2001) Necrotic cell death in *C. elegans* requires the function of calreticulin and regulators of Ca^{2+} release from the endoplasmic reticulum. Neuron 31:957–971
65. Zwaal RR, Van Baelen K, Groenen JT, van Geel A, Rottiers V, Kaletta T, Dode L, Raeymaekers L, Wuytack F, Bogaert T (2001) The sarco-endoplasmic reticulum Ca^{2+} ATPase is required for development and muscle function in *Caenorhabditis elegans*. J Biol Chem 276: 43557–43563
66. Lee D, Singaravelu G, Park BJ, Ahnn J (2007) Differential requirement of unfolded protein response pathway for calreticulin expression in *Caenorhabditis elegans*. J Mol Biol 372:331–340
67. Yan X, Xing J, Lorin-Nebel C, Estevez AY, Nehrke K, Lamitina T, Strange K (2006) Function of a STIM1 homologue in *C. elegans*: evidence that store-operated Ca^{2+} entry is not essential for oscillatory Ca^{2+} signaling and ER Ca^{2+} homeostasis. J Gen Physiol 128:443–459
68. Zheng Z, Wu K, Ruan Q, Li D, Liu W, Wang M, Li Y, Xia J, Yang D, Guo J (2022) Suppression of selective voltage-gated calcium channels alleviates neuronal degeneration and dysfunction through glutathione S-transferase-mediated oxidative stress resistance in a *Caenorhabditis elegans* model of Alzheimer's disease. Oxidative Med Cell Longev 2022: 8287633

69. Sun J, Liu Y, Aballay A (2012) Organismal regulation of XBP-1-mediated unfolded protein response during development and immune activation. EMBO Rep 13:55–60
70. Waldherr SM, Strovas TJ, Vadset TA, Liachko NF, Kraemer BC (2019) Constitutive XBP-1s-mediated activation of the endoplasmic reticulum unfolded protein response protects against pathological tau. Nat Commun 10:4443
71. Calfon M, Zeng H, Urano F, Till JH, Hubbard SR, Harding HP, Clark SG, Ron D (2002) IRE1 couples endoplasmic reticulum load to secretory capacity by processing the XBP-1 mRNA. Nature 415:92–96
72. Shen X, Ellis RE, Lee K, Liu CY, Yang K, Solomon A, Yoshida H, Morimoto R, Kurnit DM, Mori K, Kaufman RJ (2001) Complementary signaling pathways regulate the unfolded protein response and are required for *C. elegans* development. Cell 107:893–903
73. Richardson CE, Kinkel S, Kim DH (2011) Physiological IRE-1-XBP-1 and PEK-1 signaling in *Caenorhabditis elegans* larval development and immunity. PLoS Genet 7:e1002391
74. Shen X, Ellis RE, Sakaki K, Kaufman RJ (2005) Genetic interactions due to constitutive and inducible gene regulation mediated by the unfolded protein response in *C. elegans*. PLoS Genet 1:e37
75. Li D, Chen D, Li W, Ou G (2024) Inhibition of a cyclic nucleotide-gated channel on neuronal cilia activates unfolded protein response in intestinal cells to promote longevity. Proc Natl Acad Sci USA 121:e2321228121
76. Zhou H, Shen Y, Dong C, Feng W, Tian Y, Xiao Y (2025) Asperuloside promotes innate immunity via IRE-1/XBP-1 mediated unfolded protein response. Bioorg Chem 157:108318
77. Tong Y, Mukhamejanova Z, Zheng Y, Wen T, Xu F, Pang J (2021) Marine-derived xyloketal compound ameliorates MPP(+)-induced neuronal injury through regulating of the IRE1/XBP1 signaling pathway. ACS Chem Neurosci 12:3101–3111
78. Serot C, Scarcelli V, Pouget A, Largeau C, Sagot A, El-Hachami K, Dupuy D, Culetto E, Lefebvre C, Legouis R (2025) Reticulon-dependent ER-phagy mediates adaptation to heat stress in *C. elegans*. Curr Biol 35:2365–2378
79. Donahue EKF, Hepowit NL, Ruark EM, Mulligan AG, Keuchel B, Urban ND, Peng L, Stephens S, Johnson DJ, Wallace NS, Jackson LP, Ellisman MH, Arrojo E, Drigo R, Folkmann AW, Truttmann MC, MacGurn JA, Burkewitz K (2026) ER remodelling is a feature of ageing and depends on ER-phagy. Nat Cell Biol. https://doi.org/10.1038/s41556-025-01860-1
80. Shi H, Huang X, Chen X, Yang Y, Wang Z, Yang Y, Wu F, Zhou J, Yao C, Ma G, Du A (2021) Acyl-CoA oxidase ACOX-1 interacts with a peroxin PEX-5 to play roles in larval development of *Haemonchus contortus*. PLoS Pathog 17:e1009767
81. Brown LA, Baker A (2003) Peroxisome biogenesis and the role of protein import. J Cell Mol Med 7:388–400
82. Okumoto K, Tamura S, Honsho M, Fujiki Y (2020) Peroxisome: metabolic functions and biogenesis. Adv Exp Med Biol 1299:3–17
83. Tahri-Joutey M, Andreoletti P, Surapureddi S, Nasser B, Cherkaoui-Malki M, Latruffe N (2021) Mechanisms mediating the regulation of peroxisomal fatty acid beta-oxidation by PPARalpha. Int J Mol Sci 22:8969
84. Barth PG, Gootjes J, Bode H, Vreken P, Majoie CB, Wanders RJ (2001) Late onset white matter disease in peroxisome biogenesis disorder. Neurology 57:1949–1955
85. Powers JM (2001) Normal and defective neuronal membranes: structure and function: neuronal lesions in peroxisomal disorders. J Mol Neurosci 16:285–287
86. Zeng L, Li X, Preusch CB, He GJ, Xu N, Cheung TH, Qu J, Mak HY (2021) Nuclear receptors NHR-49 and NHR-79 promote peroxisome proliferation to compensate for aldehyde dehydrogenase deficiency in *C. elegans*. PLoS Genet 17:e1009635
87. Liu X, He A, Lu D, Hu D, Tan M, Abere A, Goodarzi P, Ahmad B, Kleiboeker B, Finck BN, Zayed M, Funai K, Brestoff JR, Javaheri A, Weisensee P, Mittendorfer B, Hsu FF, Van Veldhoven PP, Razani B, Semenkovich CF, Lodhi IJ (2025) Peroxisomal metabolism of branched fatty acids regulates energy homeostasis. Nature 646:1223–1231

88. Di Cara F, Savary S, Kovacs WJ, Kim P, Rachubinski RA (2023) The peroxisome: an up-and-coming organelle in immunometabolism. Trends Cell Biol 33:70–86
89. Wang Y, Li C, Zhang J, Xu X, Fu L, Xu J, Zhu H, Hu Y, Li C, Wang M, Wu Y, Zou X, Liang B (2022) Polyunsaturated fatty acids promote the rapid fusion of lipid droplets in *Caenorhabditis elegans*. J Biol Chem 298:102179
90. Van Gilst MR, Hadjivassiliou H, Jolly A, Yamamoto KR (2005) Nuclear hormone receptor NHR-49 controls fat consumption and fatty acid composition in *C. elegans*. PLoS Biol 3:e53
91. Wang Y, Hua X, Wang D (2023) Exposure to 6-PPD quinone enhances lipid accumulation through activating metabolic sensors of SBP-1 and MDT-15 in *Caenorhabditis elegans*. Environ Pollut 333:121937
92. Wang Y, Wu J, Wang D (2025) 6-PPD quinone causes lipid accumulation across multiple generations differentially affected by metabolic sensors and components of COMPASS complex in *Caenorhabditis elegans*. Environ Pollut 366:125539
93. Navarro-Herrera D, Aranaz P, Eder-Azanza L, Zabala M, Romo-Hualde A, Hurtado C, Calavia D, López-Yoldi M, Martínez JA, González-Navarro CJ, Vizmanos JL (2018) Borago officinalis seed oil (BSO), a natural source of omega-6 fatty acids, attenuates fat accumulation by activating peroxisomal beta-oxidation both in *C. elegans* and in diet-induced obese rats. Food Funct 9:4340–4351
94. Navarro-Herrera D, Aranaz P, Eder-Azanza L, Zabala M, Hurtado C, Romo-Hualde A, Martínez JA, González-Navarro CJ, Vizmanos JL (2018) Dihomo-gamma-linolenic acid induces fat loss in *C. elegans* in an omega-3-independent manner by promoting peroxisomal fatty acid beta-oxidation. Food Funct 9:1621–1637
95. Petriv OI, Rachubinski RA (2004) Lack of peroxisomal catalase causes a progeric phenotype in *Caenorhabditis elegans*. J Biol Chem 279:19996–20001
96. Sampayo JN, Olsen A, Lithgow GJ (2003) Oxidative stress in *Caenorhabditis elegans*: protective effects of superoxide dismutase/catalase mimetics. Aging Cell 2:319–326
97. Sampayo JN, Gill MS, Lithgow GJ (2003) Oxidative stress and aging—the use of superoxide dismutase/catalase mimetics to extend lifespan. Biochem Soc Trans 31:1305–1307
98. Zhou L, Liu J, Bu LL, Liao DF, Cheng SW, Zheng XL (2021) Curcumin acetylsalicylate extends the lifespan of *Caenorhabditis elegans*. Molecules 26:6609
99. Dolese DA, Junot MP, Ghosh B, Butsch TJ, Johnson AE, Bohnert KA (2022) Degradative tubular lysosomes link pexophagy to starvation and early aging in *C. elegans*. Autophagy 18:1522–1533
100. Bard JAM, Goodall EA, Greene ER, Jonsson E, Dong KC, Martin A (2018) Structure and function of the 26S proteasome. Annu Rev Biochem 87:697–724
101. Budenholzer L, Cheng CL, Li Y, Hochstrasser M (2017) Proteasome structure and assembly. J Mol Biol 429:3500–3524
102. Hochstrasser M (1996) Ubiquitin-dependent protein degradation. Annu Rev Genet 30:405–439
103. Orlowski M, Wilk S (2003) Ubiquitin-independent proteolytic functions of the proteasome. Arch Biochem Biophys 415:1–5
104. Matilainen O, Jha S, Holmberg CI (2016) Fluorescent tools for in vivo studies on the ubiquitin-proteasome system. Methods Mol Biol 1449:215–222
105. Hamer G, Matilainen O, Holmberg CI (2010) A photoconvertible reporter of the ubiquitin-proteasome system in vivo. Nat Methods 7:473–478
106. Shimada M, Kanematsu K, Tanaka K, Yokosawa H, Kawahara H (2006) Proteasomal ubiquitin receptor RPN-10 controls sex determination in *Caenorhabditis elegans*. Mol Biol Cell 17:5356–5371
107. Zhao JM, Nguyen DAH, Cervantes D, Vong B, Phillips CM (2026) The ubiquitin-proteasome pathway mediates selective degradation of unloaded Argonaute proteins in *C. elegans*. Genetics 232:iyag007
108. Keith SA, Maddux SK, Zhong Y, Chinchankar MN, Ferguson AA, Ghazi A, Fisher AL (2016) Graded proteasome dysfunction in *Caenorhabditis elegans* activates an adaptive response

involving the conserved SKN-1 and ELT-2 transcription factors and the autophagy-lysosome pathway. PLoS Genet 12:e1005823
109. Dubey AA, Sarkar A, Milcz K, Szulc NA, Thapa P, Piechota W, Serwa RA, Pokrzywa W (2024) Floxuridine supports UPS independent of germline signaling and proteostasis regulators via involvement of detoxification in *C. elegans*. PLoS Genet 20:e1011371
110. Papaevgeniou N, Sakellari M, Jha S, Tavernarakis N, Holmberg CI, Gonos ES, Chondrogianni N (2016) 18alpha-Glycyrrhetinic acid proteasome activator decelerates aging and Alzheimer's disease progression in *Caenorhabditis elegans* and neuronal cultures. Antioxid Redox Signal 25:855–869
111. Sedlacek J, Smahelova Z, Adamek M, Subova D, Svobodova L, Kadlecova A, Majer P, Machara A, Grantz Saskova K (2025) Small-molecule activators of NRF1 transcriptional activity prevent protein aggregation. Biomed Pharmacother 183:17864
112. Guo X, Yuan J, Song X, Wang X, Sun Q, Tian J, Li X, Ding M, Liu Y (2020) Bacteria metabolites from *Peganum harmala* L. polysaccharides inhibits polyQ aggregation through proteasome-mediated protein degradation in *C. elegans*. Int J Biol Macromol 161:681–691
113. Broekhuis JR, Rademakers S, Burghoorn J, Jansen G (2013) SQL-1, homologue of the Golgi protein GMAP210, modulates intraflagellar transport in *C. elegans*. J Cell Sci 126:1785–1795
114. Cao WX, Merritt DM, Pe K, Cesar M, Hobert O (2024) Comparative analysis of new mScarlet-based red fluorescent tags in *Caenorhabditis elegans*. Genetics 228:iyae126
115. Kulkarni-Gosavi P, Makhoul C, Gleeson PA (2019) Form and function of the Golgi apparatus: scaffolds, cytoskeleton and signalling. FEBS Lett 593:2289–2305
116. Duden R (2003) ER-to-Golgi transport: COP I and COP II function (review). Mol Membr Biol 20:197–207
117. Liu J, Huang Y, Li T, Jiang Z, Zeng L, Hu Z (2021) The role of the Golgi apparatus in disease (review). Int J Mol Med 47:38
118. Tang VT, Ginsburg D (2023) Cargo selection in endoplasmic reticulum-to-Golgi transport and relevant diseases. J Clin Invest 133:e163838
119. Stanley P (2011) Golgi glycosylation. Cold Spring Harb Perspect Biol 3:a005199
120. Moremen KW, Tiemeyer M, Nairn AV (2012) Vertebrate protein glycosylation: diversity, synthesis and function. Nat Rev Mol Cell Biol 13:448–462
121. Kubota Y, Sano M, Goda S, Suzuki N, Nishiwaki K (2006) The conserved oligomeric Golgi complex acts in organ morphogenesis via glycosylation of an ADAM protease in *C. elegans*. Development 133:263–273
122. Wang H, Spang A, Sullivan MA, Hryhorenko J, Hagen FK (2005) The terminal phase of cytokinesis in the *Caenorhabditis elegans* early embryo requires protein glycosylation. Mol Biol Cell 16:4202–4213
123. Stevens J, Spang A (2013) N-glycosylation is required for secretion and mitosis in *C. Elegans*. PLoS One 8:e63687
124. Stalder D, Gershlick DC (2020) Direct trafficking pathways from the Golgi apparatus to the plasma membrane. Semin Cell Dev Biol 107:112–125
125. Liu JJ (2016) Retromer-mediated protein sorting and vesicular trafficking. J Genet Genomics 43:165–177
126. Frappaolo A, Karimpour-Ghahnavieh A, Sechi S, Giansanti MG (2020) The close relationship between the Golgi trafficking machinery and protein glycosylation. Cells 9:2652
127. Struwe WB, Reinhold VN (2012) The conserved oligomeric Golgi complex is required for fucosylation of N-glycans in *Caenorhabditis elegans*. Glycobiology 22:863–875
128. Zhang Z, Bai M, Barbosa GO, Chen A, Wei Y, Luo S, Wang X, Wang B, Tsukui T, Li H, Sheppard D, Kornberg TB, Ma DK (2020) Broadly conserved roles of TMEM131 family proteins in intracellular collagen assembly and secretory cargo trafficking. Sci Adv 6:eaay7667
129. Sumakovic M, Hegermann J, Luo L, Husson SJ, Schwarze K, Olendrowitz C, Schoofs L, Richmond J, Eimer S (2009) UNC-108/RAB-2 and its effector RIC-19 are involved in dense core vesicle maturation in *Caenorhabditis elegans*. J Cell Biol 186:897–914

130. McGough IJ, de Groot REA, Jellett AP, Betist MC, Varandas KC, Danson CM, Heesom KJ, Korswagen HC, Cullen PJ (2018) SNX3-retromer requires an evolutionary conserved MON2: DOPEY2:ATP9A complex to mediate Wntless sorting and Wnt secretion. Nat Commun 9: 3737
131. Liu S, Wei J, Zhong L, Hai S, Song S, Xie C, Huang Z, Cheng Z, Zhang J, Du A, Zhang P, Yan Y, Shi A (2025) RAB-10 cooperates with EHBP-1 to capture vesicular carriers during post-Golgi exocytic trafficking. J Cell Biol 224:e202410003
132. Schuske K, Palfreyman MT, Watanabe S, Jorgensen EM (2007) UNC-46 is required for trafficking of the vesicular GABA transporter. Nat Neurosci 10:846–853
133. Joo JH, Wang B, Frankel E, Ge L, Xu L, Iyengar R, Li-Harms X, Wright C, Shaw TI, Lindsten T, Green DR, Peng J, Hendershot LM, Kilic F, Sze JY, Audhya A, Kundu M (2016) The noncanonical role of ULK/ATG1 in ER-to-Golgi trafficking is essential for cellular homeostasis. Mol Cell 62:491–506
134. Kim JD, Chun AY, Mangan RJ, Brown G, Mourao Pacheco B, Doyle H, Leonard A, El Bejjani R (2019) A conserved retromer-independent function for RAB-6.2 in *C. elegans* epidermis integrity. J Cell Sci 132:jcs223586
135. Silhankova M, Port F, Harterink M, Basler K, Korswagen HC (2010) Wnt signalling requires MTM-6 and MTM-9 myotubularin lipid-phosphatase function in Wnt-producing cells. EMBO J 29:4094–4105
136. Fanelli MJ, Welsh CM, Lui DS, Smulan LJ, Walker AK (2023) Immunity-linked genes are stimulated by a membrane stress pathway linked to Golgi function and the ARF-1 GTPase. Sci Adv 9:eadi5545
137. Meng X, Wijaya CS, Shao Q, Xu S (2023) Triggered Golgi membrane enrichment promotes PtdIns(4,5)P2 generation for plasma membrane repair. J Cell Biol 222:e202303017
138. Lacroix E, Audas TE (2022) Keeping up with the condensates: the retention, gain, and loss of nuclear membrane-less organelles. Front Mol Biosci 9:998363
139. Robinson AO, Lee J, Cameron A, Keating CD, Adamala KP (2024) Cell-free expressed membraneless organelles inhibit translation in synthetic cells. ACS Biomater Sci Eng 10: 773–781
140. Chen B, Lyssiotis CA, Shah YM (2025) Mitochondria-organelle crosstalk in establishing compartmentalized metabolic homeostasis. Mol Cell 85:1487–1508
141. Zheng S, Wang X, Zhao D, Liu H, Hu Y (2023) Calcium homeostasis and cancer: insights from endoplasmic reticulum-centered organelle communications. Trends Cell Biol 33:312–323
142. Domingues N, Pires J, Milosevic I, Raimundo N (2025) Role of lipids in interorganelle communication. Trends Cell Biol 35:46–58
143. Zhu D, Li X, Tian Y (2022) Mitochondrial-to-nuclear communication in aging: an epigenetic perspective. Trends Biochem Sci 47:645–659

Target Molecular Pharmacology 9

Abstract

In the cells, compounds and drugs will target to different molecules to exert their pharmacological effects. The potential pharmacological effects of compounds and drugs on G protein-coupled receptors (GPCRs), nuclear hormone receptors (NHRs), ion channels, cytoplasmic molecular signals, and enzymes were introduced and discussed, which is helpful for understanding underlying molecular mechanisms for observed pharmacological effects.

Keywords

Targeted molecules · Compounds · *Caenorhabditis elegans*

9.1 Introduction

After the compounds and the drugs are absorbed by the organisms and enter the cells, on the one hand, they will target to different organelles to exert their pharmacological effects. On the other hand, the compounds and the drugs will target to different proteins and/or enzymes to impose their pharmacological effects by affecting activities and functions of proteins and/or enzymes. For the water-soluble and lipid-soluble compounds and drugs, it is assumed that they can target to G protein-coupled receptors (GPCRs), nuclear hormone receptors (NHRs), ion channels, certain cytoplasmic proteins, and/or enzymes. For compounds and drugs at the nano size, their interaction with these proteins and enzymes may also exist. In the Chap. 8, the target organelle pharmacology has been introduced. In this chapter, we will further introduce and discuss the pharmacological effects of compounds and drugs on GPCRs, NHRs, cytoplasmic molecular signals, and enzymes.

D. Wang, *Pharmacology in Caenorhabditis elegans*,
https://doi.org/10.1007/978-981-92-5166-7_9

9.2 Pharmacological Effects on G Protein-Coupled Receptors and Nuclear Hormone Receptors

9.2.1 Pharmacological Effects on G Protein-Coupled Receptors

The GPCR refer to polypeptide chain that winds back and forth through the cell membrane seven times, which are also called 7-transmembrane receptor. GPCRs are the largest and most diverse family of membrane receptors responsible for detecting molecules outside the cells and environmental stimuli and activating internal signal transduction pathways in eukaryotes [1, 2]. For the GPCRs, their N-terminus is extracellular, and their C-terminus is intracellular [1, 2]. The ligands for GPCR are hydrophilic and large molecules, which usually cannot cross the membrane and bind to GPCRs externally [3, 4]. After the binding, the ligands will cause the conformational change of GPCRs in the intracellular part [5, 6], which further activates the G proteins (trimeric GTP-binding proteins) on the inside of the membrane [7, 8]. The activated G protein subunits can interact with the downstream effectors (e.g., adenylyl cyclase and phospholipase C) to lead to cellular responses by further activating the second messengers (e.g., cAMP, IP3, DAG, and calcium ion) and the intracellular signaling cascades [9–12]. A ligand-bound GPCR can activate multiple G proteins, each of which can further affect multiple second messengers, leading to a massive cellular response from the initial signal, which suggests the signal amplification [13, 14]. Considering the crucial role for signal transduction and signal amplification, the GPCRs act as the security guards at the main gate. It is predicted that the GPCRs are the targets for over 30% of modern drugs.

To assess the pharmacological effects of compounds and drugs on GPCRs, the following several aspects of assays are suggested:

1. Conformation of binding potential of compounds and drugs to GPCRs. This can be directly examined by radioligand binding assay or label-free technology. This can also be indirectly assessed by molecular docking analysis.
2. Examination of conformational change of GPCRs before and after addition of compounds and drugs. This aspect of analysis aims to determine the possible allosteric modulation of GPCRs by compounds and drugs.
3. Assay of activities of G protein-dependent signals, such as cAMP, calcium ion, and GTP.
4. Analysis of expression and functions of GPCRs before and after addition of compounds and drugs. This will provide the response and the functional evidence for the role of GPCRs as the targets of compounds and drugs.

In *C. elegans*, some GPCRs were shown to function as the possible targets of compounds and drugs:

1. Insulin Receptors: DAF-2 is the insulin receptor in the insulin signaling pathway. The antipsychotic drug olanzapine decreased lifespan was prevented by *daf-*

2 mutation, suggesting the requirement of DAF-2 for the pharmacological effect of this antipsychotic drug [15].
2. Melatonin Receptors: Among melatonin receptors in mammals, MT1 and MT2 are GPCRs. Based on MT1/2 receptor antagonist assay, the melatonin signaling might regulate locomotion through MT1-like receptor [16]. The G-protein α subunits GPA-7 and EGL-30 were required for the function of melatonin signaling [16], which further suggested the role of GPCRs.
3. Dopamine Receptors: Pepper component 7-ethoxy-4-methylcoumarin could ameliorate the phenotypes of Parkinson's disease by acting as agonist of dopamine D2 receptors [17].
4. Glutamate Receptors: The allosteric drug binding sites of orthosteric to metabotropic glutamate receptor MGL-2 was identified [18]. D-Serine regulated the adaptive foraging behavior by acting as coagonist for NMDA-type glutamate receptors [19]. ZA-II-05 functioned as NMDA-receptor antagonist to inhibit the vanadium neurotoxicity [20].
5. Serotonin Receptors: In Machado-Joseph disease (MJD) model, as the agonists, function of befiradol required serotonin receptor of SER-4, and function of tandospirone required serotonin receptors of SER-4, SER-5, and SER-7 [21]. Withanone had high binding affinities to serotonin receptors of SER-1, SER-4, SER-7, and MOD-5, and upregulated their expressions to ameliorate the stress symptoms [22].
6. Tyramine Receptors: The phentolamine showed the high affinity to tyramine receptor TYRA-2 [23].
7. Neuropeptide Receptors: Under the radiation stress condition, treatment with eleutheroside E could elevate long-term memory by activating neuropeptide receptors of NPR-2 and NPR-11 [24].

In *C. elegans*, with the aid of transgenic strains expressing human GPCRs, the potential pharmacological effects of compounds and drugs on human GPCRs can be further determined. Using transgenic strain expressing human chemokine receptor 5 (CCR5) in nociceptive neurons, among the library of 68 MIP-1alpha variants, 13 agonist residues were identified with the function of CCR5 activation [25]. Treatment with corresponding ligands could cause the receptor desensitization in transgenic strains expressing human somatostatin receptor 2 (Sstr2) and CCR5 [26].

9.2.2 Pharmacological Effects on Nuclear Hormone Receptors

The NHRs refer to the essentially ligand-activated transcription factors governing the transcriptional expression of targeted genes [27, 28]. Some small, hydrophobic (lipid-soluble) molecules act as the ligands for NHRs, such as steroid hormones, thyroid hormones, retinoids, and Vitamin D [29, 30]. Structurally, the NHRs contain: (1) N-terminal domain involved in the gene activation, (2) DNA-binding domain containing the zinc finger motifs to recognize specific DNA sequences, and (3) ligand-binding domain with the function to bind specific ligands, such as the hormones [31–33].

After the binding by ligands, the conformational change of NHRs will be formed [34, 35], and the translocation into the nucleus will occur [36, 37]. For the steroid receptors, the ligand binding can cause them to release chaperone proteins to be translocated in the nucleus [38, 39]. For the thyroid/retinoid receptors, they will bind to DNA in the nucleus by acting as the repressors or activators [40, 41]. Acting as the ligand-activated transcription factors, the function of NHRs is like the commanding officers inside the main building.

To examine the pharmacological effects of compounds and drugs on NHRs, the following several aspects of assays are suggested:

1. Assay of binding of compounds and drugs to NHRs.
2. Assay of coregulator interaction: The pharmacological effect (agonist vs. antagonist) can be defined by whether the compound and drugs can promote the recruitment of coactivators (for activation) or corepressors (for repression).
3. Analysis of expression and functions of NHRs before and after addition of compounds and drugs.
4. Assay of expression and functions of downstream targeted genes.

In *C. elegans*, some NHRs act as the possible targets of compounds and drugs:

1. NHR-49: NHR-49 is the mammalian ortholog PPARα. Betulinic acid and *Baccharis trimera* leaf extract reduced lipid accumulation by inhibiting NHR-49 [42, 43]. Lipid-lowering fibrates and omega-3 fatty acid alpha-linolenic acid extended the lifespan in the NHR-49-dependent manner [44, 45]. The fenofibrate inhibited the Abeta-induced paralysis by acting as the NHR-49/PPARalpha agonist [46]. Some compounds were identified as immune modulators to enhance the resistance of *C. elegans* to pathogen infections, which was dependent of NHR-49 function [47].
2. DAF-12: In *C. elegans*, dafachronic acids (the metabolites of distinct 3-keto-cholestenoic acid) were identified as the ligands of DAF-12 with the potential to bind to and to activate it [48, 49]. The 24-difluoromethyl analogue could retain the DAF-12 antagonist activity of 24-hydroxy-4-cholen-3-one and exhibit an improved activity [50]. The mitophagy-inducing coumarin screened from the natural product library regulated mitophagy and extended lifespan by engaging the DAF-12 [51].
3. NHR-8: During the longevity control, NHR-8 was identified as the receptor of vitamin D3 [52]. The beneficial effects of isotschimgine, such as its neuroprotective effect, could be suppressed by RNAi of *nhr-8*, suggesting the requirement of NHR-8 for the formation of beneficial effects of isotschimgine [53].

9.3 Pharmacological Effects on Ion Channels

Ion channels refer to the proteins that create openings in cell membranes, which can allow specific ions to pass through [54, 55]. Follow this, the ion channels can help the cells to establish the resting membrane potential, to shape the electrical signals, and to gate the flow of ions across the cell membrane [56–58].

To determine the pharmacological effects of compounds and drugs on ion channels, the following several aspects of assays are suggested:

1. Binding assay of compounds and drugs to ion channels.
2. Patch clamp electrophysiology assay to assess activity and physiological state of ion channels.
3. Examination of conformational change of ion channels before and after addition of compounds and drugs.
4. Analysis of expression and functions of ion channels before and after addition of compounds and drugs.

In the *C. elegans* genome, the genes can encode multiple superfamilies of ion channels [59–64]. These *C. elegans* ion channel superfamilies can include:

1. Voltage-Gated Channels: The calcium, potassium, and sodium channels will open or close in response to changes in membrane potential.
2. Cys-Loop Ligand-Gated Ion Channels: These pentameric channels can be gated by neurotransmitters, including both cation-selective (excitatory) and anion-selective (inhibitory) members.
3. Ionotropic Glutamate Receptors: They can mediate the fast excitatory neurotransmission.
4. Sensory Transduction Channels: These channels function to convert sensory stimuli (touch, smell, or temperature) into electrical signals. This superfamily includes CNG (cyclic nucleotide-gated), TRPV (transient receptor potential vanilloid), and DEG/ENaC (degenerin/epithelial sodium channel) families.
5. Acetylcholine-Gated Anion Channels: These belong to the inhibitory channels. This superfamily includes the channels gated by acetylcholine.

In *C. elegans*, some ion channels function as the possible targets of compounds and drugs:

1. Calcium Channels: Based on screen from small molecules library, nemadipine-A was identified to cause defects in morphology and egg-laying by acting as the calcium channel antagonist [65]. Ethylene glycol tetraacetic acid (EGTA) and nimodipine reduced Aβ 1-42 aggregations by suppressing selective voltage-gated calcium channels [66]. In amyotrophic lateral sclerosis model, a class of neuroleptics and nifedipine were identified to restore the motility by blocking calcium channels [67, 68].

2. Sodium Channels: Using transgenic strain expressing modified human ether-a-go-go-related gene (hERG) potassium channels as the tool, prostratin and ingenol-3,20-dibenzoate were identified as the channel modulators to activate the PKCε signaling [69]. BK channel is a large-conductance potassium channel, and melatonin could promote the sleep by activating BK channel SLO-1, a target of melatonin receptor PCDR-1 [70]. The 2-aminoethoxydiphenyl borate (2-ABP), nifedipine, nimodipine, flunarizine and ethoxzolamide were screened and identified to inhibit the abnormal movements in *C elegans* with gain-of-function mutation in *unc-77* encoding NALCN (Na leak-current channel) [71].
3. Potassium Channels: The cyclooctadepsipeptide emodepside could directly open the Ca^{2+}-gated and voltage-dependent K^+-channel SLO-1 [72].
4. Chloride Channels: The piperazine and muscimol were identified as allosteric inhibitors of serotonin-gated chloride channel MOD-1 to reduce the serotonin-elicited currents [73]. The L-glutamate-gated chloride channel GLC-3 was sensitive to 3,3-bis-trifluoromethyl-bicyclo[2,2,1]heptane-2,2-dicarbonitrile (BIDN) and fipronil, which were identified as the channel blockers [74]. Ivermectin, a member of the avermectin family, could act at the GABA-gated chloride channel [75].
5. CNG and TRPV Channels: The chemosensation to isoamyl alcohol (3-methyl-1-butanol) and 2-methyl-1-butanol produced by microbial fermentation was found to rely on CNG channel TAX-4 and TRPV channel OSM-9 [76]. The ammonium sensation depended on the CNG channels of TAX-2 and TAX-4 [77]. The chemosensation to Argentine ant extract was dependent on TRPV channel OSM-9 [78].
6. Acetylcholine-Gated Anion Channels: On acetylcholine receptor ACR-23, betaine is bound to classical neurotransmitter sites in inter-subunit interfaces of extracellular domain to lead to pore closure, and monepantel is bound to allosteric sites in inter-subunit interfaces of transmembrane domain to cause an open channel [79]. The bephenium was identified as an agonist of nicotinic acetylcholine receptors and a blocker of open-channels [80].

9.4 Pharmacological Effects on Cytoplasmic Signals

9.4.1 G Proteins

Upon the GPCRs activation, G proteins will dissociate to modulate their targets (such as adenylyl cyclase) that can produce the second messengers. The G proteins act as gatekeepers for most cytoplasmic signals [81, 82]. In *C. elegans*, accompanied with the resistance to small molecule compounds, the identified corresponding genetic mutants affected the members of the regulators of G-protein signaling (RGS) protein/G-protein complex [83]. Besides serotonin receptors SER-1 and SER-7, the pharmacological effect of gliotoxin on egg-laying behavior also required the Gq ortholog EGL-30 [84].

9.4.2 Protein Kinases

Protein kinases are enzymes with the function to phosphorylate target proteins and to alter their activity, which is the primary mechanism for the second messengers to exert their effects [85, 86]. The protein kinases contain: (1) second messenger-dependent kinases, (2) protein kinase A (PKA) activated by cAMP, (3) protein kinase C (PKC) activated by DAG and Ca^{2+}, (4) non-receptor tyrosine kinases, and (5) some kinase cascades (such as p38 MAPK, JNK MAPK, and ERK MAPK cascades) [87, 88].

In *C. elegans*, staurosporine modulated the regeneration by acting as a PKC kinase inhibitor in the neuronal type-specific manner [89]. Function of esculetin treatment in reducing lipid accumulation was partially dependent on AAK-2 a 5′-adenosine monophosphate-activated protein kinase (AMPK) [90]. Vitexin and isovitexin extended the lifespan by partially inhibiting the AKT kinases in the insulin signaling pathway [91]. Thymol treatment could enhance innate immunity against the fungal infection by activating p38 MAPK signaling [92]. Valproic acid could ameliorate the dopaminergic neurodegeneration by modulating the ERK MAPK signaling [93]. In Duchenne muscular dystrophy model, the beneficial effect of mitochondrial hydrogen sulfide in improving the health was also dependent of JNK MAPK signaling [94].

9.4.3 Cyclic AMP

Cyclic AMP (cAMP) synthesized from ATP by adenylyl cyclase regulates different biological processes via affecting the PKA [95–97]. In radiation-damaged nematodes, eleutheroside E improved the long-term memory by modulating CREB activity via cAMP-PKA signaling [24].

9.4.4 Diacylglycerol

Diacylglycerol (DAG) can produce the alongside IP3 by phospholipase C and activate the protein kinase C [98, 99]. In *C. elegans*, phorbol 12-myristate 13-acetate could rescue the activity of egg-laying circuit and the behavior deficit in Gαq signaling mutants by acting as the mimics of 1,2-diacylglycerol [100].

9.4.5 Cyclic GMP

Cyclic GMP (cGMP), synthesized from GTP by guanylyl cyclase, is involved in the control of multiple processes, such as smooth muscle relaxation and phototransduction [101–104]. In *C. elegans*, the role of bacterial metabolites in activating chemosensory neuron gene expression was controlled by distinct cGMP-dependent pathways [105].

9.4.6 Calcium Ions

Calcium ions, released from intracellular stores (e.g., endoplasmic reticulum) via IP3 receptors or generated through certain ion channels, can triggers various biological processes, such as muscle contraction and neurotransmitter release [106–108]. In *C. elegans*, the gut-derived metabolite urolithin A promoted health span by modulating inter-organellar communication via calcium-dependent mitophagy [109].

9.4.7 Regulatory Non-coding RNAs

Regulatory non-coding RNAs are a large and diverse group of non-coding RNA molecules that play crucial roles in fine-tuning how genes are expressed [110–112]. In *C. elegans*, the regulatory RNAs mainly include microRNAs, long non-coding RNAs, and circular RNAs. In *C. elegans*, betulinic acid inhibited lipid accumulation by affecting some microRNAs (such as *mir-60*, *lin-4*, *let-7*, and *mir-786*) associated with control of lipid metabolism [42]. Ginsenoside Rf could suppress the human tau proteotoxicity, which might be linked to changes of some microRNAs and long non-coding RNAs [42, 113].

9.5 Pharmacological Effects on Enzymes

Enzymes represent one of the most classical classes of targets for pharmacological compounds and drugs. The enzymes function to catalyze certain biochemical reactions to regulate different biological processes.

To examine the pharmacological effects of compounds and drugs on enzymes, the following several aspects of assays are suggested:

1. Binding assay of compounds and drugs to enzymes.
2. Examination of conformational change of enzymes before and after addition of compounds and drugs.
3. Assay of enzyme activity before and after addition of compounds and drugs.
4. Analysis of expression and functions of enzyme genes before and after addition of compounds and drugs.

9.5.1 Pharmacological Effects on Different Types of Enzymes

The International Union of Biochemistry and Molecular Biology (IUBMB) has classified the enzymes into six main families based on their chemical reactions they can catalyze. For the pharmacological effects on enzymes, we can first understand them on different types of enzymes:

1. Oxidoreductases: Function of these enzymes is oxidation-reduction by transferring the electrons. In the *C. elegans* SDH model, 19 compounds were identified to show the inhibitory activity against succinate dehydrogenase [114]. Treatment with Se compound could increase activity of thioredoxin reductase (TXNRD) and change expression of antioxidant proteins [115].
2. Transferases: Function of transferases is to transfer functional groups (e.g., methyl and phosphate). Curcumin acetylsalicylate could extend lifespan, inhibit oxidative stress, and cause the response of glutathione S-transferase [116].
3. Hydrolases: Function of hydrolases is the cleavage of bonds. In *C. elegans*, JZL184 had the neuroprotective effect on dopaminergic neurons exposed to 6-hydroxydopamine by acting as the inhibitor of fatty acid amide hydrolase-4 [117].
4. Lyases: Function of lyases is the addition or removal of groups to form double bonds.
5. Isomerases: These enzymes have the function of geometric or structural rearrangements.
6. Ligases: Function of ligases is the joining of two molecules with ATP hydrolysis. Treatment with cisplatin (CisPt) caused the inhibition in pharyngeal pumping, which could be aggravated by RNAi of *gcs-1* encoding glutamate-cysteine ligase [118].

9.5.2 Pharmacological Stimulating or Inhibitory Effects on Enzymes

The pharmacological effects of compounds and drugs on enzymes can also be reflected by their stimulating effects and different inhibitory effects.

Some compounds and drugs can show the stimulating effects on certain enzyme activity. In *C. elegans* AD model, thonningianin A delayed paralysis and ameliorated food-sensing deficits by acting as the ferroptosis inhibitor via GPX4 activation [119]. Hyperoside could extend the lifespan by activating SOD-3 and GST-4 [120].

Besides this, some compounds and drugs can also inhibit certain enzyme activity. The enzyme inhibitors include irreversible Inhibitors and reversible Inhibitors. The reversible Inhibitors further contain: (1) competitive inhibitors, (2) non-competitive inhibitors, and (3) uncompetitive inhibitors. The compound was synthesized and identified as an ATP-competitive inhibitor of GSK-3β to show implications for Alzheimer's Disease treatment [121]. Tirotundin and parthenolide were identified as reversible inhibitors against acetylcholinesterase (AChE) [122].

9.5.3 Pharmacological Effects on Enzymes Governing Different Biological Processes

Pharmacological treatment with compounds and drugs can exert their effects on enzymes governing different biological processes:

1. Effects of Enzymes Governing Glucose Metabolism: Treatment with (p-ClPhSe) 2 could reduce the glucose, which required the involvement of insulin/IGF-1 like signaling, including the AGE-1, AKT-1 and AKT-2, and PFK-1 [123].
2. Effects of Enzymes Governing Lipid Metabolism: Treatment with cycloastragenol reduced lipid accumulation by inhibiting delta(9)-fatty-acid desaturase 7 (FAT-7) and activating 5'-AMP-activated protein kinase catalytic subunit alpha-2 (AAK-2) [124].
3. Effects of Enzymes Governing Electron Transfer Chain: Treatment with rosmarinic acid enhanced the resistance of *C. elegans* to bacterial pathogen infection by upregulating expressions of *mev-1* and *gas-1* encoding components of mitochondrial complexes I and II [125].
4. Effects of Enzymes Governing Phospholipid Metabolisms: Treatment with 2-undecanone affected the sphingomyelin metabolism by decreasing acid sphingomyelinase activity [126].
5. Effects of Enzymes Governing Protein Synthesis: Based on bis(phenylmethylen)-cycloalkanones and heterocyclic analogues, some small-molecule compounds were prepared to show the effects in decreasing size and number of protein aggregates by increasing the proteasome activity [127].
6. Effects of Enzymes Governing Amino Acid Metabolisms: Ziziphi spinosae semen treatment could delay Aβ1–42-induced paralysis and reduce amyloid accumulation, which were associated with change of metabolisms of some amino acids, such as tryptophan metabolism [128].

9.6 Perspectives

We here introduced and discussed the possible pharmacological effects of compounds and drugs on GPCRs, NHRs, ion channels, cytoplasmic molecular signals, and enzymes in the cells (Fig. 9.1). However, besides these, some other molecular molecules may also act as the targets of compounds and drugs. In cells, for the epigenetic regulation, besides the introduced regulatory non-coding RNAs, molecular signals required for histone/DNA methylation and histone acetylation can also function as the targets of compounds and drugs to mediate formation of their pharmacological effects. Some protective response proteins, such as those reflecting innate immune response, mitochondrial UPR, ER UPR, and autophagy, might also be targeted by certain pharmacological compounds and drugs. Moreover, besides corresponding receptors, some secreted ligands and neurotransmitters may also function as the targets of certain pharmacological compounds and drugs.

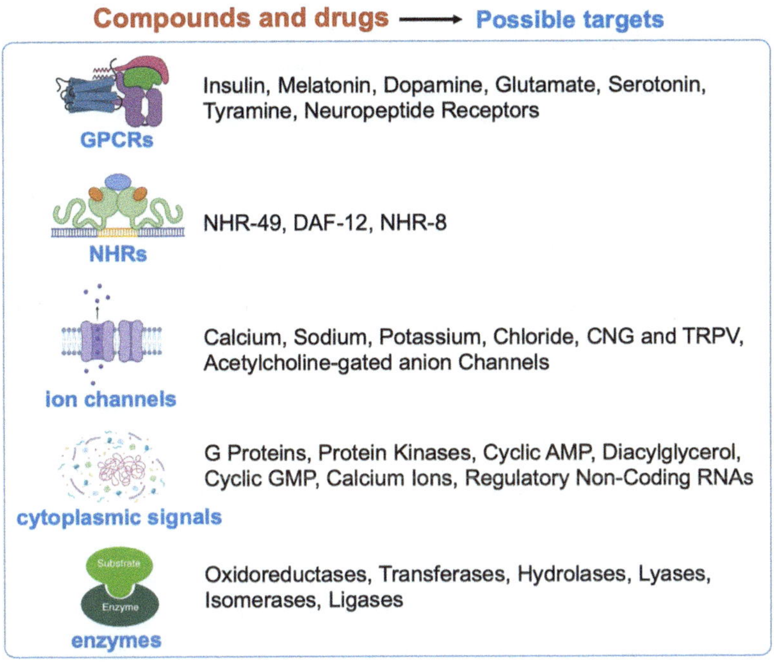

Fig. 9.1 Pharmacological effects of compounds and drugs on molecules in *C. elegans* cells

References

1. Lameh J, Cone RI, Maeda S, Philip M, Corbani M, Nádasdi L, Ramachandran J, Smith GM, Sadée W (1990) Structure and function of G protein coupled receptors. Pharm Res 7:1213–1221
2. Weis WI, Kobilka BK (2018) The molecular basis of G protein-coupled receptor activation. Annu Rev Biochem 87:897–919
3. Katritch V, Cherezov V, Stevens RC (2013) Structure-function of the G protein-coupled receptor superfamily. Annu Rev Pharmacol Toxicol 53:531–556
4. Nakliang P, Lazim R, Chang H, Choi S (2020) Multiscale molecular modeling in G protein-coupled receptor (GPCR)-ligand studies. Biomolecules 10:631
5. Latorraca NR, Venkatakrishnan AJ, Dror RO (2017) GPCR dynamics: structures in motion. Chem Rev 117:139–155
6. Latek D, Modzelewska A, Trzaskowski B, Palczewski K, Filipek S (2012) G protein-coupled receptors–recent advances. Acta Biochim Pol 59:515–529
7. Manglik A, Kruse AC (2017) Structural basis for G protein-coupled receptor activation. Biochemistry 56:5628–5634
8. Christopoulos A, Kenakin T (2002) G protein-coupled receptor allosterism and complexing. Pharmacol Rev 54:323–374
9. Krupinski J (1991) The adenylyl cyclase family. Mol Cell Biochem 104:73–79
10. Freissmuth M, Casey PJ, Gilman AG (1989) G proteins control diverse pathways of transmembrane signaling. FASEB J 3:2125–2131

11. English D (1996) Phosphatidic acid: a lipid messenger involved in intracellular and extracellular signalling. Cell Signal 8:341–347
12. Huisamen B, Lochner A (1996) Inositolpolyphosphates and their binding proteins–a short review. Mol Cell Biochem 157:229–232
13. Ross EM (1989) Signal sorting and amplification through G protein-coupled receptors. Neuron 3:141–152
14. Lamb TD, Pugh EN Jr (1992) G-protein cascades: gain and kinetics. Trends Neurosci 15:291–298
15. Weeks KR, Dwyer DS, Aamodt EJ (2010) Antipsychotic drugs activate the *C. elegans* Akt pathway via the DAF-2 insulin/IGF-1 receptor. ACS Chem Neurosci 1:463–473
16. Tanaka D, Furusawa K, Kameyama K, Okamoto H, Doi M (2007) Melatonin signaling regulates locomotion behavior and homeostatic states through distinct receptor pathways in *Caenorhabditis elegans*. Neuropharmacology 53:157–168
17. Li H, Feng Y, Chen Z, Jiang X, Zhou Z, Yuan J, Li F, Zhang Y, Huang X, Fan S, Wu X, Huang C (2021) Pepper component 7-ethoxy-4-methylcoumarin, a novel dopamine D2 receptor agonist, ameliorates experimental Parkinson's disease in mice and *Caenorhabditis elegans*. Pharmacol Res 163:105220
18. Tharmalingam S, Burns AR, Roy PJ, Hampson DR (2012) Orthosteric and allosteric drug binding sites in the *Caenorhabditis elegans mgl-2* metabotropic glutamate receptor. Neuropharmacology 63:667–674
19. Saitoh Y, Katane M, Miyamoto T, Sekine M, Sakai-Kato K, Homma H (2020) D-serine and D-alanine regulate adaptive foraging behavior in *Caenorhabditis elegans* via the NMDA receptor. J Neurosci 40:7531–7544
20. Ladagu A, Olopade F, Chazot P, Elufioye T, Luong T, Fuller M, Halprin E, Mckay J, Ates-Alagoz Z, Gilbert T, Adejare A, Olopade J (2024) ZA-II-05, a novel NMDA-receptor antagonist reverses vanadium-induced neurotoxicity in *Caenorhabditis elegans* (*C. elegans*). BMC Neurosci 25:56
21. Pereira-Sousa J, Ferreira-Lomba B, Bellver-Sanchis A, Vilasboas-Campos D, Fernandes JH, Costa MD, Varney MA, Newman-Tancredi A, Maciel P, Teixeira-Castro A (2021) Identification of the 5-HT1A serotonin receptor as a novel therapeutic target in a *C. elegans* model of Machado-Joseph disease. Neurobiol Dis 152:105278
22. Naß J, Efferth T (2021) Withanone ameliorates stress symptoms in *Caenorhabditis elegans* by acting through serotonin receptors. Pharmacopsychiatry 54:215–223
23. Smith KA, Rex EB, Komuniecki RW (2007) Are *C. elegans* receptors useful targets for drug discovery: pharmacological comparison of tyramine receptors with high identity from *Caenorhabditis elegans* (TYRA-2) and *Brugia malayi* (Bm4). Mol Biochem Parasitol 154: 52–61
24. Liu M, Xiong Y, Shan S, Zhu Y, Zeng D, Shi Y, Zhang Y, Lu W (2020) Eleutheroside E enhances the long-term memory of radiation damaged *C. elegans* through G-protein-coupled receptor and neuropeptide signaling pathways. J Nat Prod 83:3315–3323
25. Teng MS, Shadbolt P, Fraser AG, Jansen G, McCafferty J (2008) Control of feeding behavior in *C. elegans* by human G protein-coupled receptors permits screening for agonist-expressing bacteria. Proc Natl Acad Sci USA 105:14826–14831
26. Teng MS, Dekkers MP, Ng BL, Rademakers S, Jansen G, Fraser AG, McCafferty J (2006) Expression of mammalian GPCRs in *C. elegans* generates novel behavioural responses to human ligands. BMC Biol 4:22
27. Ribeiro RC, Kushner PJ, Baxter JD (1995) The nuclear hormone receptor gene superfamily. Annu Rev Med 46:443–453
28. Hu X, Lazar MA (2000) Transcriptional repression by nuclear hormone receptors. Trends Endocrinol Metab 11:6–10
29. Moras D, Gronemeyer H (1998) The nuclear receptor ligand-binding domain: structure and function. Curr Opin Cell Biol 10:384–391

30. Weatherman RV, Fletterick RJ, Scanlan TS (1999) Nuclear-receptor ligands and ligand-binding domains. Annu Rev Biochem 68:559–581
31. Kumar R, Thompson EB (1999) The structure of the nuclear hormone receptors. Steroids 64: 310–319
32. Giguère V (1990) Structure and function of the nuclear receptor superfamily for steroid, thyroid hormone and retinoic acid. Genet Eng 12:183–200
33. Parker MG (1990) Structure and function of nuclear hormone receptors. Semin Cancer Biol 1: 81–87
34. Weikum ER, Liu X, Ortlund EA (2018) The nuclear receptor superfamily: a structural perspective. Protein Sci 27:1876–1892
35. Frank F, Ortlund EA, Liu X (2021) Structural insights into glucocorticoid receptor function. Biochem Soc Trans 49:2333–2343
36. Yang Y, Guo L, Chen L, Gong B, Jia D, Sun Q (2023) Nuclear transport proteins: structure, function, and disease relevance. Signal Transduct Target Ther 8:425
37. Hager GL, Lim CS, Elbi C, Baumann CT (2000) Trafficking of nuclear receptors in living cells. J Steroid Biochem Mol Biol 74:249–254
38. Pratt WB, Galigniana MD, Morishima Y, Murphy PJ (2004) Role of molecular chaperones in steroid receptor action. Essays Biochem 40:41–58
39. Heinlein CA, Chang C (2001) Role of chaperones in nuclear translocation and transactivation of steroid receptors. Endocrine 14:143–149
40. Tambones I, le Maire A (2024) Structural insights into thyroid hormone receptors. Endocrinology 166:bqae154
41. Glass CK, Rose DW, Rosenfeld MG (1997) Nuclear receptor coactivators. Curr Opin Cell Biol 9:222–232
42. Savova MS, Todorova MN, Apostolov AG, Yahubyan GT, Georgiev MI (2022) Betulinic acid counteracts the lipid accumulation in *Caenorhabditis elegans* by modulation of *nhr-49* expression. Biomed Pharmacother 156:113862
43. Souza FRM, Silva GMM, Cadavid COM, Lisboa LDS, Silva MMCL, Paiva WS, Ferreira MJP, de Paula Oliveira R, Rocha HAO (2022) Antioxidant *Baccharis trimera* leaf extract suppresses lipid accumulation in *C. elegans* dependent on transcription factor NHR-49. Antioxidants 11:1913
44. Brandstädt S, Schmeisser K, Zarse K, Ristow M (2013) Lipid-lowering fibrates extend *C. elegans* lifespan in a NHR-49/PPARalpha-dependent manner. Aging 5:270–275
45. Qi W, Gutierrez GE, Gao X, Dixon H, McDonough JA, Marini AM, Fisher AL (2017) The omega-3 fatty acid alpha-linolenic acid extends *Caenorhabditis elegans* lifespan via NHR-49/PPARalpha and oxidation to oxylipins. Aging Cell 16:1125–1135
46. Leiteritz A, Baumanns S, Wenzel U (2020) Amyloid-beta (Abeta(1-42))-induced paralysis in *Caenorhabditis elegans* is reduced through NHR-49/PPARalpha. Neurosci Lett 730:135042
47. Hummell NA, Revtovich AV, Kirienko NV (2021) Novel immune modulators enhance *Caenorhabditis elegans* resistance to multiple pathogens. mSphere 6:e00950–e00920
48. Motola DL, Cummins CL, Rottiers V, Sharma KK, Li T, Li Y, Suino-Powell K, Xu HE, Auchus RJ, Antebi A, Mangelsdorf DJ (2006) Identification of ligands for DAF-12 that govern dauer formation and reproduction in *C. elegans*. Cell 124:1209–1223
49. Gerisch B, Rottiers V, Li D, Motola DL, Cummins CL, Lehrach H, Mangelsdorf DJ, Antebi A (2007) A bile acid-like steroid modulates *Caenorhabditis elegans* lifespan through nuclear receptor signaling. Proc Natl Acad Sci USA 104:5014–5019
50. Rodriguez CR, del Fueyo MC, Santillán VJ, Dansey MV, Veleiro AS, Castro OA, Burton G (2019) Synthesis and biological activity of fluorinated analogues of the DAF-12 receptor antagonist 24-hydroxy-4-cholen-3-one. Steroids 151:108469
51. Chamoli M, Rane A, Foulger A, Chinta SJ, Shahmirzadi AA, Kumsta C, Nambiar DK, Hall D, Holcom A, Angeli S, Schmidt M, Pitteri S, Hansen M, Lithgow GJ, Andersen JK (2023) A drug-like molecule engages nuclear hormone receptor DAF-12/FXR to regulate mitophagy and extend lifespan. Nat Aging 3:1529–1543

52. Huggins B, Farris M (2023) Vitamin D(3) promotes longevity in *Caenorhabditis elegans*. Geroscience 45:345–358
53. Shi H, Gao X, Yu J, Zhang L, Fan B, Liu Y, Wang X, Fan S, Huang C (2024) Isotschimgine promotes lifespan, healthspan and neuroprotection of *Caenorhabditis elegans* via the activation of nuclear hormone receptors. Biogerontology 26:2
54. Catterall WA (1995) Structure and function of voltage-gated ion channels. Annu Rev Biochem 64:493–531
55. Anderson PA, Greenberg RM (2001) Phylogeny of ion channels: clues to structure and function. Comp Biochem Physiol B Biochem Mol Biol 129:17–28
56. Jentsch TJ (1993) Chloride channels. Curr Opin Neurobiol 3:316–321
57. Catterall WA (1988) Structure and function of voltage-sensitive ion channels. Science 242:50–61
58. Green WN (1999) Ion channel assembly: creating structures that function. J Gen Physiol 113: 163–170
59. Strange K (2003) From genes to integrative physiology: ion channel and transporter biology in *Caenorhabditis elegans*. Physiol Rev 83:377–3415
60. Kahn-Kirby AH, Bargmann CI (2006) TRP channels in *C. elegans*. Annu Rev Physiol 68: 719–736
61. Tavernarakis N, Driscoll M (2001) Mechanotransduction in *Caenorhabditis elegans*: the role of DEG/ENaC ion channels. Cell Biochem Biophys 35:1–18
62. Bargmann CI (1998) Neurobiology of the *Caenorhabditis elegans* genome. Science 282: 2028–2033
63. Hernando G, Turani O, Rodriguez Araujo N, Bouzat C (2023) The diverse family of Cys-loop receptors in *Caenorhabditis elegans*: insights from electrophysiological studies. Biophys Rev 15:733–750
64. Salkoff L, Wei AD, Baban B, Butler A, Fawcett G, Ferreira G, Santi CM (2005) Potassium channels in *C. elegans*. WormBook, Pasadena. https://doi.org/10.1895/wormbook.1.42.1
65. Kwok TCY, Ricker N, Fraser R, Chan AW, Burns A, Stanley EF, McCourt P, Cutler SR, Roy PJ (2006) A small-molecule screen in *C. elegans* yields a new calcium channel antagonist. Nature 441:91–95
66. Zheng Z, Wu K, Ruan Q, Li D, Liu W, Wang M, Li Y, Xia J, Yang D, Guo J (2022) Suppression of selective voltage-gated calcium channels alleviates neuronal degeneration and dysfunction through glutathione S-transferase-mediated oxidative stress resistance in a *Caenorhabditis elegans* model of Alzheimer's disease. Oxidative Med Cell Longev 2022: 8287633
67. Patten SA, Aggad D, Martinez J, Tremblay E, Petrillo J, Armstrong GA, La Fontaine A, Maios C, Liao M, Ciura S, Wen XY, Rafuse V, Ichida J, Zinman L, Julien JP, Kabashi E, Robitaille R, Korngut L, Parker JA, Drapeau P (2017) Neuroleptics as therapeutic compounds stabilizing neuromuscular transmission in amyotrophic lateral sclerosis. JCI Insight 2:e97152
68. Ikenaka K, Tsukada Y, Giles AC, Arai T, Nakadera Y, Nakano S, Kawai K, Mochizuki H, Katsuno M, Sobue G, Mori I (2019) A behavior-based drug screening system using a *Caenorhabditis elegans* model of motor neuron disease. Sci Rep 9:10104
69. Jiang Q, Li K, Lu W, Li S, Chen X, Liu X, Yuan J, Ding Q, Lan F, Cai S (2018) Identification of small-molecule ion channel modulators in *C. elegans* channelopathy models. Nat Commun 9:3941
70. Niu L, Li Y, Zong P, Liu P, Shui Y, Chen B, Wang Z (2020) Melatonin promotes sleep by activating the BK channel in *C. elegans*. Proc Natl Acad Sci USA 117:25128–25137
71. Kasap M, Aamodt EJ, Sagrera CE, Dwyer DS (2020) Novel pharmacological modulation of dystonic phenotypes caused by a gain-of-function mutation in the Na^+ leak-current channel. Behav Pharmacol 31:465–476
72. Kulke D, von Samson-Himmelstjerna G, Miltsch SM, Wolstenholme AJ, Jex AR, Gasser RB, Ballesteros C, Geary TG, Keiser J, Townson S, Harder A, Krücken J (2014) Characterization

of the Ca^{2+}-gated and voltage-dependent K^{+}-channel Slo-1 of nematodes and its interaction with emodepside. PLoS Negl Trop Dis 8:e3401
73. Rodriguez Araujo N, Hernando G, Corradi J, Bouzat C (2022) The nematode serotonin-gated chloride channel MOD-1: a novel target for anthelmintic therapy. J Biol Chem 298:102356
74. Horoszok L, Raymond V, Sattelle DB, Wolstenholme AJ (2001) GLC-3: a novel fipronil and BIDN-sensitive, but picrotoxinin-insensitive, L-glutamate-gated chloride channel subunit from *Caenorhabditis elegans*. Br J Pharmacol 132:1247–1254
75. Cully DF, Paress PS (1991) Solubilization and characterization of a high affinity ivermectin binding site from *Caenorhabditis elegans*. Mol Pharmacol 40:326–332
76. Logan-Garbisch T, Fryer E, Seyahi LS, Rogel-Hernandez L, Rhee SY, Goodman MB (2025) Satiety, TAX-4, and OSM-9 tune the attraction of *Caenorhabditis elegans* nematodes to microbial fermentation products. G3 15:jkaf245
77. Frøkjaer-Jensen C, Ailion M, Lockery SR (2008) Ammonium-acetate is sensed by gustatory and olfactory neurons in *Caenorhabditis elegans*. PLoS One 3:e2467
78. Alfonso SA, Arango Sumano D, Bhatt DA, Cullen AB, Hajian CM, Huang W, Jaeger EL, Li E, Maske AK, Offenberg EG, Ta V, Whiting WW, Adebogun GT, Bachmann AE, Callan AA, Khan U, Lewis AR, Pollock AC, Ramirez D, Bradon N, Fiocca K, Cote LE, Sallee MD, McKinney J, O'Connell LA (2023) Argentine ant extract induces an *osm-9* dependent chemotaxis response in *C. elegans*. MicroPubl Biol 2023. https://doi.org/10.17912/micropub.biology.000745
79. Liu F, Li T, Gong H, Tian F, Bai Y, Wang H, Yang C, Li Y, Guo F, Liu S, Chen Q (2024) Structural insights into the molecular effects of the anthelmintics monepantel and betaine on the *Caenorhabditis elegans* acetylcholine receptor ACR-23. EMBO J 43:3787–3806
80. Turani O, Hernando G, Corradi J, Bouzat C (2018) Activation of *Caenorhabditis elegans* levamisole-sensitive and mammalian nicotinic receptors by the antiparasitic bephenium. Mol Pharmacol 94:1270–1279
81. Syrovatkina V, Alegre KO, Dey R, Huang XY (2016) Regulation, signaling, and physiological functions of G-proteins. J Mol Biol 428:3850–3868
82. Calebiro D, Koszegi Z, Lanoiselée Y, Miljus T, O'Brien S (2021) G protein-coupled receptor-G protein interactions: a single-molecule perspective. Physiol Rev 101:857–906
83. Fitzgerald K, Tertyshnikova S, Moore L, Bjerke L, Burley B, Cao J, Carroll P, Choy R, Doberstein S, Dubaquie Y, Franke Y, Kopczynski J, Korswagen H, Krystek SR, Lodge NJ, Plasterk R, Starrett J, Stouch T, Thalody G, Wayne H, van der Linden A, Zhang Y, Walker SG, Cockett M, Wardwell-Swanson J, Ross-Macdonald P, Kindt RM (2006) Chemical genetics reveals an RGS/G-protein role in the action of a compound. PLoS Genet 2:e57
84. Chen YC, Seyedsayamdost MR, Ringstad N (2020) A microbial metabolite synergizes with endogenous serotonin to trigger *C. elegans* reproductive behavior. Proc Natl Acad Sci USA 117:30589–30598
85. Krebs EG, Beavo JA (1979) Phosphorylation-dephosphorylation of enzymes. Annu Rev Biochem 48:923–959
86. Graves JD, Krebs EG (1999) Protein phosphorylation and signal transduction. Pharmacol Ther 82:111–121
87. Hanks SK, Hunter T (1995) Protein kinases 6. The eukaryotic protein kinase superfamily: kinase (catalytic) domain structure and classification. FASEB J 9:576–596
88. Hunter T (1991) Protein kinase classification. Methods Enzymol 200:3–37
89. Samara C, Rohde CB, Gilleland CL, Norton S, Haggarty SJ, Yanik MF (2010) Large-scale in vivo femtosecond laser neurosurgery screen reveals small-molecule enhancer of regeneration. Proc Natl Acad Sci USA 107:18342–18347
90. Kim AT, Park Y (2025) Esculetin inhibits fat accumulation through insulin/insulin-like growth factor- and AMP-activated protein kinase-dependent pathways in *Caenorhabditis elegans*. Nutrients 17:1565

91. Tao M, Li R, Zhang Z, Wu T, Xu T, Zogona D, Huang Y, Pan S, Xu X (2022) Vitexin and isovitexin act through inhibition of insulin receptor to promote longevity and fitness in *Caenorhabditis elegans*. Mol Nutr Food Res 66:e2100845
92. Shu C, Sun L, Zhang W (2016) Thymol has antifungal activity against Candida albicans during infection and maintains the innate immune response required for function of the p38 MAPK signaling pathway in *Caenorhabditis elegans*. Immunol Res 64:1013–1024
93. Kautu BB, Carrasquilla A, Hicks ML, Caldwell KA, Caldwell GA (2013) Valproic acid ameliorates *C. elegans* dopaminergic neurodegeneration with implications for ERK-MAPK signaling. Neurosci Lett 541:116–119
94. Ellwood RA, Hewitt JE, Torregrossa R, Philp AM, Hardee JP, Hughes S, van de Klashorst D, Gharahdaghi N, Anupom T, Slade L, Deane CS, Cooke M, Etheridge T, Piasecki M, Antebi A, Lynch GS, Philp A, Vanapalli SA, Whiteman M, Szewczyk NJ (2021) Mitochondrial hydrogen sulfide supplementation improves health in the *C. Elegans* Duchenne muscular dystrophy model. Proc Natl Acad Sci USA 118:e2018342118
95. Seino S, Shibasaki T (2005) PKA-dependent and PKA-independent pathways for cAMP-regulated exocytosis. Physiol Rev 85:1303–1342
96. Feliciello A, Gottesman ME, Avvedimento EV (2005) cAMP-PKA signaling to the mitochondria: protein scaffolds, mRNA and phosphatases. Cell Signal 17:279–287
97. Beebe SJ (1994) The cAMP-dependent protein kinases and cAMP signal transduction. Semin Cancer Biol 5:285–294
98. Newton AC (1997) Regulation of protein kinase C. Curr Opin Cell Biol 9:161–167
99. Huang KP (1989) The mechanism of protein kinase C activation. Trends Neurosci 12:425–432
100. Dhakal P, Chaudhry SI, Signorelli R, Collins KM (2022) Serotonin signals through postsynaptic Galphaq, Trio RhoGEF, and diacylglycerol to promote *Caenorhabditis elegans* egg-laying circuit activity and behavior. Genetics 221:iyac084
101. Goldberg ND, O'Dea RF, Haddox MK (1973) Cyclic GMP. Adv Cyclic Nucleotide Res 3: 155–223
102. Waldman SA, Murad F (1987) Cyclic GMP synthesis and function. Pharmacol Rev 39:163–196
103. Waldman SA, Murad F (1988) Biochemical mechanisms underlying vascular smooth muscle relaxation: the guanylate cyclase-cyclic GMP system. J Cardiovasc Pharmacol Suppl 5:S115–S118
104. Stryer L (1986) Cyclic GMP cascade of vision. Annu Rev Neurosci 9:87–119
105. Park J, Meisel JD, Kim DH (2020) Immediate activation of chemosensory neuron gene expression by bacterial metabolites is selectively induced by distinct cyclic GMP-dependent pathways in *Caenorhabditis elegans*. PLoS Genet 16:e1008505
106. Bygrave FL, Benedetti A (1996) What is the concentration of calcium ions in the endoplasmic reticulum? Cell Calcium 19:547–551
107. Perez-Zoghbi JF, Karner C, Ito S, Shepherd M, Alrashdan Y, Sanderson MJ (2009) Ion channel regulation of intracellular calcium and airway smooth muscle function. Pulm Pharmacol Ther 22:388–397
108. Augustine GJ (2001) How does calcium trigger neurotransmitter release? Curr Opin Neurobiol 11:320–326
109. Roussos A, Kitopoulou K, Borbolis F, Ploumi C, Gianniou DD, Li Z, He H, Tsakiri E, Borland H, Kostakis IK, Samiotaki M, Trougakos IP, Bohr VA, Palikaras K (2025) Urolithin alpha modulates inter-organellar communication via calcium-dependent mitophagy to promote healthy ageing. Autophagy 21:3097–3122
110. Arraiano CM (2021) Regulatory noncoding RNAs: functions and applications in health and disease. FEBS J 288:6308–6309
111. Chen LL, Kim VN (2024) Small and long non-coding RNAs: past, present, and future. Cell 187:6451–6485
112. Yang L, Wilusz JE, Chen LL (2022) Biogenesis and regulatory roles of circular RNAs. Annu Rev Cell Dev Biol 38:263–289

113. Zhang S, Wang H, Wang J, Jin W, Yan X, Chen X, Wang D, Zhao D, Wang Y, Cong D, Sun L (2022) Ginsenoside Rf inhibits human tau proteotoxicity and causes specific LncRNA, miRNA and mRNA expression changes in *Caenorhabditis elegans* model of tauopathy. Eur J Pharmacol 922:174887
114. Xu Q, Zhao Z, Liang P, Wang S, Li F, Jin S, Zhang J (2023) Identification of novel nematode succinate dehydrogenase inhibitors: virtual screening based on ligand-pocket interactions. Chem Biol Drug Des 101:9–23
115. Rohn I, Raschke S, Aschner M, Tuck S, Kuehnelt D, Kipp A, Schwerdtle T, Bornhorst J (2019) Treatment of *Caenorhabditis elegans* with small selenium species enhances antioxidant defense systems. Mol Nutr Food Res 63:e1801304
116. Zhou L, Liu J, Bu LL, Liao DF, Cheng SW, Zheng XL (2021) Curcumin acetylsalicylate extends the lifespan of *Caenorhabditis elegans*. Molecules 26:6609
117. Estrada-Valencia R, Túnez I, Tinkov AA, Aschner M, López-Goerne T, Pedraza-Chaverrí J, Santamaría A (2025) The inhibition of fatty acid amide hydrolase-4 affords neuroprotection in a toxic model induced by 6-hydroxydopamine in *Caenorhabditis elegans* nematodes. Mol Neurobiol 62:12984–12999
118. Wellenberg A, Weides L, Kurzke J, Hennecke T, Bornhorst J, Crone B, Karst U, Brinkmann V, Fritz G, Honnen S (2021) Use of *C. elegans* as a 3R-compliant in vivo model for the chemoprevention of cisplatin-induced neurotoxicity. Exp Neurol 341:113705
119. Yong Y, Yan L, Wei J, Feng C, Yu L, Wu J, Guo M, Fan D, Yu C, Qin D, Zhou X, Wu A (2024) A novel ferroptosis inhibitor, Thonningianin A, improves Alzheimer's disease by activating GPX4. Theranostics 14:6161–6184
120. Yu XT, Shi L, Huang Q, Li B, Li YQ, Ou LL, Luo HR, Wu GS (2025) Hyperoside extends lifespan in *Caenorhabditis elegans* through SEK-1/PMK-1/SKN-1 pathway. Biogerontology 26:180
121. Sukanya S, Bellver-Sanchis A, Choudhary BS, Kumar S, Pérez B, Rodríguez ALM, Brea J, Escolano C, Griñán-Ferré C, Malik R (2025) Exploring 2-thioxo-1,2,3,4-tetrahydropyrimidines as dual acting GSK-3beta/Abeta aggregation inhibitors: implications for Alzheimer's disease treatment. Drug Dev Res 86:e70112
122. Lan M, Gao X, Duan X, Li H, Yu H, Li J, Zhao Y, Hao X, Zhao Y, Ding X, Wu G (2022) Nematicidal activity of tirotundin and parthenolide isolated from Tithonia diversifolia and *Chrysanthemum parthenium*. J Environ Sci Health B 57:54–61
123. de Oliveira Pereira FS, Santos AG, Neto JSS, Silva GMM, Pinton S, Zeni GR, Nogueira CW, Ávila DS, Quines CB (2024) (p-ClPhSe)(2) modulation on carbohydrate and lipid metabolism requires the insulin-like signaling in *Caenorhabditis elegans*. Biochem Biophys Res Commun 696:149514
124. Mihaylova LV, Savova MS, Todorova MN, Tonova V, Binev BK, Georgiev MI (2026) Cycloastragenol improves fatty acid metabolism through NHR-49/FAT-7 suppression and potent AAK-2 activation in *Caenorhabditis elegans* obesity model. Int J Mol Sci 27:772
125. Zhang Y, Shu C, Li Z, Qu M, Wang C, Nie S, Xu Q, Gu S, Fan M, Shi B, Fu L, Zhang F (2025) The protective effects and mechanisms of rosmarinic acid against *Pseudomonas aeruginosa* infection in *Caenorhabditis elegans*. Front Pharmacol 16:1701885
126. Dai W, Zhai Y, Yang F, Chen W, Liu C, Tian Y, Huang F, Cai M, Zheng L, Cheng W, Chen W, Zhang J (2025) Different concentrations of 2-Undecanone triggers repellent and nematicidal responses in *Caenorhabditis elegans*. Sci Rep 15:14186
127. Sedlacek J, Smahelova Z, Adamek M, Subova D, Svobodova L, Kadlecova A, Majer P, Machara A, Grantz Saskova K (2025) Small-molecule activators of NRF1 transcriptional activity prevent protein aggregation. Biomed Pharmacother 183:117864
128. Liu J, Zhang Y, Zhang T, Zhao X, Wang M, Yang H, Lai CJ, Xie J (2026) Metabolomic insights into the neuroprotective actions of Ziziphi spinosae semen and jujuboside B against Aβ-induced toxicity. J Ethnopharmacol 360:121109

Pharmacological Effects on Gut Pathogens and Microbes

10

Abstract

Caenorhabditis elegans gut is a specific target site for compounds and drugs. In the gut, compounds and drugs will meet pathogens and microbes and exert the effects on them. The pharmacological effects of compounds and drugs on bacterial and fungal pathogens were first introduced. Moreover, the potential pharmacological effects of compounds and drugs on gut microbes were discussed.

Keywords

Gut pathogen · Gut microbe · Compound · *Caenorhabditis elegans*

10.1 Introduction

In Chap. 6, we have introduced and discussed pharmacological bioavailability, translocation, and metabolism. Moreover, we have introduced and discussed target organ pharmacology, target organelle pharmacology, and target molecular pharmacology in Chaps. 7, 8, and 9, respectively. Besides these aspects, the pharmacological compounds and drugs can also target to gut bacterial and fungal pathogens, and meanwhile modulate the innate immune response of the host of *C. elegans*. That is, the main aim of this aspect of efforts is to develop more effective and targeted therapies that consider the complex reality of the pathogen–host interaction. In this chapter, we first introduced and discussed the pharmacological effects of compounds and drugs on gut bacterial and fungal pathogens. In addition, the pharmacological effects of compounds and drugs on or from gut microbes were also introduced and discussed.

D. Wang, *Pharmacology in Caenorhabditis elegans*,
https://doi.org/10.1007/978-981-92-5166-7_10

10.2 Pharmacological Effects on Gut Bacterial Pathogens

For the bacterial pathogen–host interaction, the behaviors of at least following bacterial pathogens have been examined in *C. elegans*:

1. *Pseudomonas aeruginosa*, the gram-negative bacterial pathogen [1, 2].
2. *Salmonella enterica*, the gram-negative bacterial pathogen [3].
3. *Staphylococcus aureus*, the gram-positive bacterial pathogen [4].
4. *Enterococcus faecalis*, the gram-positive bacterial pathogen [5].
5. *Klebsiella aerogenes*, the gram-negative bacterial pathogen [6].
6. *Vibrio cholerae*, the gram-negative bacterial pathogen [7].
7. *Vibrio alginolyticus*, the gram-negative bacterial pathogen [8].
8. *Serratia marcescens*, the gram-negative bacterial pathogen [9].

10.2.1 Pharmacological Effects on Bacterial Pathogens

In the clinical, it is still an urgent need to combat the increasing antibiotic resistance of *P. aeruginosa*. In *C. elegans*, the *P. aeruginosa* is a classic model for bacterial pathogens to examine the bacterial pathogen–host interaction. The efforts on pharmacological studies focus on not only the identification of novel compounds and drugs with more promising pharmacological effects, but also the finding new ways to enhance the existing drugs. With *P. aeruginosa* as the example of bacterial pathogens, we next introduced and discussed the pharmacological effects of compounds and drugs on bacterial pathogens. In the gut, the possible effects of compounds and drugs on bacterial pathogens may mainly include the following aspects:

1. Inhibition in biofilm formation: Based on assays of crystal violate staining and OD595 absorbance, biofilm formation of *P. aeruginosa* PA14 was significantly inhibited by 1.25–10 mg/L paeoniflorin, which was also confirmed by observation under light microscopy (Fig. 10.1) [10]. Meanwhile, expressions of virulence genes (*pelA*, *pelB*, *phzA*, *lasB*, *lasR*, *rhlA*, and *rhlC*) required for formation of *P. aeruginosa* biofilm and/or quorum sensing were decreased by 10 mg/L paeoniflorin (Fig. 10.2) [10]. In addition, levels of the other virulence factors required for biofilm formation, such as pyocyanin, elastase, and rhamnolipid, were also inhibited by 10 mg/L paeoniflorin (Fig. 10.2) [10].
2. Decrease in swimming motility: The ability of swimming motility required for formation of functional polar flagellum for *P. aeruginosa* could be suppressed by paeoniflorin treatment. The swimming diameter, the diameter of swarming motility, and the twitching diameter of *P. aeruginosa* PA14 were reduced by 2.5–10 mg/L paeoniflorin (Fig. 10.3) [10].
3. Inhibition in quorum sensing: Tirazone could disrupt the cell-to-cell communication by competitively binding to quorum sensing regulatory proteins (LasR, RhlR, and PqsR) [11]. Among FDA-approved small molecules, vorinostat and

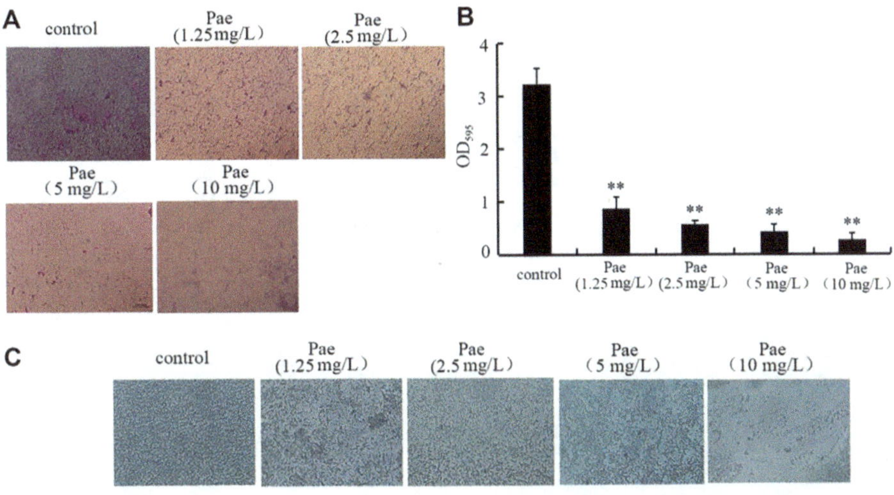

Fig. 10.1 Effect of paeoniflorin treatment on biofilm formation of *P. aeruginosa* PA14 [10]. (**a**) Crystal violate staining images. (**b**) Effect of paeoniflorin treatment on amount of biofilm formation based on OD595 absorbance analysis. (**c**) Effect of paeoniflorin treatment on biofilm formation visualized under a light microscope. Pae paeoniflorin. $^{**}P < 0.01$ vs. control

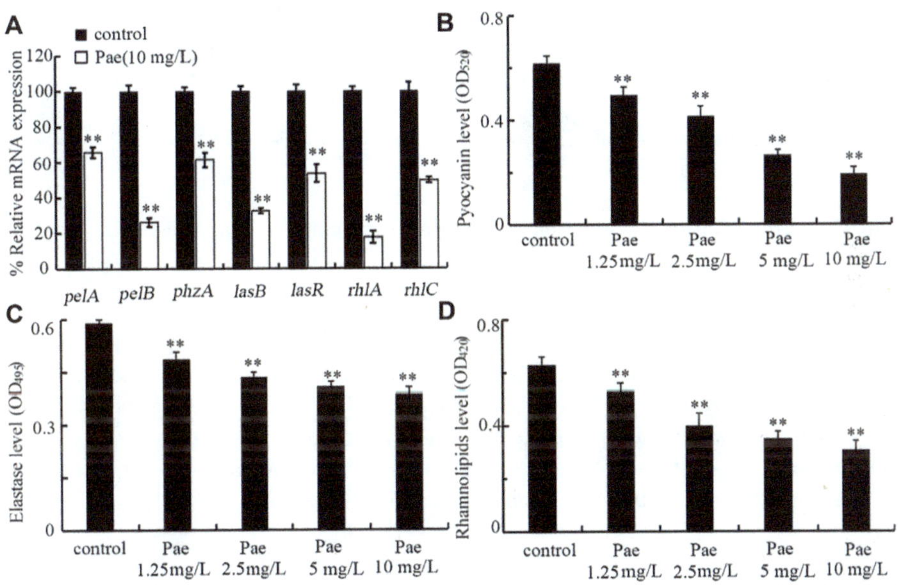

Fig. 10.2 Effect of paeoniflorin treatment on expressions of virulence genes and virulence factors of pyocyanin, elastase, and rhamnolipid [10]. (**a**) Effect of 10 mg/L paeoniflorin treatment on expressions of virulence genes required for *P. aeruginosa* biofilm formation. (**b–d**) Effect of paeoniflorin treatment on levels of pyocyanin, elastase, and rhamnolipid. Pae paeoniflorin. $^{**}P < 0.01$ vs. control

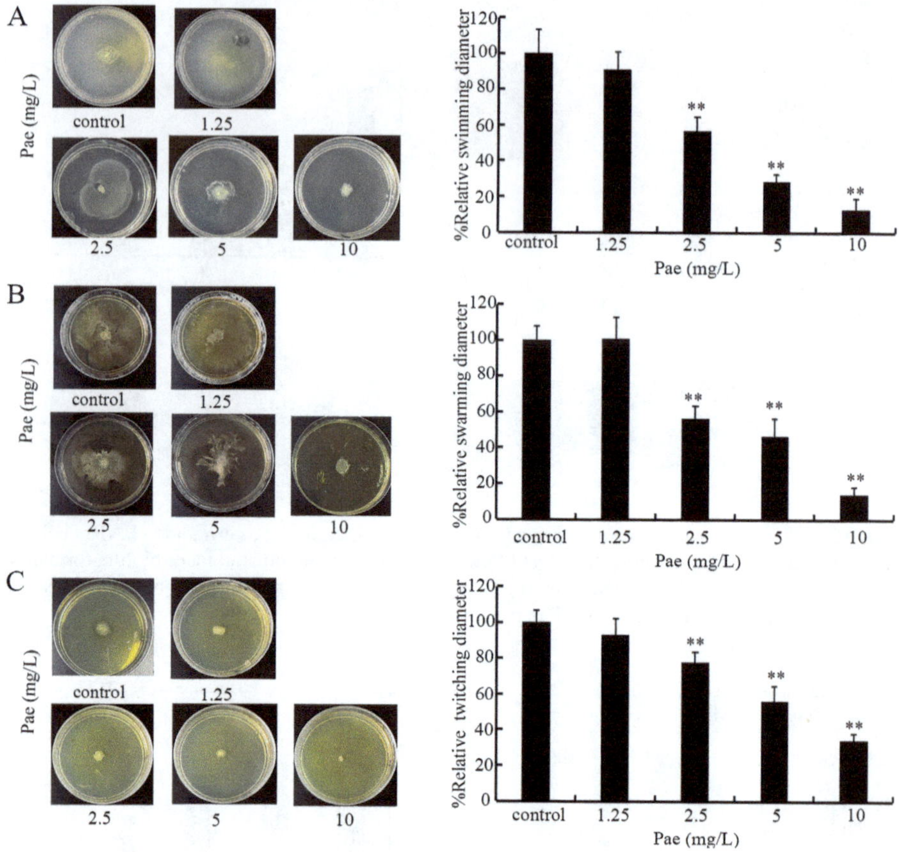

Fig. 10.3 Effect of paeoniflorin treatment on swimming motility (**a**), swarming motility (**b**), and twitching motility (**c**) of *P. aeruginosa* PA14. Pae paeoniflorin [10]. $^{**}P < 0.01$ vs control

apomorphine were identified as inhibitors for quorum sensing enzyme PqsE by binding to the active site of PqsE or disrupting the PqsE–RhlR interaction [12].

4. Overcoming resistance mechanisms: Certain β-lactam/β-lactamase inhibitors, such as ceftazidime-avibactam and ceftolozane-tazobactam, can be used to treat the multidrug-resistant *P. aeruginosa* bacteremia by overcoming specific resistance mechanisms, such as β-lactamases and efflux pumps [13, 14].

10.2.2 Pharmacological Effects on Innate Immunity Against Bacterial Pathogens

In *C. elegans*, the innate immunity and the modulating effects of compounds and drugs on innate immune response induced by bacterial pathogens can be assessed with the aid of following endpoints [15–18]:

1. Expression of antimicrobial genes: The antimicrobial genes (e.g., C-type lectins and lysozymes) act as the immune effector genes. The expression of antimicrobial genes can be assessed by transcriptional expressions or transcriptional microarray assay. In addition, some corresponding fluorescent reporters (e.g., GFP) can also be employed to assess this.
2. Survival and health span assays: The survival (lifespan) assay was used to measure the rate of death after or during the pathogen infection, which is considered as the gold standard for determining if a gene or pharmacological treatment affects susceptibility or resistance to pathogen infection. The health span analysis was used to quantify the decline in physiological state before the death, and to provide a more sensitive measurement of pathogen infection impact.
3. Pathogen loading and possible physical interaction: The live bacterial pathogen loading can be assessed by colony-forming units (CFUs) after lysing infected worms. For the pathogen colonization and the physical interaction (such as tissue invasion and biofilm formation), they can be further directly observed using fluorescent *P. aeruginosa* strains after the feeding.

Using these endpoints, some compounds and drugs were found to have modulatory effects on innate immunity in bacterial pathogen-infected nematodes [19–22]. With the paeoniflorin as the example of pharmacological compounds, *P. aeruginosa* infection induced immunosuppression reflected by decreased expressions of antimicrobial genes (*lys-1* and *lys-8*) could be inhibited by paeoniflorin treatment [23]. Treatment with paeoniflorin could extend the lifespan in *P. aeruginosa*-infected nematodes [10] and increase survival rate and locomotion behavior during immunosuppression stage in *P. aeruginosa*-infected nematodes [23]. The colonization and accumulation in intestinal lumen were also suppressed by paeoniflorin treatment (Fig. 10.4) [10].

Treatment with compounds and drugs modulate the innate immunity to bacterial pathogen infection at least through the following several molecular mechanisms:

1. Direct modulation of host immune pathways: Among molecular signals involved in control of innate immunity, treatment with paeoniflorin elevated expressions of *pmk-1*, *bar-1*, and *egl-1* in *P. aeruginosa*-infected nematodes (Fig. 10.5) [23]. Moreover, the beneficial effects of paeoniflorin treatment against the *P. aeruginosa* infection induced immunosuppression, colonization and accumulation in intestinal lumen, and decrease in survival were suppressed by RNAi of *pmk-1*, *bar-1*, and *egl-1* (Fig. 10.6) [23], suggesting that paeoniflorin modulated the innate immunity by affecting these molecular signals. The compound LK56 could confer the resistance of *C. elegans* to *Enterococcus faecalis*, which required the involvement of MDT-15/MED15 and NHR-49/HNF4 [22].
2. Activation of microbiota–brain axis: The beneficial effect of antibiotic Q203 in enhancing host survival after *P. aeruginosa* infection was dependent on the response of nervous system, as evidenced by modulation on host UNC-13 governing the head neurotransmitter release [24].

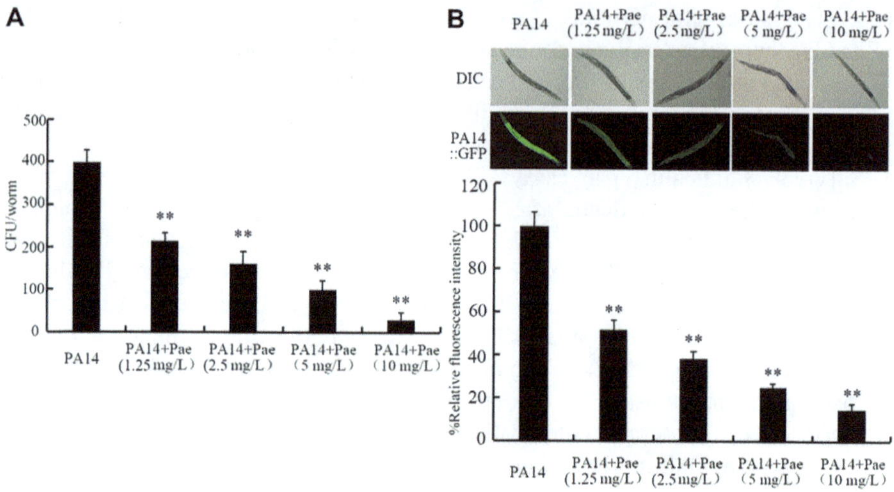

Fig. 10.4 Effect of paeoniflorin treatment on CFU of *P. aeruginosa* PA14 (**a**) and accumulatio of *P. aeruginosa* PA14:GFP (**b**) in nematodes after the infection [10]. Pae paeoniflorin. $^{**}P < 0.01$ vs. PA14

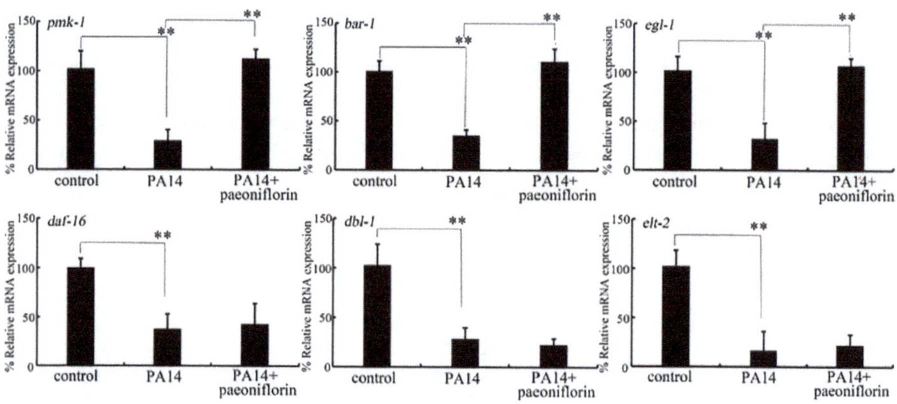

Fig. 10.5 Effect of 100 mg/L paeoniflorin treatment on expressions of *daf-16*, *bar-1*, *elt-2*, *dbl-1*, *pmk-1*, and *egl-1* in nematodes at the immunosuppression stage (day-5) after *P. aeruginosa* infection [23]. $^{**}P < 0.01$

3. Disarming the bacterial pathogens: Unlike the antibiotics that can kill the bacterial pathogens, some compounds can suppress the bacterial virulence factors to enhance the host innate immunity. Besides the beneficial effect on this aspect introduced above, treatment with Traditional Chinese Medicine Xuebijing could also inhibited *P. aeruginosa* biofilm and quorum sensing system by reducing the expression of key virulence factor genes [25].

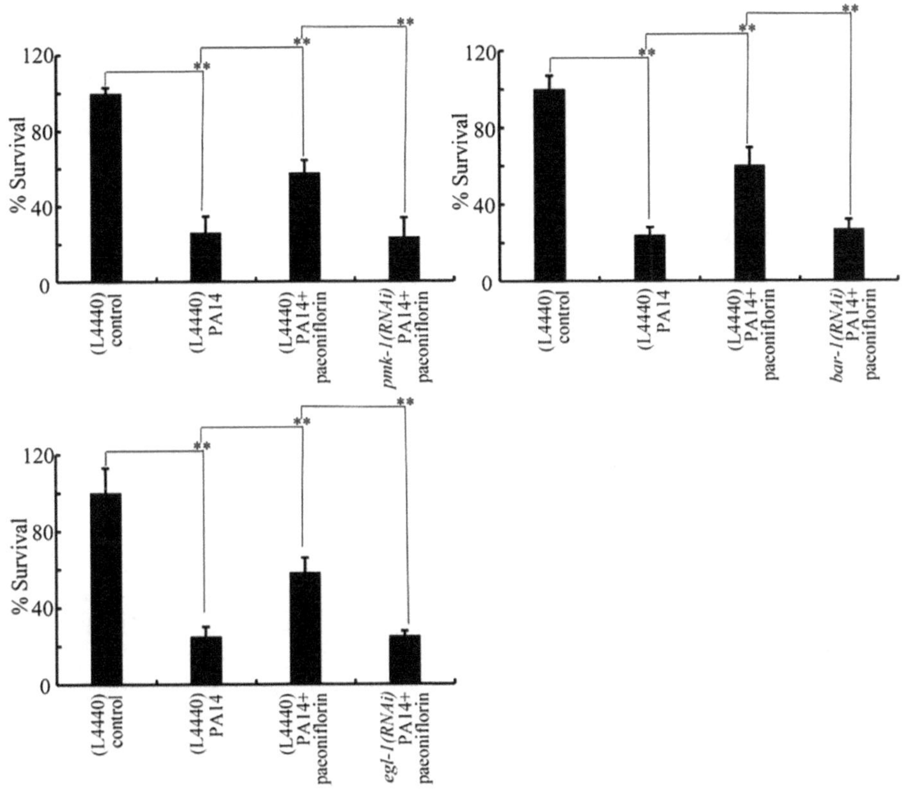

Fig. 10.6 RNAi of *bar-1*, *pmk-1*, and *egl-1* inhibited the effect of 100 mg/L paeoniflorin treatment for 1 day in increasing survival of nematodes at the immunosuppression stage (day-5) after *P. aeruginosa* infection [23]. Feeding with HT115 expressing L4440 (empty vector) was employed as the control. $^{**}P < 0.01$

10.3 Pharmacological Effects on Gut Fungal Pathogens

The endpoints used to assess innate immunity and modulating effects of compounds and drugs on innate immune response induced by fungal pathogens are similar to those for bacterial pathogens. In *C. elegans*, the behaviors of at least following fungal pathogens have been examined:

1. *Candida albicans*, the dimorphic yeast (opportunistic human pathogen) [26, 27].
2. *Candida auris*, the multidrug-resistant yeast (emerging human pathogen) [28].
3. *Cryptococcus neoformans*, the encapsulated yeast (human pathogen) [29].
4. *Aspergillus fumigatus*, the filamentous fungus [30].
5. *Drechmeria coniospora*, the nematophagous fungus [31].

6. *Harposporium* sp., the nematophagous fungus [32].
7. *Mucor hiemalis*, the filamentous fungus [33].
8. *Purpureocillium lavendulum*, the nematophagous fungus [34].
9. *Histoplasma capsulatum*, the dimorphic fungus [35].
10. *Talaromyces marneffei*, the dimorphic fungus [36].

10.3.1 Pharmacological Effects on Fungal Pathogens

In *C. elegans*, the *C. albicans* is a classic model for fungal pathogens to examine the fungal pathogen–host interaction. In the gut, the possible effects of compounds and drugs on fungal pathogens may mainly include the following aspects:

1. Disruption in cell membrane integrity: Some compounds, such as fluconazole, capsaicin, and octenidine dihydrochloride, could disrupt *C. albicans* cell membrane integrity and function by blocking ergosterol synthesis [37–39].
2. Disruption in cell wall integrity: Triazine amine compounds showed the effect to inhibit the *C. albicans* cell wall integrity [40].
3. Inhibition in biofilm formation: Based on crystal violet staining and analysis of OD600 absorbance, the biofilm formation of *C. albicans* SC5314 was reduced by 10–100 μM procatechuic acid and protocatechuic aldehyde, which was associated with decrease in expression of biofilm-related gene *ALS3* (Fig. 10.7) [41].
4. Inhibition in hyphal growth: The *C. albicans* hyphal growth could be inhibited by procatechuic acid and protocatechuic aldehyde (Fig. 10.8) [41]. Meanwhile, among genes governing hyphal growth, treatment with procatechuic acid reduced expressions of *Vac1*, *Vam7*, *CaVps34*, and *HWP1*, and treatment with protocatechuic aldehyde decreased expressions of *Vac1*, *Vam7*, and *CaVps34* (Fig. 10.8) [41].

10.3.2 Pharmacological Effects on Innate Immunity Against Fungal Pathogens

In *C. elegans*, some compounds have been shown to modulate the innate immunity after fungal pathogen infection [42–45]. Treatment with compounds and drugs modulate the innate immunity to fungal pathogen infection at least through the following several molecular mechanisms:

1. Modulation of host immune signaling: In *C. albicans*-infected *C. elegans*, treatment with procatechuic acid and protocatechuic aldehyde could extend the lifespan (Fig. 10.9) [41] and reduce the *C. albicans* colonization and accumulation in the body (Fig. 10.10) [41]. Moreover, these aspects of beneficial effects of procatechuic acid and protocatechuic aldehyde treatment were suppressed by RNAi of *daf-16* and *pmk-1* (Fig. 10.11) [41], suggesting the involvement of

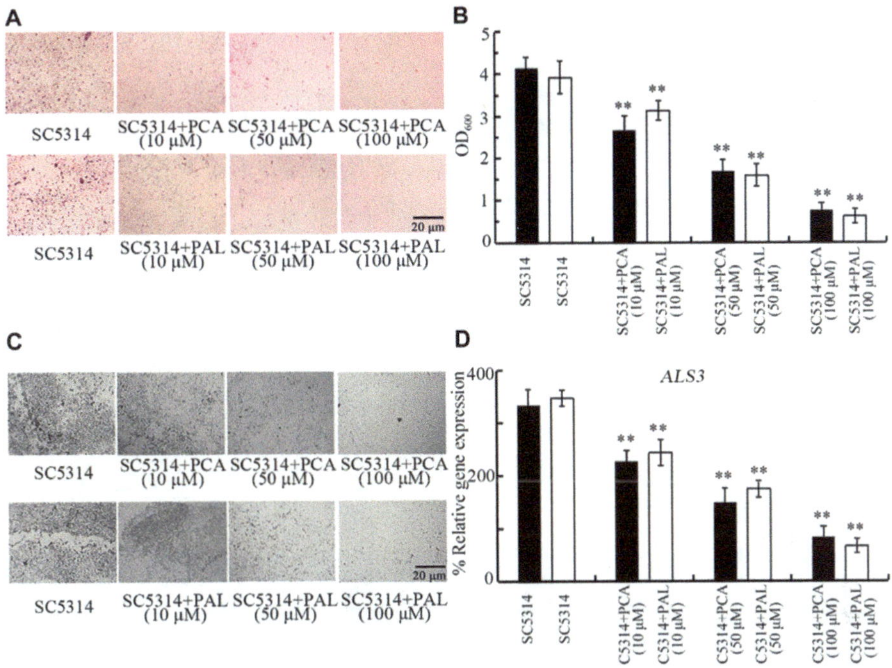

Fig. 10.7 Effect of PCA and PAL on *C. albicans* biofilm formation [41]. (**a**) Crystal violate staining images. (**b**) Effect of PCA and PAL on amount of biofilm formation based on OD600 absorbance analysis. (**c**) Effect of PCA and PAL on biofilm formation visualized under a light microscope. (**d**) Effect of PCA and PAL on expression of *ALS3*. PCA protocatechuic acid, PAL protocatechuic aldehyde. $^{**}P < 0.01$ vs. SC5314

insulin and p38 MAPK signals in regulating the pharmacological effect of procatechuic acid and protocatechuic aldehyde in modulating innate immunity.

2. Modulation of neuro-immune axis signaling: Treatment with naloxone could shorten the time of paralysis induced by fungal infection and mimics the effect of NLP-27 overexpression by acting as opioid receptor antagonist [46].
3. Disarming the fungal pathogens: As described above, treatment with procatechuic acid and protocatechuic aldehyde could inhibit the biofilm formation and the hyphal growth, which in turn enhanced the host innate immunity.

10.4 Pharmacological Effects on or from Gut Microbes

In *C. elegans*, the pharmacological effects on gut microbes can be considered as a paradigm shift from directly targeting pathogens to modulation of host's resident microbial community. That is, treatment with certain compounds and drugs can reshape composition and metabolism of the gut microbiome, leading to the further benefits for the hosts.

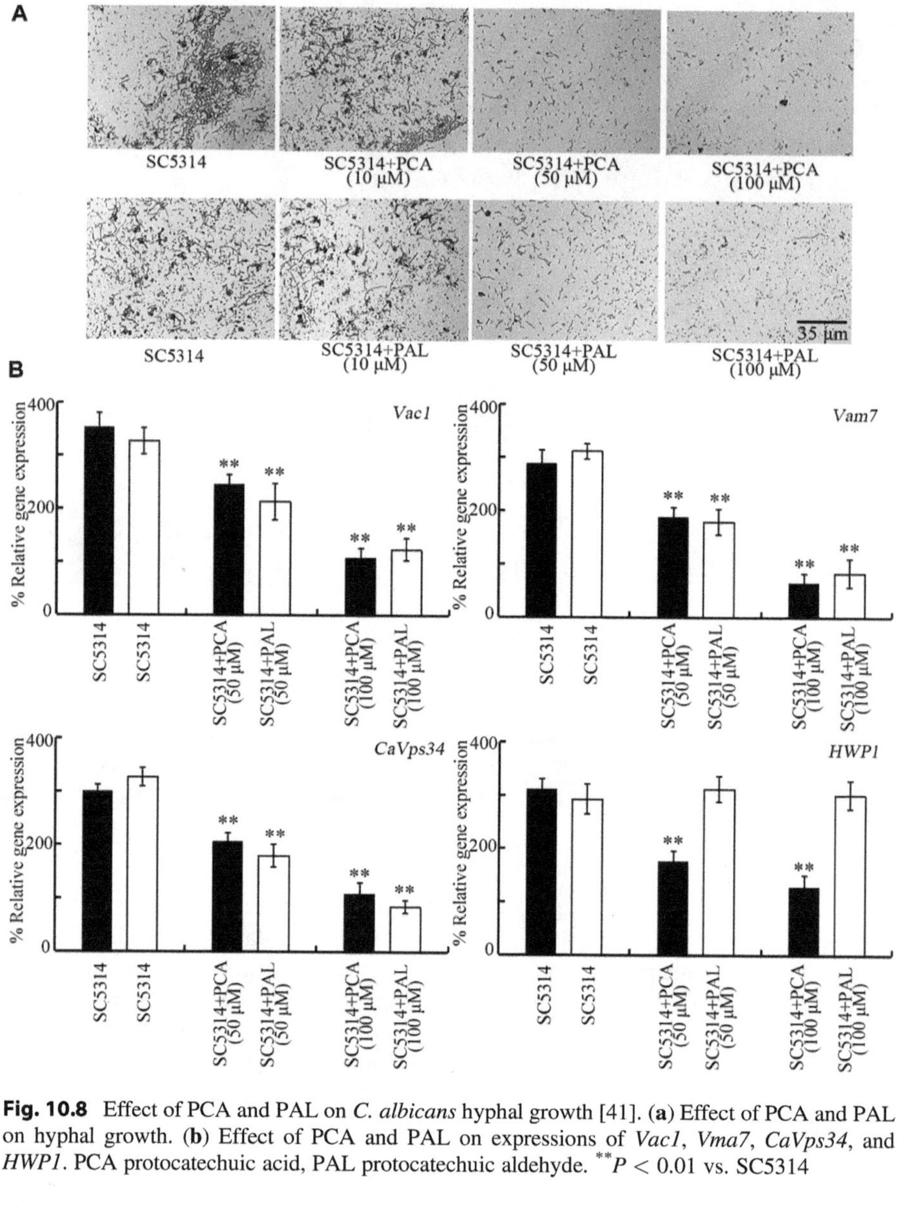

Fig. 10.8 Effect of PCA and PAL on *C. albicans* hyphal growth [41]. (**a**) Effect of PCA and PAL on hyphal growth. (**b**) Effect of PCA and PAL on expressions of *Vac1*, *Vma7*, *CaVps34*, and *HWP1*. PCA protocatechuic acid, PAL protocatechuic aldehyde. $^{**}P < 0.01$ vs. SC5314

The main pharmacological effects of compounds and drugs on *C. elegans* gut microbiota include the following several aspects:

1. Effects on Gut Microbiome
   Probiotics have been widely examined for their possible beneficial effects in modulating gut microbes in hosts, including in *C. elegans*. Besides the nutritional

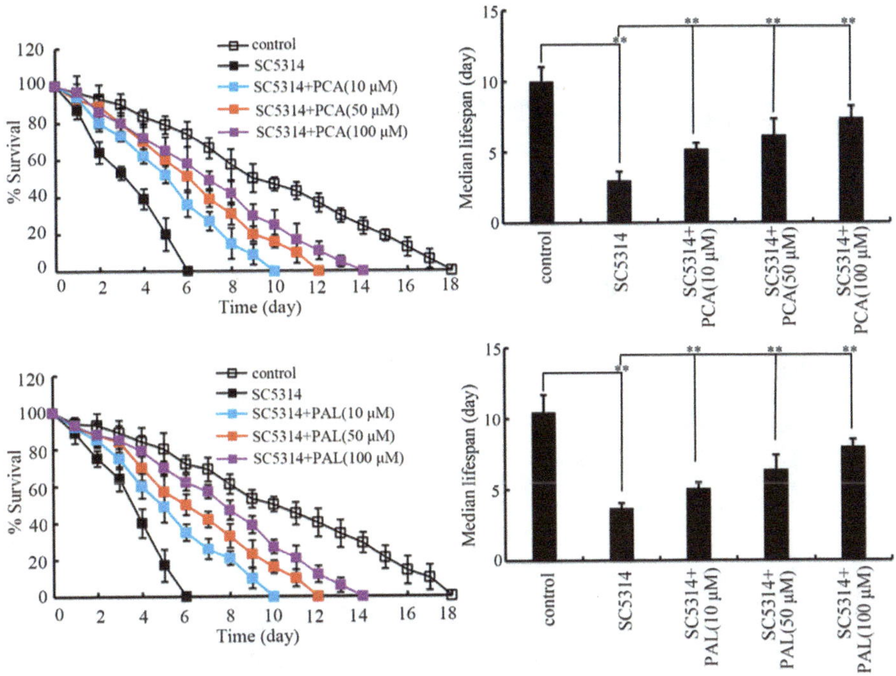

Fig. 10.9 Effect of PCA and PAL treatment on lifespan of nematodes after *C. albicans* infection [41]. PCA protocatechuic acid, PAL protocatechuic aldehyde. $^{**}P < 0.01$. Lifespan curves of SC4314 showed a significant difference ($P < 0.01$) compared to control. Lifespan curves of SC5314 + PCA (10 μM), SC5314 + PCA (50 μM), SC5314 + PCA (100 μM), SC5314 + PAL (10 μM), SC5314 + PAL (50 μM), and SC5314 + PAL (100 μM) showed a significant difference ($P < 0.01$) compared to SC5314

role, the probiotics can further actively modulate the physiology and the gut ecosystem of hosts. In *C. elegans*, it was reported that feeding with *Lacticaseibacillus casei*, IDCC 3451 could alleviate cognitive and behavioral deficits under the chronic stress condition by regulating intestinal barrier function and reshaping gut microbiome [47]. Feeding with *Lacticaseibacillus casei*, HY2782 could also extend lifespan by improving intestinal barrier function and tract environment [48]. In addition, feeding with lactic acid bacteria could inhibit toxicity of graphene oxide by maintaining normal intestinal permeability [49], suggesting the role of probiotics in modulating intestinal physiology.

Besides the probiotics, some compounds also have the effects on gut microbiome. Treatment with *Spirulina platensis* polysaccharide showed the antioxidant effect [50]. Meanwhile, in the intestinal flora, *Spirulina platensis* polysaccharide modulated gut microbiota by increasing abundance of *Flavobacterium*, *Achromobacter*, *Empedobacter*, *Anaerolinea*, and *Pseudoalteromonas* [50].

2. Effects on Metabolisms of Gut Microbiome

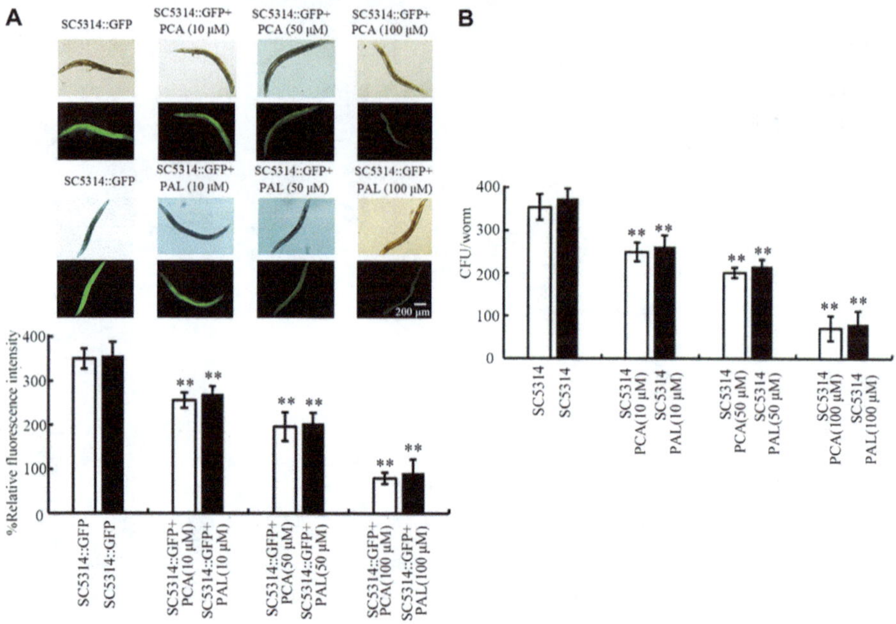

Fig. 10.10 Effect of PCA and PAL treatment on *C. albicans* accumulation in intestinal lumen of nematodes [41]. (**a**) Effect of PCA and PAL treatment on relative fluorescence intensity of SC5314: GFP in intestinal lumen of nematodes. (**b**) Effect of PCA and PAL treatment on CFU of SC5314 in infected nematodes. PCA protocatechuic acid, PAL protocatechuic aldehyde. $^{**}P < 0.01$ vs. SC5314:GFP or SC5314

In *C. elegans*, some bacterial metabolites can affect the behavior and physiology of host. For example, bacterial nitric oxide could extend the lifespan [51]. Thus, the possibility is existing that pharmacological treatment with compounds and drugs may affect the metabolisms of gut microbiome, which in turn influences the host. It was observed that treatment with metformin treatment could extend the *C. elegans* lifespan by modulating metabolisms of microbial folate and methionine [52].

3. Effects from Metabolisms of Gut Microbiome
   The gut microbiome can be considered as a "metabolic organ" after treatment with compounds. The bacteria living in the gut possess the enzymes that can chemically modify the compounds. In the gut, the compounds may be activated, inactivated, or even converted into other compounds by the gut microbiome. That is, we can dissect how gut bacteria chemically transform the compounds, thereby altering their effects on the host. In *C. elegans*, *E. coli* metabolisms, such as those from vitamin B6, B9, and nucleotide, could impact the drug efficacy of 5-fluorouracil (5-FU) to cause its stronger or weaker effect on host through metabolic drug interconversion [53, 54].

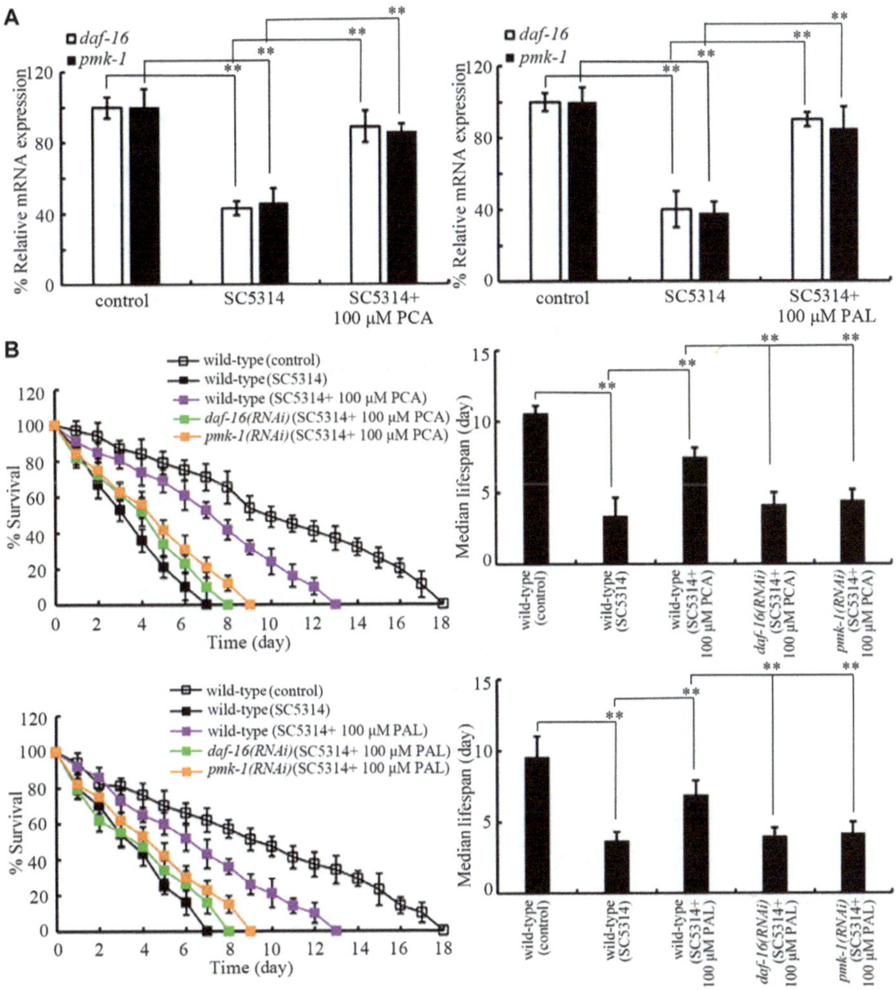

Fig. 10.11 Effect of *pmk-1* or *daf-16* RNAi on function of PCA and PAL against *C. albicans* infection [41]. (**a**) Effect of PCA and PAL treatment on expressions of *pmk-1* and *daf-16* in *C. albicans*-infected nematodes. (**b**) Effect of *pmk-1* or *daf-16* RNAi on function of PCA and PAL in increasing survival of *C. albicans*-infected nematodes. RNAi of *pmk-1* or *daf-16* was performed after *C. albicans* infection. PCA protocatechuic acid, PAL protocatechuic aldehyde. $^{**}P < 0.01$. Lifespan curves of SC4314 showed a significant difference ($P < 0.01$) compared to control. Lifespan curves of SC5314 + 100 μM PCA and SC5314 + 100 μM PAL showed a significant difference ($P < 0.01$) compared to SC5314. Lifespan curves of *daf-16(RNAi)* (SC5314 + 100 μM PCA) and *pmk-1(RNAi)* (SC5314 + 100 μM PCA) showed a significant difference ($P < 0.01$) compared to the group of SC5314 + 100 μM PCA. Lifespan curves of *daf-16(RNAi)* (SC5314 + 100 μM PAL) and *pmk-1(RNAi)* (SC5314 + 100 μM PAL) showed a significant difference ($P < 0.01$) compared to the group of SC5314 + 100 μM PAL

10.5 Perspectives

In this chapter, we first introduced and discussed pharmacological effects of compounds and drugs on gut bacterial and fungal pathogens. Besides bacterial and fungal pathogens, the virus–host interaction (such as the interaction between Orsay virus and host) has also been examined in *C. elegans* [55–58]. Compared to research on bacterial and fungal infections, the study of pharmacological effects on viral infections is a more nascent field in *C. elegans*. The pharmacological effects of compounds and drugs on gut virus, virus infection, and innate immune response to virus are also suggested to be paid attention to. For example, dietary zinc supplementation inhibited viral replication to lead to an ~1620-fold reduction in viral RNA, which was linked to the ability of zinc to reduce lipid levels [59].

We also introduced and discussed the pharmacological effects of compounds and drugs on gut microbes and metabolism of gut microbes. That is, the gut microbiome can be considered as a direct pharmacological target. More importantly, the possible effects of gut microbe on compounds and drugs were also suggested. Therefore, instead of trying to kill pathogens (which can normally lead to the resistance), we can use probiotics and compounds to steer our resident microbial community toward a state that is beneficial for the health. Besides the introduced effects of probiotics and compounds, the pharmacological effects of engineered bacterial and fungal cells and novel designed compounds and drugs that can overcome the influence of intestinal flora are also suggested to be considered.

References

1. Yu Y-L, Zhi L-T, Wu Q-L, Jing L-N, Wang D-Y (2018) NPR-9 regulates innate immune response in *Caenorhabditis elegans* by antagonizing activity of AIB interneurons. Cell Mol Immunol 15:27–37
2. Zhi L-T, Yu Y-L, Li X-Y, Wang D-Y, Wang D-Y (2017) Molecular control of innate immune response to *Pseudomonas aeruginosa* infection by intestinal *let-7* in *Caenorhabditis elegans*. PLoS Pathog 13:e1006152
3. Ding AJ, Zhang WM, Tao J, Chen B, Liu XC, Dong Y, Ma HJ, Pan SD, He JB, Zeng WK (2023) *Salmonella enterica* serovar Paratyphi A-induced immune response in *Caenorhabditis elegans* depends on MAPK pathways and DAF-16. Front Immunol 14:1118003
4. Schifano E, Ficociello G, Vespa S, Ghosh S, Cipollo JF, Talora C, Lotti LV, Mancini P, Uccelletti D (2019) *Pmr-1* gene affects susceptibility of *Caenorhabditis elegans* to *Staphylococcus aureus* infection through glycosylation and stress response pathways' alterations. Virulence 10:1013–1025
5. Dasgupta M, Shashikanth M, Gupta A, Sandhu A, De A, Javed S, Singh V (2020) NHR-49 transcription factor regulates Immunometabolic response and survival of *Caenorhabditis elegans* during *Enterococcus faecalis* infection. Infect Immun 88:e00130–e00120
6. Gowripriya T, Yashwanth R, James PB, Suresh R, Balamurugan K (2025) Dopaminergic neuronal regulation determines innate immunity of *Caenorhabditis elegans* during *Klebsiella aerogenes* infection. Microbes Infect 27:105430
7. List C, Grutsch A, Radler C, Cakar F, Zingl FG, Schild-Prüfert K, Schild S (2018) Genes activated by *Vibrio cholerae* upon exposure to *Caenorhabditis elegans* reveal the mannose-sensitive hemagglutinin to be essential for colonization. mSphere 3:e00238-18

8. Mir DA, Balamurugan K (2019) *In vitro* and *in vivo* efficacy of *Caenorhabditis elegans* recombinant antimicrobial protein against gram-negative bacteria. Biofouling 35:900–921
9. White PS, Penley MJ, Tierney ARP, Soper DM, Morran LT (2019) Dauer life stage of *Caenorhabditis elegans* induces elevated levels of defense against the parasite *Serratia marcescens*. Sci Rep 9:11575
10. Wang Y-X, Zhang L, Yuan X-A, Wang D-Y (2023) Treatment with paeoniflorin increases lifespan of Pseudomonas aeruginosa infected Caenorhabditis elegans by inhibiting bacterial accumulation in intestinal lumen and biofilm formation. Front Pharmacol 14:1114219
11. Feng M, Wu X, Hu X, Wu Y, Gou S, Ran Q, Yuan Y, Huang T, Dan L, Chu Y, Zhou X, Zhao K, Du L (2026) Repurposing Tirazone as an effective quorum-sensing inhibitor against *Pseudomonas aeruginosa* virulence and biofilm formation. J Antibiot. https://doi.org/10.1038/s41429-026-00901-7
12. Jones HA, Baxter MJ, Zimmermann N, Li A, Uptmor KAP, Taylor IR (2026) Screening of FDA-approved small molecules to discover inhibitors of the *Pseudomonas aeruginosa* quorum-sensing enzyme, PqsE. Biochemistry 65:263–269
13. Yahav D, Giske CG, Grāmatniece A, Abodakpi H, Tam VH, Leibovici L (2020) New beta-lactam-beta-lactamase inhibitor combinations. Clin Microbiol Rev 34:e00115–e00120
14. Shah S, Kline EG, Haidar G, Squires KM, Pogue JM, McCreary EK, Ludwig J, Clarke LG, Stellfox M, Van Tyne D, Shields RK (2025) Rates of resistance to ceftazidime-avibactam and ceftolozane-tazobactam among patients treated for multidrug-resistant *Pseudomonas aeruginosa* bacteremia or pneumonia. Clin Infect Dis 80:24–28
15. Kim DH, Ewbank JJ (2018) Signaling in the innate immune response. WormBook. https://doi.org/10.1895/wormbook.1.83.2
16. Hajdú G, Szathmári C, Sőti C (2024) Modeling host-pathogen interactions in *C. elegans*: lessons learned from *Pseudomonas aeruginosa* infection. Int J Mol Sci 25:7034
17. Martineau CN, Kirienko NV, Pujol N (2021) Innate immunity in *C. elegans*. Curr Top Dev Biol 144:309–351
18. Peterson ND, Pukkila-Worley R (2018) *Caenorhabditis elegans* in high-throughput screens for anti-infective compounds. Curr Opin Immunol 54:59–65
19. Zhou H, Shen Y, Dong C, Feng W, Tian Y, Xiao Y (2025) Asperuloside promotes innate immunity via IRE-1/XBP-1 mediated unfolded protein response. Bioorg Chem 157:108318
20. Zhu X, Liu F, Wu Q, Li S, Ruan G, Yang J, Yu C, Jiang N, Xiao Y, Liu Y (2022) Brevilin A enhances innate immunity and the resistance of oxidative stress in *Caenorhabditis elegans* via p38 MAPK pathway. Int Immunopharmacol 113:109385
21. Xiao Y, Zhou H, Cui Y, Zhu X, Li S, Yu C, Jiang N, Liu L, Liu F (2024) Schisandrin A enhances pathogens resistance by targeting a conserved p38 MAPK pathway. Int Immunopharmacol 128:111472
22. Hummell NA, Revtovich AV, Kirienko NV (2021) Novel immune modulators enhance *Caenorhabditis elegans* resistance to multiple pathogens. mSphere 6:e00950-20
23. Zhang L, Wang Y-X, Wang D-Y (2023) Paeoniflorin increases the survival of *Pseudomonas aeruginosa* infected *Caenorhabditis elegans* at the immunosuppression stage by activating PMK-1, BAR-1, and EGL-1 signals. Arch Pharm Res 46:616–628
24. Wu Y, Li G, Tang H (2024) Antibiotics trigger host innate immune response via microbiota-brain communication in *C. elegans*. Int J Mol Sci 25:8866
25. Zhang L, Wang Y-X, Cao C, Zhu Y-K, Huang W, Yang Y, Qiu H-B, Liu S-Q, Wang D-Y (2022) Beneficial effect of Xuebijing against *Pseudomonas aeruginosa* infection in *Caenorhabditis elegans*. Front Pharmacol 13:949608
26. Li D, Deng Y-J, Wang S-T, Du H-H, Xiao G-S, Wang D-Y (2020) Assessment of nanopolystyrene toxicity under fungal infection condition in *Caenorhabditis elegans*. Ecotoxicol Environ Saf 197:110625
27. Sun L-M, Zhi L-T, Shakoor S, Liao K, Wang D-Y (2016) microRNAs involved in the control of innate immunity in *Candida* infected *Caenorhabditis elegans*. Sci Rep 6:36036

28. Hernando-Ortiz A, Mateo E, Perez-Rodriguez A, de Groot PWJ, Quindós G, Eraso E (2021) Virulence of *Candida auris* from different clinical origins in *Caenorhabditis elegans* and Galleria mellonella host models. Virulence 12:1063–1075
29. Kitisin T, Muangkaew W, Sukphopetch P (2022) *Caenorhabditis elegans* DAF-16 regulates lifespan and immune responses to *Cryptococcus neoformans* and Cryptococcus gattii infections. BMC Microbiol 22:162
30. Ahamefule CS, Qin Q, Odiba AS, Li S, Moneke AN, Ogbonna JC, Jin C, Wang B, Fang W (2020) *Caenorhabditis elegans*-based *Aspergillus fumigatus* infection model for evaluating pathogenicity and drug efficacy. Front Cell Infect Microbiol 10:320
31. Lebrigand K, He LD, Thakur N, Arguel MJ, Polanowska J, Henrissat B, Record E, Magdelenat G, Barbe V, Raffaele S, Barbry P, Ewbank JJ (2016) Comparative genomic analysis of *Drechmeria coniospora* reveals core and specific genetic requirements for fungal endoparasitism of nematodes. PLoS Genet 12:e1006017
32. Chen M, Chavez L, Kang J, Hu J, Dong J, Tan Y, Chen Z, Huang B, Li C, Sun C, Hywel-Jones N, Liu X, Li Z (2024) Harposporium incensis sp. nov. a South American cordycipitoid species exhibiting inter-phylum host-jumping and having potential as a biological control agent for pest management. Mycology 16:238–249
33. Ni J, Sowa JN (2025) Novel infection by *Mucor hiemalis* kills *Caenorhabditis* hosts through intestinal perforation. bioRxiv. https://doi.org/10.1101/2025.06.04.657490
34. Zhuang XM, Guo ZY, Zhang M, Chen YH, Qi FN, Wang RQ, Zhang L, Zhao PJ, Lu CJ, Zou CG, Ma YC, Xu J, Zhang KQ, Cao YR, Liang LM (2023) Ethanol mediates the interaction between *Caenorhabditis elegans* and the nematophagous fungus *Purpureocillium lavendulum*. Microbiol Spectrum 11:e0127023
35. Costa ADC, Fernandes MR, Nobre AFD, Rocha MG, Mesquita JRL, Freire RS, Monteiro AJ, Silveira Vieira R, Brilhante RSN (2025) *In vitro* study of essential oils encapsulated in chitosan microparticles against histoplasma capsulatum and their pathogenicity in *Caenorhabditis elegans*. Biofouling 41:181–196
36. Huang XW, Lu S, Pan W, Zhong MZ, Chai JW, Liu YH, Zeng K, Xi LY (2023) Autophagy benefits the in vitro and *in vivo* clearance of *Talaromyces marneffei*. Microb Pathog 180:106146
37. Suchodolski J, Muraszko J, Bernat P, Krasowska A (2021) Lactate like fluconazole reduces ergosterol content in the plasma membrane and synergistically kills *Candida albicans*. Int J Mol Sci 22:5219
38. Behbehani JM, Irshad M, Shreaz S, Karched M (2023) Anticandidal activity of cpsaicin and is effect on ergosterol biosynthesis and membrane integrity of *Candida albicans*. Int J Mol Sci 24: 1046
39. Fang T, Xiong J, Wang L, Feng Z, Hang S, Yu J, Li W, Feng Y, Lu H, Jiang Y (2023) Unexpected inhibitory effect of octenidine dihydrochloride on *Candida albicans* filamentation by impairing ergosterol biosynthesis and disrupting cell membrane integrity. Antibiotics 12: 1675
40. Jiang L, Wang M, Yang L, Yang B, Xing L, Liu J, Yan R, Han J, Li G, Wang R (2025) Triazine amine compounds synergistically enhance the efficacy against fluconazole-resistant *Candida albicans* with fluconazole by inhibiting cell wall synthesis. Biochem Pharmacol 241:117181
41. Yuan C-Y, Wang Y-X, Zhang L, Wang D-Y (2024) Procatechuic acid and protocatechuic aldehyde increase survival of Caenorhabditis elegans after fungal infection and inhibit fungal virulence. Front Pharmacol 15:1396733
42. Camaioni L, Lambert D, Sendid B, Billamboz M, Jawhara S (2023) Antifungal properties of hydrazine-based compounds against *Candida albicans*. Antibiotics 12:1043
43. Kong C, Eng SA, Lim MP, Nathan S (2016) Beyond traditional antimicrobials: a *Caenorhabditis elegans* model for discovery of novel anti-infectives. Front Microbiol 7:1956
44. Mena L, Billamboz M, Charlet R, Desprès B, Sendid B, Ghinet A, Jawhara S (2022) Two new compounds containing pyridinone or triazine heterocycles have antifungal properties against *Candida albicans*. Antibiotics 11:72

45. Shu C, Sun L, Zhang W (2016) Thymol has antifungal activity against *Candida albicans* during infection and maintains the innate immune response required for function of the p38 MAPK signaling pathway in *Caenorhabditis elegans*. Immunol Res 64:1013–1024
46. Pop M, Klemke AL, Seidler L, Wernet N, Steudel PL, Baust V, Wohlmann E, Fischer R (2024) *Caenorhabditis elegans* neuropeptide NLP-27 enhances neurodegeneration and paralysis in an opioid-like manner during fungal infection. iScience 27:109484
47. Kang A, Eor JY, Lee J, Kwak M, Lee DJ, Seo E, Lee WJ, Son S, Song M, Kim J, Kim HW, Yang J, Oh S, Kim Y (2025) *Lacticaseibacillus casei* IDCC 3451 alleviates cognitive and behavioral functions by reshaping the gut microbiome and regulating intestinal barrier integrity in chronic stress animal models. Curr Res Food Sci 10:101051
48. Bae WY, Nguyen UTT, Le TAN, Tran SH, Lee S, Hong SC, Cha KH, Choi ID, Kang K (2025) *Lacticaseibacillus casei* HY2782 improves the intestinal barrier and tract environment and ultimately prolongs the lifespan of *Caenorhabditis elegans*. Food Funct 16:5871–5884
49. Zhao Y-L, Yu X-M, Jia R-H, Yang R-L, Rui Q, Wang D-Y (2015) Lactic acid bacteria protects *Caenorhabditis elegans* from toxicity of graphene oxide by maintaining normal intestinal permeability under different genetic backgrounds. Sci Rep 5:17233
50. Chen Y, Wan X, Wu D, Ouyang Y, Gao L, Chen Z, El-Seedi HR, Wang M, Chen X, Zhao C (2020) Characterization of the structure and analysis of the anti-oxidant effect of microalga *Spirulina platensis* polysaccharide on *Caenorhabditis elegans* mediated by modulating microRNAs and gut microbiota. Int J Biol Macromol 163(2020):2295–2305
51. Gusarov I, Gautier L, Smolentseva O, Shamovsky I, Eremina S, Mironov A, Nudler E (2013) Bacterial nitric oxide extends the lifespan of *C. elegans*. Cell 152:818–830
52. Cabreiro F, Au C, Leung K, Vergara-Irigaray N, Cocheme HM, Noori T, Weinkove D, Schuster E, Greene NDE, Gems D (2013) Metformin retards aging in *C. elegans* by altering microbial folate and methionine metabolism. Cell 153:228–239
53. Scott TA, Quintaneiro LM, Norvaisas P, Lui PP, Wilson MP, Leung KY, Herrera-Dominguez L, Sudiwala S, Pessia A, Clayton PT, Bryson K, Velagapudi V, Mills PB, Typas A, Greene NDE, Cabreiro F (2017) Host-microbe co-metabolism dictates cancer drug efficacy in *C. elegans*. Cell 169:442–456
54. García-González AP, Ritter AD, Shrestha S, Andersen EC, Yilmaz LS, Walhout AJM (2017) Bacterial metabolism affects the *C. elegans* response to cancer chemotherapeutics. Cell 169: 431–441
55. Batachari LE, Sarmiento MB, Wernet N, Troemel ER (2024) Orsay virus infection in *Caenorhabditis elegans*. Curr Protoc 4:e1098
56. Zhou Y, Chen H, Zhong W, Tao YJ (2024) Collagen and actin network mediate antiviral immunity against Orsay virus in *C. elegans* intestinal cells. PLoS Pathog 20:e1011366
57. Kalugotla G, Marmerstein V, Schriefer LA, Wang L, Morrison SA, Perez LC, Silverman GA, Schedl T, Pak SC, Baldridge MT (2025) ATG-3 limits Orsay virus infection in *C. elegans* and regulates collagen pathways. PLoS Pathog 21:e1012900
58. Jiang H, Chen K, Sandoval LE, Leung C, Wang D (2017) An evolutionarily conserved pathway essential for Orsay virus infection of *Caenorhabditis elegans*. MBio 8:e00940–e00917
59. Casorla-Perez LA, Guennoun R, Cubillas C, Peng B, Kornfeld K, Wang D (2022) Orsay virus infection of *Caenorhabditis elegans* is modulated by zinc and dependent on lipids. J Virol 96: e0121122

Underlying Mechanisms for Pharmacological Effects of Compounds 11

Abstract

In *Caenorhabditis elegans*, the underling molecular mechanisms for pharmacological effects of compounds have been widely determined. We first introduced and discussed underling molecular mechanisms for antioxidation, anti-stress, and anti-aging effects and activation of protective responses by compounds. After that, the underling molecular mechanisms for promotion in growth, increase in reproduction, modulation of feeding and food intake, maintenance of functional state of biological barriers, improvement in regeneration and wounding healing, and anti-tumor by compounds were introduced and discussed. Moreover, we introduced the underling molecular mechanisms for anti-neurodegeneration-related diseases, anti-seizure-like, addiction, anxiety-like, and autism-like phenotypes, modulation of sleep, and improvement in muscle function by compounds. Finally, the underling molecular mechanisms for modulation of metabolic processes (such as reduction in fat storage) by compounds were discussed.

Keywords

Molecular mechanisms · Pharmacological effects · Compounds · *Caenorhabditis elegans*

11.1 Introduction

In Chap. 3, we have introduced the classification of examined compounds and drugs in *Caenorhabditis elegans*. Moreover, in Chap. 4, we have introduced the physiological and pathological models for pharmacological study in *C. elegans*. Using these models, together with the different research aims, different pharmacological effects of compounds and drugs can be observed in *C. elegans*. After the observation

D. Wang, *Pharmacology in Caenorhabditis elegans*,
https://doi.org/10.1007/978-981-92-5166-7_11

of pharmacological effects of compounds and drugs, the further pharmacological mechanisms are needed to be elucidated. The pharmacological mechanisms for observed pharmacological effects of compounds and drugs that we usually discuss refer to the pharmacological molecular mechanisms. Besides this, the interactions between compounds or drugs can provide the corresponding chemical mechanisms to a certain degree, which will be introduced and discussed in Chap. 12. Moreover, for plant extracts and Traditional Chinese Medicine formulas, their pharmacological effects are largely due to the existence of active compounds and their combinations, which will be introduced and discussed in Chap. 13. In this chapter, we focus on introduction and discussion of molecular mechanisms for observed different pharmacological effects of compounds and drugs in *C. elegans*.

11.2 Underlying Mechanisms for Antioxidation Effect of Compounds

In *C. elegans*, the oxidative stress models can be generated by chemical treatments with paraquat, juglone, or hydrogen peroxide (H_2O_2). Using these models or certain stress models, the potential pharmacological effect of compounds against oxidative stress can be determined [1]. For the underlying mechanisms, the compounds can exert antioxidation effects primarily through a conserved network organized by multiple stress response pathways. To be against the oxidative stress, the compounds can not only act as the chemical scavengers, but also function as hormetic agents.

11.2.1 Modulation of Insulin Signaling Pathway

In *C. elegans*, insulin receptor DAF-2 inhibit expression and function of transcriptional factor DAF-16 by activating downstream kinase cascade [2]. Once activated or not phosphorylated by kinases of AKT-1/2, DAF-16 will be translocated to nucleus to activate the transcription of targeted genes. Thus, the activated DAF-16 by compounds can function to reduce intracellular ROS accumulation by driving expression of antioxidant enzymes, such as SOD-2 and SOD-3, two mitochondria Mn-SODs [3]. Isoxanthohumol could show the antioxidation effect by activating DAF-16 and increasing its nuclear translocation, and the stress resistance induced by isoxanthohumol treatment could be abolished by *daf-16* mutation [4]. Amentoflavone, a component from Selaginella extract, could also confer the resistance to oxidative stress by promoting DAF-16 nuclear translocation and upregulating some antioxidant genes, such as *sod-3* and *gst-3/4* [5].

11.2.2 Modulation of SKN-1/Nrf Signaling

SKN-1, the homologue of transcriptional factor Nrf2, is considered as the primary antioxidant response regulator. As the transcriptional factor, SKN-1 can control the phase II detoxification and antioxidant gene expressions so as to be against the oxidative stress [6]. Another important molecular mechanism for pharmacological effect of compounds to inhibit oxidative stress is through activation of SKN-1. For example, flavonoids from *Folium Artemisiae* Argyi exhibited the antioxidation effect by activating both SKN-1 and its targets, such as GST-4 [7]. Coniferaldehyde treatment could also inhibit generation of reactive oxygen species (ROS) via promoting SKN-1 translocation into the nucleus and enhancing GST-4 expression [8].

In *C. elegans*, the SKN-1 could also be indirectly activated by compounds via activating its upstream regulators. The normally examined upstream regulator is PMK-1, the p38 MAPK. Hyperoside treatment could induce resistance to oxidative stress by activating SEK-1/PMK-1/SKN-1 pathway [9]. In addition, chaetoglobosin F could reduce intracellular ROS level, which was linked to upregulation of PMK-1 and SKN-1 [10]. Besides these, it was also observed that the obvious ROS generation could be caused by berberine, and this was followed by activation of PMK-1/SKN-1 signaling and improved health span [11]. That is, certain compounds can somctimcs furthcr act as thc hormetic agents.

11.2.3 Modulation of Mitochondrial Signals

At least change in three aspects of mitochondrial signals can mediate the pharmacological effects of compounds and drugs against the oxidative stress. The first group of mitochondrial signals are genes (such as *mev-1*, *gas-1*, *clk-1*, and *isp-1*) encoding the components of different mitochondrial complexes on the electron transfer chain, which governs the mitochondrial function and are also involved in control of oxidative stress [12]. The second group of mitochondrial signals are genes encoding mitochondrial Mn-SODs (SOD-2 and SOD-3) as described above, and these mitochondrial Mn-SODs act as the targets of DAF-16. The third group of mitochondrial signals are genes encoding proteins governing the mitochondrial unfolded protein response (mt UPR), and its marker HSP-6 can acts as the target of both DAF-16 and SKN-1 [13, 14]. Eleutheroside B treatment could combat the oxidative stress via activating SKN-1 and HSP-6 [15].

11.3 Underlying Mechanisms for Anti-Stress Effect of Compounds

Besides the oxidative stress, treatment with compounds can also induce the resistance to other forms of stresses. Actually, alteration in above-described three aspects of molecular signals (insulin signaling, SKN-1/Nrf signaling, and mitochondrial signals) also definitely mediate the pharmacological effects of compounds and drugs against the other forms of stresses. Nevertheless, in addition to these three aspects of molecular signals, treatment with compounds may also exert their pharmacological effects by modulating other aspects of molecular signals. The following will introduce and discuss this.

11.3.1 Underlying Mechanisms for Anti-Heat Stress Effect of Compounds

Relatively specific molecular mechanisms for anti-heat stress effect of compounds mainly contain three aspects:

1. Modulation of Heat Shock Factor (HSF-1) pathway
   The master regulator of this response is the transcription factor HSF-1. Heat stress would cause the protein misfolding, which titrates away chaperones from HSF-1 to allow its activation [16, 17]. That is, heat stress can activate HSF-1, leading to increased expression of small HSPs (such as HSP-16.2 and HSP-12.6) to stabilize the cellular proteostasis [16, 17]. Treatment with alpha-lipoic acid could increase survival rate under heat stress condition by activating HSF-1 [18]. The *C. elegans* lifespan under heat stress condition could be extended by water-soluble polysaccharides extracted from *Pueraria lobata* via activation of HSF-1 and its targets, such as HSP-16.2 [19].
2. Modulation of Dietary Restriction Pathway
   Dietary restriction is a well-established strategy that enhances stress resistance, including thermotolerance [20]. EAT-2, a nicotinic receptor subunit, is involved in controlling the dietary restriction process [21]. Genistein treatment conferred the resistance to heat stress by partially modulating dietary restriction signaling [22]. The pharmacological effect of tectochrysin against the heat stress could not be observed in *eat-2* mutants [23].
3. Modulation of Guanylyl Cyclase Signaling
   C. elegans can detect the temperature primarily through specialized sensory neurons (such as AFD sensory neurons) and AIY interneurons [24]. In AFD sensory neurons, guanylyl cyclases (GCY-8, GCY-18, and GCY-23) are critical for thermosensation, which function redundantly to produce cGMP and to open TAX-2/TAX-4 ion channels [25, 26]. It is supposed that compounds can also induce the resistance to heat stress by at least partially affecting the neuronal circuit and related molecular signals in this neuronal circuit for thermosensation.

11.3.2 Underlying Mechanisms for Anti-Irradiation Stress Effect of Compounds

The pharmacological effects of compounds against the toxicity of both ionizing (X-ray, γ-ray) irradiation and non-ionizing (UV) irradiation have been examined in *C. elegans*. Relatively specific molecular mechanisms for anti-irradiation stress effect of compounds mainly contain two aspects:

1. Modulation of DNA Damage Signaling
   In DNA damage signaling pathway, some DNA damage checkpoints (such as HUS-1) induce the DNA damage by activating CEP-1 [27]. Quercetin and its mixture could increase reproductive capacity in UV-B-irradiate nematodes by

inhibiting HUS-1 and CEP-1 [28]. Meanwhile, *C. elegans* uses the DNA repair system to counteract the DNA damage [27]. Thus, the pharmacological effect of compounds against the reproductive toxicity of irradiation stresses may also due to the activation of DNA repair system.

2. Modulation of Certain G Protein-Coupled Receptors (GPCRs)
   Administration with eleutheroside E (EE) could protect against γ-ray radiation-induced neurotoxicity, which was possibly linked to the modulation of GPCRs and its downstream signaling cascade [29].

11.3.3 Underlying Mechanisms for Anti-Osmotic Stress Effect of Compounds

Osmotic stress, especially hypertonic stress (high salt), is one of the best-characterized stress models in *C. elegans*. Relatively specific molecular mechanisms for anti-osmotic stress effect of compounds mainly contain two aspects:

1. Modulation of Osmosensation Signals
   TRPV channel OSM-9 functions in the regulation of osmosensation and related behavioral avoidance by controlling the proteostasis capacity [30]. Crotamiton derivative JM03 induced resistance to osmotic stress by inhibiting OSM-9, which further caused the aggregation of Q35 [31].
2. Modulation of Glycerol-3-phosphate Dehydrogenases
   Glycerol is the primary osmolyte in *C. elegans*, which is synthesized by GPDH-1 and GPDH-2, glycerol-3-phosphate dehydrogenases [32]. That is, GPDH-1 and GPDH-2 act as core effector enzymes to synthesize the organic osmolyte glycerol essential for the survival under osmotic stress conditions [32]. Another possible mechanism for observed pharmacological effect of compounds against the osmotic stress may be through modulation of expression and function of GPDH-1 and GPDH-2.

11.3.4 Underlying Mechanisms for Anti-Glucose Stress Effect of Compounds

In *C. elegans*, insulin and AMPK signals act as important regulators for glucose metabolism [33, 34]. Relatively specific molecular mechanisms for anti-glucose stress effect of compounds may be mainly reflected by their modulation on insulin and AMPK signals. In glucose-treated nematodes, paeoniflorin treatment increased *daf-2* expression and inhibited *daf-16* expression [35]. The pharmacological effect of paeoniflorin treatment in extending lifespan of glucose-treated nematodes was increased by *daf-2* RNAi and suppressed by *daf-16* RNAi (Fig. 11.1) [35]. DAF-2 acted upstream of DAF-16 to regulate the pharmacological effect of paeoniflorin in extending lifespan of glucose-treated nematodes [35]. Paeoniflorin potentially binds

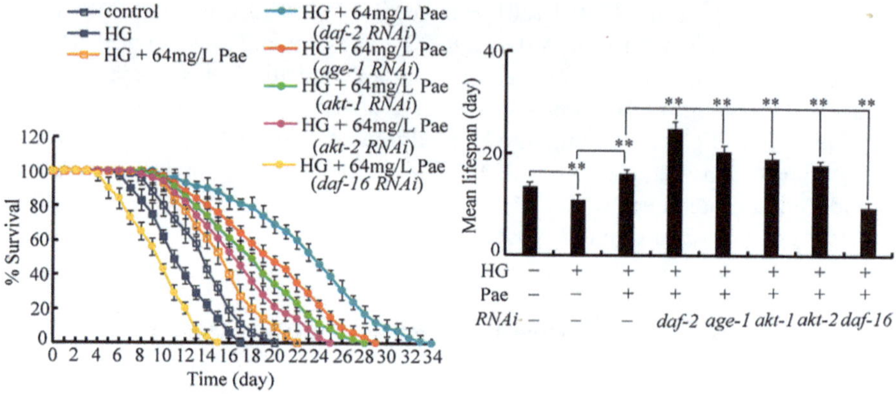

Fig. 11.1 Effect of *daf-2*, *age-1*, *akt-1*, *akt-2*, and *daf-16* RNAi on pharmacological effect of paeoniflorin (64 mg/L) in increasing lifespan in 50 mM glucose-treated nematodes [35]. Lifespan curve of HG was significantly ($P < 0.01$) different from control. Lifespan curve of HG + 64 mg/L Pae was significantly ($P < 0.01$) different from HG. Lifespan curves of HG + Pae + *daf-2(RNAi)*, HG + Pae + *age-1(RNAi)*, HG + Pae + *akt-1(RNAi)*, HG + Pae + *akt-2(RNAi)*, and HG + Pae + *daf-16(RNAi)* were significantly ($P < 0.01$) different from HG + Pae. HG high glucose (50 mM), Pae paeoniflorin. $^{**}P < 0.01$

to DAF-2, AGE-1, AKT-1, and AKT-2 [35], which provided the possible chemical basis for the observed pharmacological effect of paeoniflorin.

11.3.5 Underlying Mechanisms for Anti-Heavy Metal Stress Effect of Compounds

Relatively specific molecular mechanisms for anti-heavy metal stress effect of compounds may mainly contain two aspects:

1. Modulation of Metallothionein Defense
   Exposure to heavy metals, especially the cadmium (Cd), can trigger a multi-faceted toxic response. Nevertheless, the primary defense against heavy metals (such as Cd) is the induction of metallothioneins (MTL-1 and MTL-2), two cysteine-rich proteins [36]. The observed beneficial effect of selenium treatment in protecting heavy metal toxicity was closely related to response of MTL-1 ad MTL-2 [37].
2. Inhibition in Mitochondrial Toxicity
   Exposure to pollutants including the heavy metals can induce the severe toxicity in *C. elegans* by causing the mitochondrial toxicity [38, 39]. The Hg exposed induced oxidative damage was linked to decrease in *mev-1* and *gas-1* expressions and increase in *clk-1* and *isp-1* expression, which was protected by pretreatment with vitamin E [40]. In addition, the MeHg caused oxidative damage and

mitochondrial dysfunction could also be protected by NAD^+ supplementation [41].

11.4 Underlying Mechanisms for Anti-Aging Effect of Compounds

The main molecular signaling pathways involved in regulating the longevity contain insulin/IGF-1 signaling pathway, dietary restriction pathway, mitochondrial electron transport-related signaling, target of rapamycin (TOR) pathway, and germline signaling. Corresponding to this, the anti-aging effects of compounds and drugs in *C. elegans* are mediated by an integrated network of these molecular signaling pathways at the organismal level.

11.4.1 Modulation of Insulin/IGF-1 Signaling Pathway

The insulin signaling pathway is the core signaling pathway governing the longevity, which is the most extensively characterized molecular signaling pathway in *C. elegans*. During the control of longevity, the insulin ligands modulated DAF-2 receptor can further activate a kinase cascade that inhibits the FOXO transcriptional factor DAF-16. At the most cases, treatment with compounds can show the an-aging effect by modulating the insulin signaling pathway. This can be reflected by at least these five aspects:

1. Activation of DAF-16
   Treatment with compounds can activate DAF-16 expression and/or enhance the DAF-16 translocation into the nucleus. Accompanied with the lifespan extension by *Glycyrrhizae radix* extract, *daf-16* expression was activated [42]. In addition, the beneficial effect of *Glycyrrhizae radix* extract in extending lifespan was inhibited by *daf-16* mutation [42]. Curcumin acetylsalicylate could extend the lifespan by enhancing DAF-16 translocation into the nucleus, and this beneficial effect of curcumin acetylsalicylate was blocked by *daf-16* mutation [43]. Similarly, 2-butoxytetrahydrofuran and palmitic acid extended the lifespan by upregulating DAF-16 expression and enhancing DAF-16 translocation into the nucleus, and the beneficial effect of these compounds was also inhibited by *daf-16* mutation [44].
2. Inhibition of DAF-2 and its Downstream Kinase Cascade
   In the insulin signaling pathway, another molecular basis for beneficial effect of compounds in extending lifespan was the inhibition of DAF-2 receptor and its downstream kinase cascade. After the treatment with *Glycyrrhizae radix* extract, besides the activation of DAF-16, DAF-2, AGE-1, SGK-1, and AKT-1 were inhibited (Fig. 11.2) [42]. Besides this, the beneficial effect of chrysophanol and icariin in delaying the lifespan was also blocked by *daf-2* mutation [45, 46].
3. Modulation of Insulin Ligands

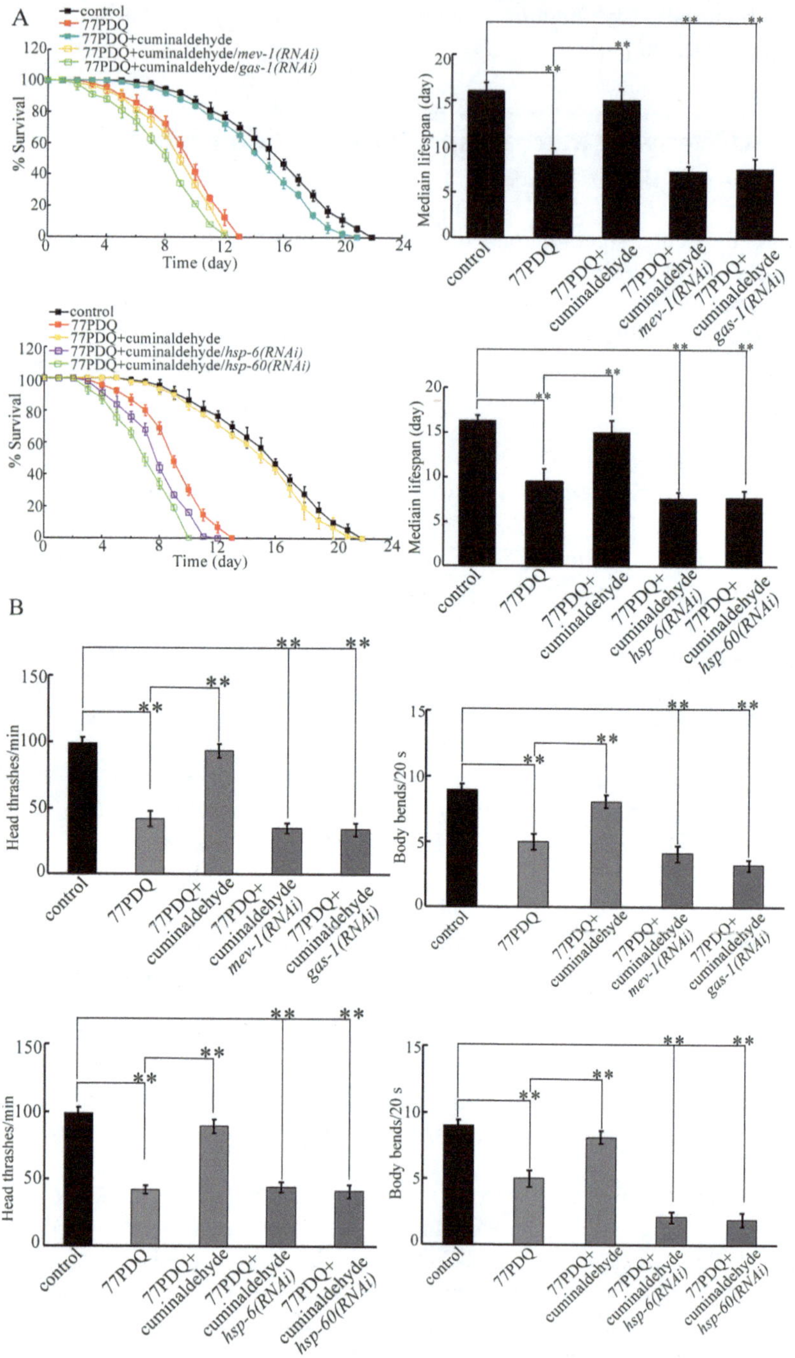

Fig. 11.2 Effect of RNAi of *mev-1*, *gas-1*, *hsp-6*, and *hsp-60* on pharmacological effect of cuminaldehyde on 6-PPDQ toxicity in reducing lifespan (**a**) and inhibiting locomotion behavior at adult day-8 (**b**) [83]. RNAi of *mev-1*, *gas-1*, *hsp-6*, and *hsp-60* was performed during the

In the insulin signaling pathway, the DAF-2 receptor can be activated by multiple corresponding insulin ligands. Among these insulin ligands, some of them (such as INS-1) can also regulate the longevity [47, 48]. Therefore, compound treatment may also extend the lifespan by modulating the insulin ligands.

4. Modulation of JNK-1/DAF-16 signaling
   In *C. elegans*, JNK-1, the JNK MAPK, can regulate the longevity by activating DAF-16 and enhance its nuclear translocation [49], suggesting the role of JNK-1 as the upregulator of DAF-16. Thus, the treatment with compounds can also extend the lifespan by modulating JNK-1/DAF-16 signaling. It was reported that treatment with lovastatin, patchouli essential oil, and black goji berry anthocyanins all could show anti-aging effect by activating the JNK-1/DAF-16 signaling [50–52].
5. Modulation of TGF-β/DAF-16 Signaling
   The TGF-β ligands, such as DAF-7, can further regulate the longevity by controlling the DAF-16 translocation and at least partially modulating the IIS signaling pathway [53]. Thus, this implies that treatment with compounds may also extend the lifespan by modulating the TGF-β/DAF-16 signaling. Administration with *Butyricicoccus pullicaecorum* and *Megasphaera elsdenii* could extend the lifespan, and this observed beneficial effect could be inhibited by mutation of *dbl-1* encoding another TGF-β ligand [54].

11.4.2 Modulation of Dietary Restriction Pathway

The dietary restriction pathway is one of the most potent and evolutionarily conserved mechanisms for extending lifespan, and the dietary restriction can extend the adult lifespan without major pleiotropic defects [55, 56]. For molecular mediators of this signaling pathway, PHA-4 acts as the master regulator, AAK-2/AMPK is the energy sensor, and SIR-2.1/Sirtuin can also be activated under dietary restriction condition to deacetylate certain targets [57–59]. Sesamin could extend the lifespan, and this beneficial effect was blocked by mutation of *aak-2* and *sir-2.1* [60]. Shatavarin IV, a steroidal saponin, could extend the lifespan and reduce the aging by-product, and these beneficial effects were dependent of EAT-2 [61]. Treatment with d-allulose, a stereoisomer of d-fructose, extended the lifespan, which was inhibited by *aak-2* mutation [62].

Fig. 11.2 (continued) cuminaldehyde treatment. Exposure concentration of 77PDQ was 10 μg/L, and treatment concentration of cuminaldehyde was 75 mg/L. Data are presented the mean ± SD. $^{**}P < 0.01$

11.4.3 Modulation of Mitochondrial Electron Transport Chain-Related Signaling

In *C. elegans*, mutation of some genes (such as *mev-1*, *clk-1*, *isp-1*, and *nuo-6*) encoding components of mitochondrial complexes on electron transfer chain can extend or reduce the lifespan [63–65]. Trigonelline and benzimidazole derivative M084 could extend the lifespan, and this anti-aging effect could not be observed in *clk-1* and *isp-1* mutants, which have the defect in mitochondrial respiratory chain complex III [66, 67]. Besides this, Benzothiazepine CGP37157 extended the lifespan, which could not be observed in *nuo-6* mutant with the defect in mitochondrial respiratory chain complex I [68].

11.4.4 Modulation of Target of Rapamycin (TOR) Pathway

Target of rapamycin (TOR) pathway is a nutrient-sensing mechanism. Inhibition in the TOR signaling can extend the lifespan, and this signaling pathway can intersect with dietary restriction signaling and insulin/IGF-1 (IIS) signaling [69, 70]. The core components of the TOR signaling pathway at least contain LET-363/mTOR, DAF-15/Raptor, RICT-1/Rictor, and RSKS-1 (substrate of TORC1) [69, 70]. Astaxanthin, a xanthophyll carotenoid, and hydrogen could extend the lifespan and decrease the *let-363* expression [71, 72]. The function of diosgenin in extending lifespan relied on the *let-363* [73]. The beneficial effect of sesamin treatment in extending lifespan could be abolished by heterozygotic mutation of *daf-15* [60]. The amino acids of serine, proline, and histidine had the effect of lifespan extension, which could be inhibited in *rsks-1* mutant nematodes [74].

11.4.5 Modulation of Germline Signaling

In *C. elegans*, removing the germline (not the somatic gonad) extends the lifespan, which suggests that the intact germline produces a shortening signal, and it can regulate the lifespan through a complex communication system between germline, somatic gonad, and the rest of the body [75]. In the gonad, the DTCs can sense absence of germline and activate KRI-1 (a CRB-like adaptor protein) that acts in the intestine to promote DAF-16 nuclear localization [75]. That is, the communication between reproductive system and intestine can be established through lipophilic-hormone signaling. The mutation of *glp-1* causes germline loss phenotype, which is usually used as the germline loss model [76]. The carnitine, aspirin, and Aiweixin (a traditional Uyghur medicine prescription) could extend the lifespan, and their anti-aging effects could all be suppressed in *glp-1* mutant nematodes [77–79].

11.5 Underlying Mechanisms for Activation of Protective Responses by Compounds

The genes reflecting various protective responses normally act as the effectors for different signaling pathway and are activated by transcriptional factors in different signaling pathways.

11.5.1 Activation of Antioxidation System

In the antioxidation system, SOD-3 and GST-4 are two normally examined proteins. After the pharmacological treatment with compounds with antioxidation effect, these two proteins are usually activated by modulating their upstream insulin signaling and SKN-1 signaling. We have introduced and discussed this in Sects. 11.2.1 and 11.2.2 in this chapter.

11.5.2 Activation of Mitochondrial Unfolded Protein Response

In *C. elegans*, the mitochondrial unfolded protein response (mt UPR) can be reflected by the markers of HSP-6 and HSP-60 [80]. Some transcriptional factors, such as ATFS-1, DVE-1, and UBL-5, govern this biological process [81, 82]. Besides this, some mitochondrial signals, such as LONP-1, are also required for mt UPR activation [81, 82]. Some compounds have been shown to have the effect in activating mt UPR. Exposure to 77PD quinone resulted in both reduction in lifespan and suppression in mt UPR, which could be reversed by cuminaldehyde treatment [83]. Moreover, the beneficial effect of cuminaldehyde treatment in extending the lifespan and improving the health span could be suppressed by RNAi of *hsp-6* and *hsp-60* (Fig. 11.2) [83], which suggested the link of activation of mt UPR with the pharmacological effect of cuminaldehyde in extending the lifespan and improving the health span. In addition, caffeic acid could activate the mt UPR to be against pathogen infection via ATFS-1 [84]. Similarly, accompanied with inhibition in Pseudomonas toxin pyocyanin, chlorogenic acid treatment induced mt UPR by activating ATFS-1 [85].

11.5.3 Activation of Mitophagy

Both mt UPR and mitophagy are required for the maintenance of mitochondrial function. In *C. elegans*, mitophagy has two conserved pathways: PINK1/ Parkindependent mitophagy and receptor-mediated mitophagy [86]. In PINK1/ Parkin-dependent mitophagy pathway, PINK-1 recruits E3 ubiquitin ligase PDR-1, which can bind to mitophagy receptor SQST-1 to induce autophagosome formation [87]. Treatment with urolithin A could activate mitophagy in a PINK-1- and PDR-1-dependent manner to be against amyotrophic lateral sclerosis pathologies [88]. In receptor-mediated mitophagy pathway, outer mitochondrial membrane receptors (such as FNDC-1 and DCT-1) and inner mitochondrial membrane receptor (such as PHB-2) bind to LGG-1 to induce autophagosome [89]. It was observed that treatment with quercetin could ameliorate the decreased ovarian reserve by activating DCT-1-medaited mitophagy [90].

11.5.4 Activation of Autophagy

Treatment with drugs (such as TOR inhibitor rapamycin) or exposure to stresses (such as starvation and heat stress) can induce the autophagy. LGG-1::GFP puncta formation in body wall muscle or intestine can be used to minor autophagy induction. In the canonical core autophagy machinery, LGG-1 and LGG-2 are lipidated to form autophagosome, ATG-12, ATG-5, and EPG-3/ATG-16 organize the conjugation system, VPS-34 functions to produce PI3P, BEC-1 acts as part of VPS34 complex, EPG-9 is a scaffold, and UNC-51 is a kinase to initiate autophagy [91–93]. Treatment with beta-amyrin exhibited the anti-Parkinsonian effect, which was largely due to the autophagy induction [94]. Cannabidiol could activate the autophagy, which was helpful for improving the neuronal health [95]. Coniferaldehyde activated the autophagy to confer the resistance to oxidative stress through PAR-4/AAK-2/SKN-1 pathway [8]. In addition, activation of autophagy by chaetoglobosin F contributed to inhibition in amyloid-beta induced neurotoxicity, which was under the control of insulin and p38 MAPK signaling pathways [10].

11.5.5 Activation of Endoplasmic Reticulum Unfolded Protein Response

Endoplasmic reticulum unfolded protein response (ER UPR) is a cellular stress response activated when misfolded or unfolded proteins accumulate in the ER [96, 97]. HSP-4::GFP can be used to minor ER UPR induction. The function of ER UPR is to restore ER homeostasis by increasing ER chaperone production, enhancing ER-associated degradation, reducing protein translation, or inducing autophagy [97, 98]. For the molecular basis, the ER UPR is governed by three conserved sensors: IRE-1 (most conserved), PEK-1 (PERK homolog), and ATF-6 [99, 100]. Asperuloside could promote the innate immunity by activating ER UPR driven by IRE-1/XBP-1 signaling [101]. In addition, certain marine-derived xyloketal compound ameliorate MPP^+-induced neuronal damage by activating ER UPR through IRE-1/XBP-1 signaling [102].

11.5.6 Enhancement in Proteostasis

Proteostasis is the integrated regulation of protein folding (chaperones), stress responses (such as mt UPR and ER UPR), and degradation (ubiquitin–proteasome system and autophagy) that ensures a functional proteome during the development [103, 104]. The chaperones, mt UPR, ER UPR, and autophagy have been introduced above. The ubiquitin–proteasome system is involved in maintaining proteostasis at least by quality control of newly synthesized proteins, proteasome-mediated clearance of oxidized or damaged proteins, and ER-associated degradation [103, 104]. During this process, cytosolic chaperones (such as HSP-70 and HSP-90) work together with E3 ligases (such as CHN-1) to target misfolded proteins for ubiquitin–proteasome system degradation [105, 106]. Some identified small-molecule compounds derived from bis(phenylmethylene) cycloalkanones and heterocyclic analogues could enhance proteasome activity and reduce size and number of protein aggregates [107]. Bafilomycin A and chloroquine could reduce the amyloid-β induced toxicity, which was associated with enhancement in proteasomal activity [108].

11.5.7 Activation of Innate Immune Response

In *C. elegans*, the innate immunity refers to the defense system that uses biological barriers (such as intestinal barrier and epidermal barrier), pattern recognition receptors (such as GPCRs), and conserved signaling pathways (such as p38 MAPK, insulin, Wnt, and autophagy-related signaling pathways) to induce antimicrobial effectors (such as NLPs, CNCs, and lysozymes) against infection of bacteria, fungi, and viruses [109–113]. Treatment with brevilin A enhanced the innate immune response by activating p38 MAPK signaling [114]. Schisandrin A and thymol also conferred the resistance to pathogen infection and enhance the innate immunity by targeting to p38 MAPK signaling [115, 116]. Cranberry extract could promote the innate immune response via activating both p38 MAPK and HSF-1 signals [117]. Treatment with certain components from red seaweed *Chondrus crispus* could enhance the innate immune response by modulating insulin and p38 MAPK signaling pathways [118]. Exposure to 6-PPD quinone (6-PPDQ) could cause the immunosuppression, which was recovered by the following cuminaldehyde treatment by activating both DAF-16 and PMK-1 (Fig. 11.3) [119]. The beneficial effect of cuminaldehyde treatment in increasing innate immune response in 6-PPDQ-exposed nematodes was inhibited by RNAi of *daf-16* and *pmk-1* [119].

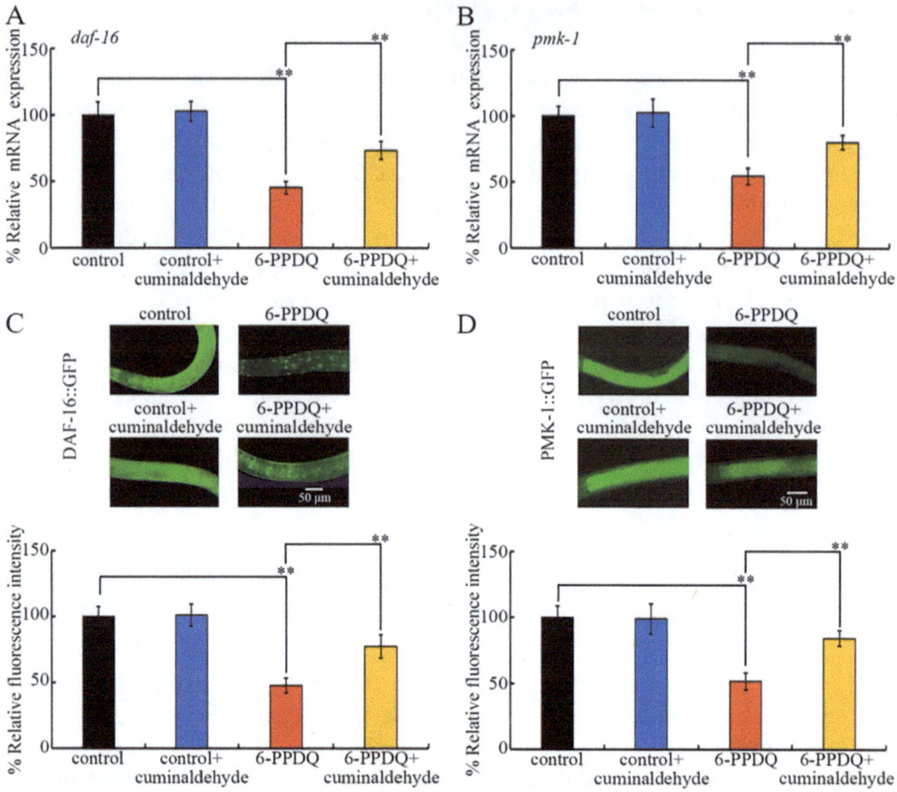

Fig. 11.3 Effect of cuminaldehyde treatment on expressions of DAF-16 and PMK-1 in 6-PPDQ exposed nematodes [119]. (**a**) Effect of cuminaldehyde treatment on *daf-16* expression in 6-PPDQ exposed nematodes. (**b**) Effect of cuminaldehyde treatment on *pmk-1* expression in 6-PPDQ exposed nematodes. (**c**) Effect of cuminaldehyde treatment on DAF-16::GFP expression in 6-PPDQ exposed nematodes. (**d**) Effect of cuminaldehyde treatment on PMK-1::GFP expression in 6-PPDQ exposed nematodes. The 6-PPDQ exposure concentration was 10 μg/L, and the treatment concentration of cuminaldehyde was 7 mg/L. $^{**}P < 0.01$

11.6 Underlying Mechanisms for Promotion in Growth and Increase in Reproduction by Compounds

11.6.1 Promotion in Growth

The *C. elegans* growth can be measured as increased body size, faster developmental rate, and enhanced biomass. Normally, the promotion in growth is linked to these three aspects of possible underlying mechanisms:

1. Activation of anabolic pathways (such as mTOR and insulin/IGF-1 signals).
2. Inhibition in certain catabolic pathways.
3. Enhancement in nutrient absorption or metabolic efficiency.

The *C. elegans clk-1* mutant shows the phenotype of delayed postembryonic growth [120]. In the *clk-1* mutant nematodes, administration with coenzyme Q10 could recover the growth retardation and restore the expressions of down-regulated genes important for growth (i.e., RNA polymerase II and eukaryotic initiation factor) [120]. Some amino acids, such as proline, can promote the growth by activating TOR signaling [121, 122]. Mobilization of cholesterol could speed the growth by inducing transition from quiescence to growth via steroid hormone and mTOR signaling [123].

11.6.2 Increase in Reproduction

The molecular mechanisms by which various pharmacological compounds increase reproduction (brood size, germline proliferation, egg-laying, or oocyte quality) contain the following several aspects:

1. Neuroendocrine control of egg-laying
   The neurotransmitter serotonin can stimulate the egg-laying [124, 125]. Moreover, fluoxetine and imipramine controlled the egg-laying by modulating serotonin reuptake transporter (SERT) and serotonin receptors, respectively [124]. In addition, another neurotransmitter dopamine could antagonize the serotonin action via serotonin-gated ion channel MOD-1, suggesting its role in coupling serotonin signaling and dopamine signaling [124]. The function of serotonin could also be antagonized by octopamine [126].
2. Increase in Germline Proliferation
   Pharmacological compounds can enhance reproductive capacity and germline proliferation by inhibiting germline apoptosis. For example, in nanoplastic-exposed nematodes, after the treatment with paeoniflorin, the brood size was increased, and the germline apoptosis was inhibited [127]. Moreover, paeoniflorin treatment inhibited DNA damage checkpoints (CLK-2, HUS-1, and MRT-2) and CEP-1 governing the germline DNA damage, and paeoniflorin has the potential to bind to these DNA damage checkpoints [127]. These observations suggest that paeoniflorin can increase the germline proliferation by inhibiting germline apoptosis through the modulation of DNA damage checkpoints.

   Besides this, some pharmacological treatment can directly cause the germline proliferation. Both folate 10-formyl-tetrahydrofolate-Glun and bacterial folate precursor dihydropteroate could stimulate the germline proliferation [128]. Moreover, the folate receptor homolog FOLR-1 was identified to be required for the observed stimulation of germline proliferation by 10-formyl-tetrahydrofolate-Glun and dihydropteroate [128].

3. Increase in Oocyte Quality
   Treatment with recapitulate, the serotonin reuptake inhibitor, could promote the healthy oocyte aging, suggesting the role of serotonin and related signals in maintaining oocyte quality [129].

11.7 Underlying Mechanisms for Modulation of Feeding and Food Intake by Compounds

11.7.1 Modulation of Feeding

Feeding involves the control of pharyngeal pumping, which is a rhythmic and myogenic behavior. Pharmacological compounds can modulate the feeding rate via specific molecular targets in different tissues.

The first well-described pharmacological mechanism is the involvement of serotonin signaling. Exogenous serotonin could increase the pumping rate. The serotonin could stimulate pharyngeal activity by inducing formation of eicosanoid 17,18-epoxyeicosatetraenoic acid via SER-7 receptor [130], which was identified to be required for stimulation of pharyngeal pumping by serotonin [131]. Moreover, a reciprocal modulation mechanism was further raised between serotonin and octopamine to regulate the pharyngeal pumping via feed forward and feedback neuronal circuits [132].

Besides the serotonin signaling, the dopamine signaling is also involved in control of the feeding behavior. Treatment with risperidone (an antagonist of dopamine D2 receptors) and aripiprazole (a partial agonist of dopamine D2 receptor) all could change the pharyngeal pumping [133]. After treatment with these compounds, the pharyngeal pumping could change to varying degrees in dopamine receptor mutants [133].

Moreover, both acetylcholine signaling and GABA signaling were required for pharyngeal pumping. Mutations in genes governing acetylcholine release and an antagonist of nicotinic acetylcholine receptor reduced the pharyngeal pumping [134], suggesting the role of acetylcholine in stimulating pharyngeal pumping. Treatment with levamisole, an agonist of acetylcholine, to selective stimulate body wall muscle receptors could inhibit the pharyngeal pumping [135], which further supported this. In contrast to these, treatment with muscimol, an agonist for GABA could inhibit the pharyngeal pumping [136], suggesting the role of GABA in inhibiting the pharyngeal pumping.

11.7.2 Modulation of Food Intake

The food intake refers to a concept that includes not just pharyngeal pumping but also food attraction/aversion and satiety signaling.

1. Modulation of Food Attraction/Aversion

Treatment with isoamyl alcohol and diacetyl represents the food attraction by acting as the attractants [137, 138]. These attractants modulate the food intake by altering the behavior to bacterial food patches to increase the time on food, which were sensed by AWA olfactory neurons [139, 140]. During this process, ODR-3 regulated the locomotion toward food by activating OSM-9, a TRPV channel [141]. In contrast, treatment with 1-octanol and 2-nonanone represents the food aversion by acting as the repellents [142, 143]. After treatment with repellents (such as 1-octanol), the ASH nociceptive neurons were activated to inhibit forward locomotion and to reduce food intake [144]. Moreover, the D2-like receptor DOP-3 was identified in the ASH neurons to dampen behavioral sensitivity to repellents [144].

2. Modulation of Food Satiety

The food satiety means the tendency to reduce the food intake. Considering that *C. elegans* quiescence can share some properties on satiety in mammals, *C. elegans* provides an important model for the study of food satiety [145]. During this process, the satiety induced quiescence was under the control of insulin peptides and TGF-β signals in certain sensory neurons (such as ASI neurons) through EGL-4 cGMP-dependent protein kinase [145]. Besides the insulin peptides, the RFamide-related peptide FLP-11 was further identified to regulate the food satiety quiescence in the RIS interneurons by modulating the cGMP signaling [146].

11.8 Underlying Mechanisms for Maintenance of Functional State of Biological Barriers by Compounds

It is assumed that certain pharmacological treatment (including the treatment with compounds) can be helpful for maintaining the functional state of biological barriers (e.g., intestinal barrier and epidermal barrier). Actually, activation of innate immune response can be helpful to enhance the functional state of biological barriers, such as the intestinal barrier [147, 148]. We have introduced and discussed the pharmacological effect of compounds on innate immunity in Chap. 10 and Sect. 11.5.7 in this chapter. Besides this, some other forms of pharmacological treatments are also helpful to enhance the functional state of biological barriers.

11.8.1 Pharmacological Effects of Phytochemicals

Treatment with jujube peel polyphenols/zein complexes was helpful to improve the biological barrier function, which was associated with the activation of innate immunity and modulation of lipid transportation mediated by altered lysosome pathway [149]. In 6-PPDQ-exposed nematodes, the intestinal barrier function was damaged, which could be inhibited by treatment with paeoniflorin, the major compound in paeony [150]. The beneficial effect of paeoniflorin treatment in

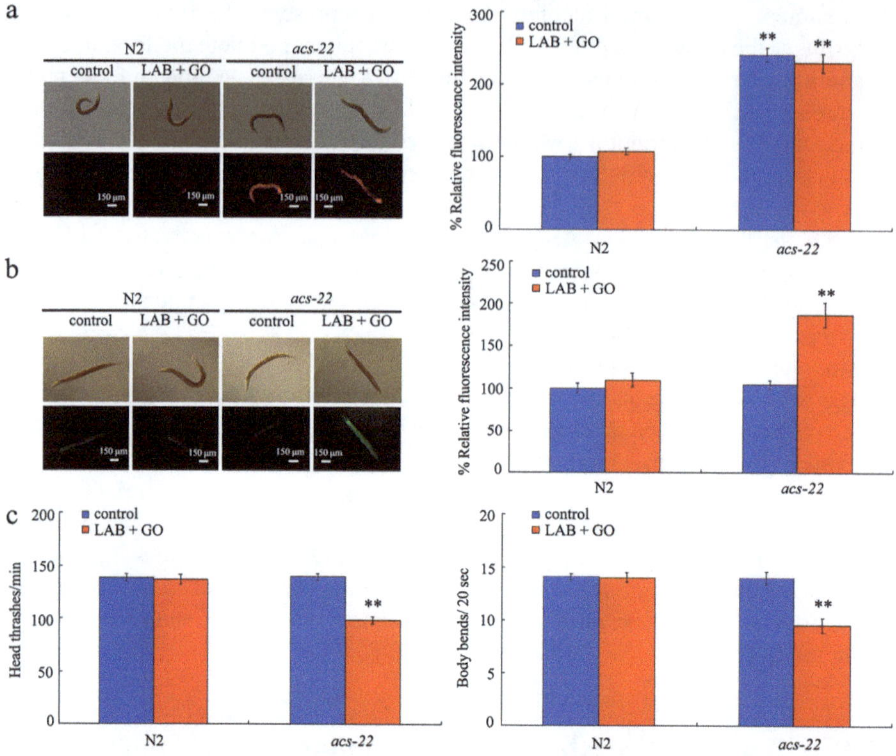

Fig. 11.4 Effects of *acs-22* mutation on beneficial effects of LAB administration against GO toxicity [151]. (**a**) Effects of *acs-22* mutation on beneficial effects of LAB administration against GO-induced enhancement of intestinal permeability as indicated by Nile Red staining results. (**b**) Effects of *acs-22* mutation on beneficial effects of LAB administration against GO-induced intestinal ROS production. (**c**) Effects of *acs-22* mutation on beneficial effects of LAB administration against GO-induced decrease in locomotion behavior. GO exposure concentration was 100 mg/L. The used LAB strain was L. bulgaricus. L4-larvae were pre-treated with LAB for 12 h, and then exposed to GO for 24 h at 20 °C. Bars represent means ± S.E.M. $^{**}P < 0.01$ vs N2

alleviating damage of 6-PPDQ on intestinal barrier function was largely due to the activation of ACS-22, a fatty acid transporter [150].

11.8.2 Pharmacological Effects of Lactic Acid Bacteria

In *C. elegans*, exposure to graphene oxide, a carbon-based nanomaterial, caused intestinal oxidative stress, enhanced intestinal permeability, and severe accumulation of graphene oxide in the body, which could be significantly suppressed by treatment with lactic acid bacteria (*Lactobacillus bulgaricus*) [151]. During this process, ACS-22 was required for the formation of beneficial effect of lactic acid bacteria, since the beneficial effects of lactic acid bacteria treatment in blocking intestinal oxidative damage and decreasing intestinal permeability were suppressed by *acs-22* mutation (Fig. 11.4) [151]. Besides this, treatment with *Lacticaseibacillus casei* HY2782 also improved the intestinal barrier function, which was associated with the extension of lifespan via DAF-16 signaling and SKN-1 signaling [152].

11.9 Underlying Mechanisms for Anti-Neurodegeneration-Related Diseases by Compounds

For the neurodegeneration-related diseases, we have introduced the Alzheimer' disease (AD) models, Parkinson's disease (PD) models, Huntington's disease (HD) models, amyotrophic lateral sclerosis (ALS) models, Machado–Joseph disease (MJD/SCA3) models, Friedreich Ataxia (FRDA) models, and Adult Neuronal Ceroid Lipofuscinosis (ANCL) models in Chap. 4. To a great degree, these generated disease models can also be considered as specific stresses. For this reason, the stress response-related signaling pathways (such as insulin signaling pathway, SKN-1/Nrf signaling, and mitochondrial signals) described in Sect. 11.2 in this chapter can definitely mediate the pharmacological effects of compounds and drugs in alleviating or blocking neurodegeneration-related diseases. Moreover, the activation of certain protective response (such as antioxidation, mt UPR, mitophagy, autophagy, ER UPR, proteostasis, and innate immunity) described in Sect. 11.5 in this chapter will also mediate the observed beneficial effects of compounds and drugs against the neurodegeneration-related diseases. We next introduced and discussed the relatively specific underlying mechanisms for pharmacological treatment with compounds against different neurodegeneration-related diseases.

11.9.1 Underlying Mechanisms for Anti-Alzheimer's Disease by Compounds

Using transgenic strains expressing human amyloid-beta (Aβ) and tau proteins, pharmacological effects of compounds and drugs on AD have been widely examined. The relatively specific underlying mechanisms for pharmacological treatment with compounds against the AD include:

1. Involvement of Autophagy
   Autophagy can mediate the pharmacological effect of compounds to clear the protein aggregates (such as the Aβ) by acting as a central cellular degradation pathway. Dendrobium officinale phenolic extract had the beneficial effect against the AD by activating autophagy to maintain the proteostasis [153]. Nelumbo nucifera leaf extract showed the function to be against amyloid-beta toxicity by partially activating autophagy, and the observed protective effect of *Nelumbo nucifera* leaf extract could be inhibited by RNAi of autophagy-related genes (*bec-1* and *vps-34*) [154].

2. Modulation of Neurotransmission
   Treatment with dimethyl sulfoxide (DMSO) could delay the Aβ-induced paralysis [155]. This observed protective effect of DMSO was inhibited by quisqualic acid, a glutamage agonist [155]. In addition, DMSO suppressed the acetylcholine neurotransmission as indicated by aldicarb and levamisole assays [155]. Thus, the DMSO could exhibit the protective effect against the AD by modulating glutamate/acetylcholine neurotransmission.

11.9.2 Underlying Mechanisms for Anti-Parkinson's Disease by Compounds

PD is the second most common neurodegenerative disorder. Using chemicals induced models and transgenic strains expressing human α-synuclein, pharmacological effects of compounds and drugs on PD have also been analyzed in *C. elegans*. The relatively specific underlying mechanisms for pharmacological treatment with compounds against the PD include:

1. Involvement of Autophagy
   The autophagy can also be activated by compounds to clear the α-synuclein aggregates and damaged organelles. Araloside A treatment could alleviate protein aggregation and behavioral deficits through activation of ERK-dependent autophagy [156]. The *H. scabra* compounds (HSEA-P1 and HSEA-P2) could activate autophagy and up-regulating autophagy-related genes, and the autophagy-related genes of *lgg-1* and *atg-7* were confirmed to be required for α-synuclein degradation and DAergic neuroprotection induced by HSEA-P1 and HSEA-P2 [157]. Similarly, treatment with beta-amyrin had the anti-Parkinsonian effect via LGG-1 mediated autophagy [94].
2. Modulation of Mitochondrial Signals
   Mitochondrial dysfunction caused by efficiency in mitochondrial complexes (such as the complex I) is a well-identified feature of PD pathogenesis. That is, compounds can exhibit the protective effect on the PD by modulating mitochondrial complexes, mt UPR, mitophagy, and mitochondrial biogenesis. Treatment with deoxytrillenoside CA and epitrillenoside CA could reduce the α-synuclein accumulation by activating mitophagy and promoting mitochondrial biogenesis, and the protective effect of deoxytrillenoside CA and epitrillenoside CA against the PD could be suppressed by RNAi of genes governing mitophagy and mitochondrial biogenesis [158].
3. Direct α-synuclein Aggregation Inhibition
   Shikonin (SHK), a natural naphthoquinone from *Lithospermum erythrorhizon*, was identified to inhibit the α-synuclein aggregation by binding to the C-terminus of α-synuclein to maintain the α-helical and to disorder the secondary structures, which led to the reduced β-sheet content and complexity of aggregates [159]. Trodusquemine treatment showed multistep inhibition of α-synuclein

aggregation, such as reduction in the number of α-synuclein inclusions and suppression in α-synuclein oligomers [160].

11.9.3 Underlying Mechanisms for Anti-Huntington's Disease by Compounds

Using transgenic strains expressing human Htt-Q150, polyQ40::YFP, or htt57-128Q, pharmacological effects of compounds and drugs on HD can be assessed in *C. elegans*. The relatively specific underlying mechanisms for pharmacological treatment with compounds against the HD include:

1. Involvement of Autophagy
   Autophagy disorder can be considered as a hallmark of HD pathogenesis. Thus, autophagy can be activated by compounds to clear the misfolded proteins, such as polyQ-expanded huntingtin. Perillaldehyde treatment could alleviate the polyQ induced neurodegeneration by activating autophagy, and this protective effect of perillaldehyde was inhibited by RNAi of *lgg-1* [161].
2. Modulation of Mitochondrial Signals
   Perillaldehyde could also suppress polyQ-induced neurodegeneration by activating mt UPR, and this beneficial effect of perillaldehyde could be inhibited by mt UPR-related genes [161]. During this process, synthesis and neurosecretion of serotonin was further found to be required for mt UPR activation and neuroprotective function of perillaldehyde [161]. In addition, treatment with deoxytrillenoside CA and epitrillenoside CA could also reduce poly40 aggregates and inhibit motor deficits by activating mitophagy and increasing mitochondrial biogenesis, and the function of deoxytrillenoside CA and epitrillenoside CA in reducing poly40 aggregates was inhibited by RNAi of mitophagy and mitochondrial biogenesis-related genes [158].

11.9.4 Underlying Mechanisms for Anti-Amyotrophic Lateral Sclerosis Disease by Compounds

Using mutants (such as *dnc-1* mutant) or transgenic strains expressing human mutant SOD1 or wild-type and mutant TDP-43, pharmacological effects of compounds and drugs on ALS have been assessed in *C. elegans*. The relatively specific underlying mechanisms for pharmacological treatment with compounds against the ALS include:

1. Involvement of Autophagy
   Treatment with metformin could reduce the ALS phenotypes in hSOD1 transgenic strain, which was associated with the activation of autophagy [162]. Moreover, the pharmacological effect of metformin against the ALS could be enhanced by LGG-1 overexpression, and abolished by mutation of *lgg-1* [162].

2. Modulation of Mitochondrial Signals
   Treatment with isoginkgetin, a bioflavonoid from *Ginkgo biloba*, could suppress the ALS pathologies by activating mitophagy [163]. The activation of mitophagy by isoginkgetin was due to the stabilization of the PINK1/TOM complex [163].
3. Modulation of Calcium Channels
   In the ALS, the calcium dysregulation is recognized as a contributor to motor neuron degeneration. Based on a behavior-based drug screening, nifedipine, an L-type calcium channel blocker, was identified to ameliorate the motor deficits in *dnc-1* mutant nematodes [164]. This pharmacological effect of nifedipine could be further confirmed in TDP-43 overexpressing worms [164]. In addition, among the neuroleptics, pimozide was identified to be helpful for treating ALS by blocking the T-type Ca^{2+} channels [165].

11.9.5 Underlying Mechanisms for Anti-Machado-Joseph Disease by Compounds

Using transgenic strains expressing human mutant ATXN3 with 89 glutamines pan-neuronally, pharmacological effects of compounds and drugs on MJD/SCA3 can be evaluated in *C. elegans*. The relatively specific underlying mechanisms for pharmacological treatment with compounds against the MJD/SCA3 include:

1. Involvement of autophagy
   Based on the drug screen of 3942 compounds, some small molecules were identified to potentially rescue the motor-deficient phenotype in transgenic ATXN3-CAG89 strain [166]. The observed beneficial effects including restoring the motility, inhibition in neurodegeneration, and extension of lifespan by compounds of alfacalcidol, chenodiol, cyclophosphamide, fenbufen, and sulfaphenazole were associated with the autophagy activation mediated by transcriptional factor HLH-30 [166].
2. Modulation of Neurotransmitter Receptors
   In MJD model, aripiprazole treatment could reduce the motor dysfunction [167]. Moreover, receptors of dopamine and serotonin were identified as the targets of aripiprazole, and the dopamine D2-like and serotonin 5-HT1A and 5-HT2A receptors were shown to regulate the mutant ATXN3-mediated pathogenesis [167].

11.10 Underlying Mechanisms for Anti-Seizure-Like, Addiction, Anxiety-Like, and Autism-Like Phenotypes by Compounds

11.10.1 Underlying Mechanisms for Anti-Seizure-Like Phenotypes by Compounds

The main underlying mechanisms observed for anti-seizure like phenotypes by compounds contain the following several aspects:

1. Modulation of GABAergic Signaling
   Seizure-like activity can be induced by GABA antagonist pentylenetetrazole (PTZ) or by mutations in GABA pathway genes. Based on this, it was found that the treatment with 2-[(4-methoxyphenylselenyl)methylthio]-5-phenyl-1,3,4-oxadiazole (MPMT-OX) could enhance the resistance to nematodes to PTZ-induced paralysis and aid the recovery of locomotor activity after induction of seizures [168]. This observed effects of MPMT-OX were inhibited by mutation of *unc-25* and *unc-47*, suggesting the anti-seizure effect of MPMT-OX required the UNC-25 (glutamate decarboxylase required for GABA synthesis enzyme) and UNC-47 (vesicular GABA transporter) [168]. Moreover, a high-throughput screen was performed among 1690 compounds and chlorothymol was identified as the anticipated compound [169]. Based on the further genetic analysis, GABA receptor subunit LGC-37 was identified as the target of chlorothymol against the seizures like phenotypes [169].
2. Modulation of cGMP/PKG Signaling
   The cyclic GMP-dependent protein kinase G (PKG) is a certain target for seizure therapeutics. It was shown that treatment with sildenafil citrate, a PDE-5 inhibitor, could rescue the nematodes from prolonged seizure-like movement duration following docetaxel exposure by acting as a PKG activator [170]. Besides the sildenafil citrate, 8-bromo-cGMP also acted as the pharmacological PKG activator to reduce the seizure duration [171].

11.10.2 Underlying Mechanisms for Anti-Addiction Phenotypes by Compounds

The main underlying mechanisms observed for anti-addiction like phenotypes by compounds contain the following several aspects:

1. Modulation of Opioid-Like System
   NPR-17 is a G-protein-coupled receptor sharing 68% identity with opioid receptor-like protein ORL1, and NLP-24 is its corresponding ligand [172]. Naltrexone was identified to selectively block the preference for multiple substances of abuse, such as ethanol, cocaine, and nicotine by acting as a pan-opioid receptor antagonist [173]. In addition, considering that its pretreatment did not affect

general locomotion, food chemotaxis, or benzaldehyde attraction, treatment with naltrexone may have the specific pharmacological effect on addiction [173].

2. Modulation of Dopamine Signaling
   The dopamine signaling may function as the central hub for the control of addiction, and the dopamine-deficient worms (such as *cat-1* and *cat-2* mutants) fail to develop preference for multiple addictive substances. Treatment with morphine can produce the addiction-like behaviors through NPR-17, and morphine could increase dopamine level in PDE neurons in an NPR-17-dependent manner [174]. This implies that treatment with antagonists of morphine can be helpful to treat the addiction behavior.
3. Modulation of Sigma-2 Receptor
   The sigma receptors can function as the modulators of the rewarding and reinforcing effects of alcohol. In *C. elegans*, two small molecules were identified as the selective modulators of sigma-2 receptor (σ2R/TMEM97) to reduce alcohol withdrawal behavior [175].

11.10.3 Underlying Mechanisms for Anti-Anxiety-Like Phenotypes by Compounds

The main underlying mechanisms observed for anti-anxiety like phenotypes by compounds contain the following several aspects:

1. Modulation of Serotonin Signaling
   The serotonin system plays the central role during the occurrence of anxiety disorders, and serotonin reuptake inhibitors (such as fluoxetine) are the first-line anti-anxiety compounds. Ursolic acid treatment could enhance the stress resistance and reduce the ROS accumulation by acting through serotonin receptors [176]. Similarly, withanone could reduce the ROS accumulation by activating serotonin receptors [177].
2. Modulation of dopamine signaling
   The important role of dopamine receptors in occurrence of anxiety has been suggested [178]. Ursolic acid treatment could confer the resistance to multiple stresses, including oxidative stress, osmotic stress, and heat stress [179]. This observed beneficial effect of ursolic acid might be linked to activation of dopamine receptors of DOP-1 and DOP-3 [179].

11.10.4 Underlying Mechanisms for Anti-Autism-Like Phenotypes by Compounds

The main underlying mechanisms observed for anti-autism like phenotypes by compounds contain the following several aspects:

1. Modulation of Excitatory/Inhibitory Balance

The DAF-18, a PTEN ortholog, regulates neurodevelopment by specifically influencing inhibitory GABAergic neurons, and mutation of *daf-18* causes an imbalance where excitation dominates over inhibition. Administration with ketone body β-hydroxybutyrate that can be naturally produced during fasting or ketogenic diets was shown to reverse phenotypes in *daf-18* mutants by restoring expression of genes critical for inhibitory neuron development and rebalancing excitatory/inhibitory ratio [180].

2. Modulation of Serotonin Signaling and Dopamine Signaling
   The nematodes with mutation of *nlg-1* encoding Neuroligin can be used as a useful model with the autism like phenotypes [181]. The behavioral defects in *nlg-1* mutant could be rescued by treatment with methylphenidate (dopamine reuptake inhibitor) and fluoxetine (serotonin reuptake inhibitor), which suggested the modulation of serotonin signaling and dopamine signaling [181]. Additionally, based on the screening, some monoamine-targeting drugs could improve behavioral phenotypes in nematodes with mutation of *nrx-1* encoding Neurexin [182]. For example, treatment with antagonism of serotonin receptors could be helpful to increase behavioral activity of *nrx-1* mutants [182].

11.11 Underlying Mechanisms for Modulation of Sleep by Compounds

Based on the screening from 3683 bioactive compounds, the amitriptyline was identified to modulate the sleep by acting downstream of or in parallel to activation of sleep-promoting neurons [183]. In *C. elegans*, neuropeptide FLP-11 acts together with its receptor DMSR-1 to provide a functional switch that both initiates and terminates the sleep [184]. This implies that the compounds that can enhance FLP-11 release, stabilize FLP-11 binding to DMSR-1, or act as DMSR-1 agonists would promote the sleep.

11.12 Underlying Mechanisms for Improvement in Muscle Function by Compounds

Under the stress condition, the stress response-related signaling pathways (such as insulin signaling pathway, SKN-1/Nrf signaling, and mitochondrial signals) and the activation of certain protective response (such as antioxidation, mt UPR, mitophagy, autophagy, ER UPR, proteostasis, and innate immunity) described in Sects. 11.2 and 11.5 in this chapter will mediate the pharmacological effects of compounds and drugs in improving muscle function. Besides this, the underlying mechanisms for improvement in muscle function by compounds also contain the following several aspects:

1. Metabolite Supplementation

In *C. elegans*, eicosanoids are derived from polyunsaturated fatty acids (PFUAs). During this process, cytochrome P450 enzyme CYP-13A12 functioned as a PUFA epoxygenase to produce 17,18-EEQ (17,18-epoxyeicosatetraenoic acid) from EPA and 14,15-EET from arachidonic acid [185]. Treatment with 17,18-EEQ could increase locomotion behavior by elevating muscle activity [185]. The impaired locomotion in PUFA-deficient mutants could also be rescued by 17,18-EEQ treatment [185]. In addition, treatment with picolinic acid, a tryptophan metabolite, could increase the mobility by improving the muscle function [186].

2. Modulation of Mitochondrial Na^+/Ca^{2+} Exchanger
   CGP37157 is the inhibitor of mitochondrial Na^+/Ca^{2+} exchanger. Treatment with CGP37157 could delay the development of sarcopenia in body wall muscle and increase the average and maximum moving speed by modulating cytosolic and mitochondrial Ca^{2+} homeostasis [187].
3. Modulation of Action of Neurotransmitters
   In the dystrophin-dependent muscular dystrophy model, serotonin, as well as some of its agonists, could inhibit the muscle degeneration [188]. Treatment with inhibitors of serotonin reuptake transporters had the similar effect in muscular dystrophy model by prolonging action of endogenous serotonin [188].

11.13 Underlying Mechanisms for Improvement in Regeneration and Wounding Healing by Compounds

11.13.1 Underlying Mechanisms for Improvement in Regeneration by Compounds

Based on the analysis of high-throughput microfluidics and/or femtosecond laser microsurgery, some small molecules were identified to alter the neurite regeneration [189, 190]. Among them, treatment with staurosporine, a PKC kinase inhibitor, modulated the neuronal regeneration in a neuronal type-specific manner [190]. Moreover, the neuronal regeneration could be enhanced by prostratin, a PKC activator [190].

11.13.2 Underlying Mechanisms for Improvement in Wounding Healing by Compounds

In *C. elegans*, H_2O_2 could oxidize certain cysteine residues on CDC-42 (Cys18 and Cys105), suggesting the redox-sensitive regulation of CDC-42 [191]. Moreover, the low concentrations of H_2O_2 promoted the wound healing by facilitating CDC-42 activation, while high concentrations of H_2O_2 inhibited the wound healing by promoting excessive oxidation [191]. Besides this, treatment with xestospongin C, the antagonist of inositol 1,4,5-trisphosphate, could phenocopy the defect in epidermal cell migration in *itr-1* mutant, suggesting its possible role in modulating wound healing [192]. In addition, Aloe-emodin was also found to regulate the epidermal cell migration through AMPK signaling [193]. The nicotinamide, a pyridine carboxamide oxime, was further identified to show the wound healing ability [194].

11.14 Underlying Mechanisms for Anti-Tumor by Compounds

The first aspect of underlying mechanism is the modulation of cell cycle and germline apoptosis machinery. Some small-molecule inhibitors of fatty acid synthase were identified to show anti-tumor activity by inducing cell cycle arrest and cell division inhibition [195]. Besides this, treatment with sanguinarine, a natural alkaloid, could trigger CEP-1/p53-dependent apoptosis, implying the activation of DNA damage signaling [196].

The second aspect of underlying mechanism is the modulation of certain molecular signaling, such as insulin signaling and Ras-MAPK signaling. For example, treatment with EAPB02303, a second-generation imiqualine, could exhibit the anti-tumor effects with IC50 values in the nanomolar range, and this observed anti-cancer activity of EAPB02303 was formed through dual inhibition of both insulin signaling and Ras-MAPK signaling [197]. In addition, based on anti-tumor drug screening, the anti-tumor activity of identified edelfosine was blocked by mutation of genes in insulin signaling pathway [198].

11.15 Underlying Mechanisms for Modulation of Metabolic Processes by Compounds

11.15.1 Underlying Mechanisms for Reduction in Fat Storage by Compounds

The underlying mechanisms for compounds to reduce the fat storage are mainly reflected by the following several aspects:

1. Modulation of Key Enzymes Involved in Fatty Acid Synthesis
   The fat storage is associated with activation of enzymes governing fatty acid synthesis. The related enzymes contain (1) fatty acid synthase FASN-1, (2) acetyl-CoA carboxylase POD-2, and (3) Delta-9 desaturases (FAT-5, FAT-6, and FAT-7). Piceatannol could reduce fat storage by inhibiting FASN-1 [199]. Cycloastragenol could reduce lipid accumulation by inhibiting the FAT-7.
2. Modulation of Enzymes Involved in Fatty Acid β-oxidation
   The fat storage is associated with inhibition in enzymes governing fatty acid β-oxidation. The related enzymes contain (1) enzymes governing mitochondrial fatty acid β-oxidation, (2) enzymes governing peroxisomal fatty acid β-oxidation, and (3) triglyceride lipases. Dihomo-gamma-linolenic acid inhibited lipid accumulation by promoting peroxisomal fatty acid β-oxidation [201]. Piceatannol could also reduce the fat storage in glucose-treated worms by activating HOSL-1, an ortholog of hormone-sensitive lipase [199].

3. Modulation of Metabolic Sensors

   In *C. elegans*, some transcriptional factors act as the metabolic sensors to control fat storage with genes governing fatty acid synthesis and β-oxidation as their downstream targeted genes. Among them, MDT-15 interacts SBP-1 to modulate the fatty acid synthesis and the fatty acid β-oxidation. Cranberry product could reduce the fat storage by inhibiting the SBP-1 [202]. Litchi flower essential oil reduced fat storage by inhibiting SBP-1 and MDT-15 [203].

   Besides SBP-1 and MDT-15, NHR-49, the ortholog of the mammalian PPARα, is another metabolic sensor, which mediates the breakdown of stored fat (lipolysis) and its conversion into usable energy through β-oxidation. The function of cranberry product and Litchi flower essential oil in reducing fat storage was also associated with the activation of NHR-49 and related fatty acid β-oxidation genes [202, 203].
4. Modulation of Energy Sensing Signals

   The AMP-activated protein kinase (AMPK) is a cellular energy sensor, and its activation will promote the catabolic pathways (such as fatty acid oxidation). Function of cycloastragenol treatment in reducing fat storage was linked to activation of AAK-2 [200]. Treatment with 3,3′-diindolylmethane and esculetin reduced the fat storage, and their function was partially dependent on AAK-1 [204, 205].
5. Modulation of Insulin Signaling

   In the insulin signaling pathway, DAF-2 and DAF-16 function as central and antagonistic regulators of fat storage. That is, when insulin signals are active, DAF-2 promotes fat storage by inhibiting DAF-16. The observed effect of esculetin in reducing fat storage was also dependent on DAF-2 and DAF-16 [205]. In addition, DAF-2 was required for the function of fermented barley extracts to reduce the fat storage [206].
6. Modulation of Neuronal Signaling

   In the nervous system, TPH-1 (the serotonin synthesis enzyme) and some serotonin receptors can stimulate the lipolysis and fatty acid β-oxidation. Coffee silverskin could reduce the fat storage by activating TPH-1 and serotonin receptor SER-6, which suggesting the activation of serotonin-mediated regulation of lipid metabolism [207].

11.15.2 Underlying Mechanisms for Modulation of Co-Factors by Compounds

Nicotinamide adenine dinucleotide (NAD^+) is a critical cofactor for energy metabolism, and the NAD^+ can be synthesized via three different pathways in *C. elegans*. With the NAD^+ as the example of co-factors, pharmacological restoration of NAD^+ by administration with its salvage pathway precursors nicotinamide and NAM riboside could prevent the age-associated metabolic decline and extend the lifespan [208], which suggested that the dietary NAD^+ precursors can show profound effects on biological processes, such as longevity, by modulating the cofactor availability.

11.16 Perspectives

In this chapter, we introduced and discussed the underlying molecular mechanisms of different pharmacological effects of compounds. Nevertheless, we did not cover the underlying mechanisms for all of the observed pharmacological effects of compounds in *C. elegans*. Among the introduced and discussed underlying mechanisms, it was observed that the compounds usually exhibited their beneficial effects by modulating stress response-related signaling pathways (such as insulin signaling pathway, SKN-1/Nrf signaling, and mitochondrial signals) and activating certain protective response (such as antioxidation, mt UPR, mitophagy, autophagy, ER UPR, proteostasis, and innate immunity). Nevertheless, it's not all like this.

In *C. elegans*, at the most cases, determination of underlying molecular mechanisms of pharmacological effects of compounds is based on the integration of gene expression assay and genetic analysis. Normally, the evidence to indicate the direct binding of compounds to targeted proteins or enzymes is lacking or limited. In addition, most of the obtained data suggested that treatment with compounds can show pharmacological effects by modulating certain molecular signaling. More efforts on the targeting of compounds to candidate G protein-coupled proteins, ion channels, nuclear hormone receptors, and rate-limiting enzymes are suggested to be further made and paid attention to.

References

1. Hernández-Cruz EY, Eugenio-Pérez D, Ramírez-Magaña KJ, Pedraza-Chaverri J (2023) Effects of vegetal extracts and metabolites against oxidative stress and associated diseases: studies in *Caenorhabditis elegans*. ACS Omega 8:8936–8959
2. Murphy CT, Hu PJ (2013) Insulin/insulin-like growth factor signaling in *C. elegans*. WormBook. https://doi.org/10.1895/wormbook.1.164.1
3. Rodriguez-Colman MJ, Dansen TB, Burgering BMT (2024) FOXO transcription factors as mediators of stress adaptation. Nat Rev Mol Cell Biol 25:46–64
4. Büchter C, Havermann S, Koch K, Wätjen W (2016) Isoxanthohumol, a constituent of hop (*Humulus lupulus* L.), increases stress resistance in *Caenorhabditis elegans* dependent on the transcription factor DAF-16. Eur J Nutr 55:257–265
5. Chen X, Yan B, Wu Y, Quan W, Liu Y, Huang Y, Shao Y, Wang Y, Zhou Q, Liu Y, Liu S, Wang J, Li P, Shi Z, Shu P, Gou J (2025) *Selaginella* extracts extend lifespan and mitigate oxidative stress in *Caenorhabditis elegans*. Front Pharmacol 16:1658991
6. Loboda A, Damulewicz M, Pyza E, Jozkowicz A, Dulak J (2016) Role of Nrf2/HO-1 system in development, oxidative stress response and diseases: an evolutionarily conserved mechanism. Cell Mol Life Sci 73:3221–3247
7. Hu Q, Liu Z, Guo Y, Lu S, Du H, Cao Y (2021) Antioxidant capacity of flavonoids from *Folium Artemisiae* Argyi and the molecular mechanism in *Caenorhabditis elegans*. J Ethnopharmacol 279:114398
8. Ramatchandirane M, Rajendran P, Athira MP, Suchiang K (2024) Coniferaldehyde activates autophagy and enhances oxidative stress resistance and lifespan of *Caenorhabditis elegans* via *par-4/aak-2/skn-1* pathway. Biogerontology 26:25
9. Yu XT, Shi L, Huang Q, Li B, Li YQ, Ou LL, Luo HR, Wu GS (2025) Hyperoside extends lifespan in *Caenorhabditis elegans* through SEK-1/PMK-1/SKN-1 pathway. Biogerontology 26:180

10. Lu X, Xu K, Zhou Z, Ding G, Zhang Y, Liang Y, Zhou J, Jia M, Zhang Y, Shen L, Li H (2026) Chaetoglobosin F attenuates amyloid-beta-induced neurotoxicity in *Caenorhabditis elegans* by regulating autophagy and oxidative stress via the insulin/IGF-1 and p38 MAPK pathways. Neurochem Res 51:80
11. Xiao Y, Zhang L, Zhou H, Cui Y, Chen K, Zhang H, Wu Q, Liu F (2025) Berberine extends healthspan and delays neurodegenerative diseases in *Caenorhabditis elegans* through ROS-dependent PMK-1/SKN-1 activation. Arch Gerontol Geriatr 128:105644
12. Wang D-Y (2019) Molecular toxicology in *Caenorhabditis elegans*. Springer Nature Singapore Pte Ltd.
13. Wang D-Y (2022) Toxicology at environmentally relevant concentrations in *Caenorhabditis elegans*. Springer Nature Singapore Pte Ltd.
14. Liu H-L, Zhao Y-L, Hua X, Wang D-Y (2022) Induction of transgenerational toxicity is associated with the activated germline insulin signals in nematodes exposed to nanoplastic at predicted environmental concentrations. Ecotoxicol Environ Saf 243:114022
15. Fu X, Han J, Xia Y, Wang W, Wen Y, Zhang B, Zhang X, Wang X, Han S, Zhang C (2026) Eleutheroside B ameliorates AD-like pathological features in *Caenorhabditis elegans* by inducing autophagy and combating oxidative stress. Free Radic Biol Med 245:43–55
16. Kyriakou E, Taouktsi E, Syntichaki P (2022) The thermal stress coping network of the nematode *Caenorhabditis elegans*. Int J Mol Sci 23:14907
17. Rubio-Tomás T, Alegre-Cortés E, Lionaki E, Fuentes JM, Tavernarakis N (2024) Heat shock and thermotolerance in *Caenorhabditis elegans*: an overview of laboratory techniques. Methods Cell Biol 185:1–17
18. You L, Huang Z, He W, Zhang L, Yu H, Zeng Y, Huang Y, Zeng S, Zheng L (2025) Dietary alpha-lipoic acid alleviates heat stress by modulating insulin-like signaling to maintain homeostasis in *C. elegans*. Food Funct 16:2824–2839
19. Shao XY, He T, Lai YL, Chen M, Tong ZH (2022) Water-soluble polysaccharides extracted from *Pueraria lobata* delay aging of *Caenorhabditis elegans* under heat stress. Plant Foods Hum Nutr 77:220–225
20. Cornwell AB, Zhang Y, Thondamal M, Johnson DW, Thakar J, Samuelson AV (2024) The C. elegans Myc-family of transcription factors coordinate a dynamic adaptive response to dietary restriction. Geroscience 46:4827–4854
21. Kozlova AA, Lotfi M, Okkema PG (2019) Cross talk with the GAR-3 receptor contributes to feeding defects in *Caenorhabditis elegans eat-2* mutants. Genetics 212:231–243
22. Zhang SY, Qin ZC, Sun YY, Chen YS, Chen WB, Wang HG, An D, Sun D, Liu YQ (2023) Genistein promotes anti-heat stress and antioxidant effects via the coordinated regulation of IIS, HSP, MAPK, DR, and mitochondrial pathways in *Caenorhabditis elegans*. Antioxidants 12:125
23. Lu M, Tan L, Zhou XG, Yang ZL, Zhu Q, Chen JN, Luo HR, Wu GS (2020) Tectochrysin increases stress resistance and extends the lifespan of *Caenorhabditis elegans* via FOXO/DAF-16. Biogerontology 21:669–682
24. Matsuyama HJ, Mori I (2020) Neural coding of thermal preferences in the nematode *Caenorhabditis elegans*. eNeuro 7:ENEURO.0414-19.2020
25. Huang TT, Matsuyama HJ, Tsukada Y, Singhvi A, Syu RT, Lu Y, Shaham S, Mori I, Pan CL (2020) Age-dependent changes in response property and morphology of a thermosensory neuron and thermotaxis behavior in *Caenorhabditis elegans*. Aging Cell 19:e13146
26. Zhou Y, Loeza-Cabrera M, Liu Z, Aleman-Meza B, Nguyen JK, Jung SK, Choi Y, Shou Q, Butcher RA, Zhong W (2017) Potential nematode alarm pheromone induces acute avoidance in *Caenorhabditis elegans*. Genetics 206:1469–1478
27. Gartner A, Engebrecht J (2022) DNA repair, recombination, and damage signaling. Genetics 220:iyab178
28. Li S, Liu D, Liu Y, Liu B, Chen X (2020) Quercetin and its mixture increase the stress resistance of *Caenorhabditis elegans* to UV-B. Int J Environ Res Public Health 17:1572

29. Liu M, Xiong Y, Shan S, Zhu Y, Zeng D, Shi Y, Zhang Y, Lu W (2020) Eleutheroside E enhances the long-term memory of radiation-damaged *C. elegans* through G-protein-coupled receptor and neuropeptide signaling pathways. J Nat Prod 83:3315–3323
30. Lee EC, Kim H, Ditano J, Manion D, King BL, Strange K (2016) Abnormal osmotic avoidance behavior in *C. elegans* is associated with increased hypertonic stress resistance and improved proteostasis. PLoS One 11:e0154156
31. Bao K, Liu W, Song Z, Feng J, Mao Z, Bao L, Sun T, Hu Z, Li J (2022) Crotamiton derivative JM03 extends lifespan and improves oxidative and hypertonic stress resistance in *Caenorhabditis elegans* via inhibiting OSM-9. elife 11:e72410
32. Davis M, Montalbano A, Wood MP, Schisa JA (2017) Biphasic adaptation to osmotic stress in the *C. elegans* germ line. Am J Physiol Cell Physiol 312:C741–C748
33. Liu Z-J, Li Y-H, Wang D-Y (2024) Exposure to 6-PPD quinone disrupts glucose metabolism associated with lifespan reduction by affecting insulin and AMPK signals in *Caenorhabditis elegans*. Chemosphere 363:142975
34. Liu Z-J, Li Y-H, Wang D-Y (2024) Transgenerational response of glucose metabolism in *Caenorhabditis elegans* exposed to 6-PPD quinone. Chemosphere 367:143653
35. Liu T-W, Zhuang Z-H, Wang D-Y (2023) Paeoniflorin mitigates high glucose-induced lifespan reduction by inhibiting insulin signaling in *Caenorhabditis elegans*. Front Pharmacol 14:1202379
36. de Oliveira NR, Siqueira AS, Bueno PSA, Gonçalves EC, Zanette J (2026) Metal-coordination specificity and structural dynamics of *C. elegans* metallothionein I: insights from 3D modeling and MD simulations. Proteins 94:620–632
37. Zhao Y, Ni S, Pei C, Sun L, Wu L, Xu A, Nie Y, Liu Y (2024) Parental treatment with selenium protects *Caenorhabditis elegans* and their offspring against the reproductive toxicity of mercury. Sci Total Environ 912:169461
38. Martins AC, Virgolini MB, Ávila DS, Scharf P, Li J, Tinkov AA, Skalny AV, Bowman AB, Rocha JBT, Aschner M (2023) Mitochondria in the spotlight: *C. elegans* as a model organism to evaluate xenobiotic-induced dysfunction. Cells 12:2124
39. Akinrinde A, Sanchez LA, Maglioni S, Ventura N (2026) *Caenorhabditis elegans* as a model system for environmental mitotoxicants. Annu Rev Pharmacol Toxicol 66:261–284
40. Wu Q-L, He K-W, Liu P-D, Li Y-X, Wang D-Y (2011) Association of oxidative stress with the formation of reproductive toxicity from mercury exposure on hermaphrodite nematode *Caenorhabditis elegans*. Environ Toxicol Pharmacol 32:175–184
41. Caito SW, Aschner M (2016) NAD^+ supplementation attenuates methylmercury dopaminergic and mitochondrial toxicity in *Caenorhabditis elegans*. Toxicol Sci 151:139–149
42. Ruan Q-L, Qiao Y, Zhao Y-L, Xu Y, Wang M, Duan J-A, Wang D-Y (2016) Beneficial effects of *Glycyrrhizae radix* extract in preventing oxidative damage and extending the lifespan of *Caenorhabditis elegans*. J Ethnopharmacol 177:101–110
43. Zhou L, Liu J, Bu LL, Liao DF, Cheng SW, Zheng XL (2021) Curcumin acetylsalicylate extends the lifespan of *Caenorhabditis elegans*. Molecules 26:6609
44. Jattujan P, Srisirirung S, Watcharaporn W, Chumphoochai K, Kraokaew P, Sanguanphun T, Prasertsuksri P, Thongdechsri S, Sobhon P, Meemon K (2022) 2-Butoxytetrahydrofuran and palmitic acid from *Holothuria scabra* enhance *C. elegans* lifespan and healthspan via DAF-16/FOXO and SKN-1/NRF2 signaling pathways. Pharmaceuticals 15:1374
45. Zhang H, Xiong J, Wang Q, Song Q, Meng L, Zhang H, Bao Y, Liu F, Xiao Y (2025) Chrysophanol delays aging via insulin/IGF-1 signaling pathway. Free Radic Biol Med 232: 269–278
46. Todorova MN, Savova MS, Mihaylova LV, Georgiev MI (2023) Icariin improves stress resistance and extends lifespan in *Caenorhabditis elegans* through *hsf-1* and *daf-2*-driven hormesis. Int J Mol Sci 25:352
47. Zhang Y, Zhang W, Dong M (2018) The *miR-58* microRNA family is regulated by insulin signaling and contributes to lifespan regulation in *Caenorhabditis elegans*. Sci China Life Sci 61:1060–1070

48. Zheng S, Chiu H, Boudreau J, Papanicolaou T, Bendena W, Chin-Sang I (2018) A functional study of all 40 *Caenorhabditis elegans* insulin-like peptides. J Biol Chem 293:16912–16922
49. Oh SW, Mukhopadhyay A, Svrzikapa N, Jiang F, Davis RJ, Tissenbaum HA (2005) JNK regulates lifespan in *Caenorhabditis elegans* by modulating nuclear translocation of forkhead transcription factor/DAF-16. Proc Natl Acad Sci USA 102:4494–4499
50. Jahn A, Scherer B, Fritz G, Honnen S (2020) Statins induce a DAF-16/Foxo-dependent longevity phenotype via JNK-1 through mevalonate depletion in *C. elegans*. Aging Dis 11: 60–72
51. Xu R, Li AP, Tan X, Tang X, He XP, Wang LX, Kang JJ, Li SH, Liu Y (2025) Patchouli essential oil extends the lifespan and healthspan of *Caenorhabditis elegans* through JNK-1/DAF-16. Life Sci 360:123270
52. Xu Q, Zheng B, Li T, Liu RH (2025) Black goji berry anthocyanins extend lifespan and enhance the antioxidant defenses in *Caenorhabditis elegans* via the JNK-1 and DAF-16/FOXO pathways. J Sci Food Agric 105:2282–2293
53. Shaw WM, Luo S, Landis J, Ashraf J, Murphy CT (2007) The *C. elegans* TGF-beta dauer pathway regulates longevity via insulin signaling. Curr Biol 17:1635–1645
54. Kwon G, Lee J, Koh JH, Lim YH (2018) Lifespan extension of *Caenorhabditis elegans* by *Butyricicoccus pullicaecorum* and *Megasphaera elsdenii* with probiotic potential. Curr Microbiol 75:557–564
55. Loo J, Shah Bana MAF, Tan JK, Goon JA (2023) Effect of dietary restriction on health span in *Caenorhabditis elegans*: a systematic review. Exp Gerontol 182:112294
56. Lee GD, Wilson MA, Zhu M, Wolkow CA, de Cabo R, Ingram DK, Zou S (2006) Dietary deprivation extends lifespan in *Caenorhabditis elegans*. Aging Cell 5:515–524
57. Kapahi P, Kaeberlein M, Hansen M (2017) Dietary restriction and lifespan: lessons from invertebrate models. Ageing Res Rev 39:3–14
58. Walker G, Houthoofd K, Vanfleteren JR, Gems D (2005) Dietary restriction in *C. elegans*: from rate-of-living effects to nutrient sensing pathways. Mech Ageing Dev 126:929–937
59. Bamps S, Wirtz J, Savory FR, Lake D, Hope IA (2009) The *Caenorhabditis elegans* sirtuin gene, *sir-2.1*, is widely expressed and induced upon caloric restriction. Mech Ageing Dev 130: 762–770
60. Nakatani Y, Yaguchi Y, Komura T, Nakadai M, Terao K, Kage-Nakadai E, Nishikawa Y (2018) Sesamin extends lifespan through pathways related to dietary restriction in *Caenorhabditis elegans*. Eur J Nutr 57:1137–1146
61. Smita SS, Trivedi S, Pandey T, Trivedi M, Pandey R (2020) A bioactive compound Shatavarin IV-mediated longevity as revealed by dietary restriction-induced autophagy in *Caenorhabditis elegans*. Biogerontology 21:827–844
62. Shintani T, Sakoguchi H, Yoshihara A, Izumori K, Sato M (2017) d-Allulose, a stereoisomer of d-fructose, extends *Caenorhabditis elegans* lifespan through a dietary restriction mechanism: a new candidate dietary restriction mimetic. Biochem Biophys Res Commun 493:1528–1533
63. Adachi H, Fujiwara Y, Ishii N (1998) Effects of oxygen on protein carbonyl and aging in *Caenorhabditis elegans* mutants with long (*age-1*) and short (*mev-1*) life spans. J Gerontol A Biol Sci Med Sci 53:B240–B244
64. Khan MH, Ligon M, Hussey LR, Hufnal B, Farber R 2nd, Munkácsy E, Rodriguez A, Dillow A, Kahlig E, Rea SL (2013) TAF-4 is required for the life extension of *isp-1*, *clk-1* and *tpk-1* Mit mutants. Aging 5:741–758
65. Yang W, Hekimi S (2010) A mitochondrial superoxide signal triggers increased longevity in *Caenorhabditis elegans*. PLoS Biol 8:e1000556
66. Zeng WY, Tan L, Han C, Zheng ZY, Wu GS, Luo HR, Li SL (2021) Trigonelline extends the lifespan of *C. elegans* and delays the progression of age-related diseases by activating AMPK, DAF-16, and HSF-1. Oxid Med Cell Longev 2021:7656834

67. Ding AJ, Wu GS, Tang B, Hong X, Zhu MX, Luo HR (2017) Benzimidazole derivative M084 extends the lifespan of *Caenorhabditis elegans* in a DAF-16/FOXO-dependent way. Mol Cell Biochem 426:101–109
68. García-Casas P, Arias-Del-Val J, Alvarez-Illera P, Wojnicz A, de Los Ríos C, Fonteriz RI, Montero M, Alvarez J (2019) The neuroprotector benzothiazepine CGP37157 extends lifespan in *C. elegans* worms. Front Aging Neurosci 10:440
69. Blackwell TK, Sewell AK, Wu Z, Han M (2019) TOR signaling in *Caenorhabditis elegans* development, metabolism, and aging. Genetics 213:329–360
70. Lapierre LR, Hansen M (2012) Lessons from *C. elegans*: signaling pathways for longevity. Trends Endocrinol Metab 23:637–644
71. Ding F, Zhao Y (2025) Astaxanthin promotes the longevity of *Caenorhabditis elegans* via modulation of the intracellular redox status and PHA-4-mediated autophagy. Food Funct 16: 617–627
72. Zhang M, Li Z, Gao D, Gong W, Gao Y, Zhang C (2020) Hydrogen extends *Caenorhabditis elegans* longevity by reducing reactive oxygen species. PLoS One 15:e0231972
73. Govindhan T, Amirthalingam M, Govindan S, Duraisamy K, Cho JH, Tawata S, Periyakali SB, Palanisamy S (2024) Diosgenin intervention: targeting lipophagy to counter high glucose diet-induced lipid accumulation and lifespan reduction. 3 Biotech 14:171
74. Edwards C, Canfield J, Copes N, Brito A, Rehan M, Lipps D, Brunquell J, Westerheide SD, Bradshaw PC (2015) Mechanisms of amino acid-mediated lifespan extension in *Caenorhabditis elegans*. BMC Genet 16:8
75. Berman JR, Kenyon C (2006) Germ-cell loss extends *C. elegans* life span through regulation of DAF-16 by *kri-1* and lipophilic-hormone signaling. Cell 124:1055–1068
76. TeKippe M, Aballay A (2010) *C. elegans* germline-deficient mutants respond to pathogen infection using shared and distinct mechanisms. PLoS One 5:e11777
77. Liu D, Zeng X, Li L, Ou ZL (2020) Carnitine promotes recovery from oxidative stress and extends lifespan in *C. elegans*. Aging 13:813–830
78. Huang XB, Mu XH, Wan QL, He XM, Wu GS, Luo HR (2017) Aspirin increases metabolism through germline signalling to extend the lifespan of *Caenorhabditis elegans*. PLoS One 12: e0184027
79. Zhu B, Jo K, Yang P, Tohti J, Fei J, Abudukerim K (2019) Aiweixin, a traditional Uyghur medicinal formula, extends the lifespan of *Caenorhabditis elegans*. Evid Based Complement Alternat Med 2019:3684601
80. Hua X, Wang D-Y (2024) 6-PPD quinone at environmentally relevant concentrations induced damage on longevity in *C. elegans*: mechanistic insight from inhibition in mitochondrial UPR response. Sci Total Environ 954:176275
81. Pellegrino MW, Nargund AM, Haynes CM (2013) Signaling the mitochondrial unfolded protein response. Biochim Biophys Acta 1833:410–416
82. Haynes CM, Hekimi S (2022) Mitochondrial dysfunction, aging, and the mitochondrial unfolded protein response in *Caenorhabditis elegans*. Genetics 222:iyac160
83. Shu C-J, Hu D-Y, Cao S-Q, Hou B-W, Gu S-L, Fu L, Zhang F-L, Wang D-Y (2026) Exposure to 77PD quinone inhibits longevity and healthspan via affecting mitochondrial signals in *Caenorhabditis elegans*. Ecotoxicol Environ Saf 310:119767
84. Xiao Y, Hong C, Liu F, Shi D, Zhu X, Yu C, Jiang N, Li S, Liu Y (2024) Caffeic acid activates mitochondrial UPR to resist pathogen infection in *Caenorhabditis elegans* via the transcription factor ATFS-1. Infect Immun 92:e0049423
85. Xiao Y, Li L, Han C, Huang T, Ren S, Wang X, Xing Q, Liu F (2025) Chlorogenic acid inhibits pseudomonas toxin pyocyanin and activates mitochondrial UPR to protect host against pathogen infection. Sci Rep 15:5508
86. Hua X, Wang D-Y (2025) Environmentally relevant concentration of 6-PPD quinone inhibits two types of mitophagy to cause mitochondrial dysfunction and lifespan reduction in *Caenorhabditis elegans*. Environ Sci Process Impacts 27:1928–1940

87. Yi J, Wang HL, Lu G, Zhang H, Wang L, Li ZY, Wang L, Wu Y, Xia D, Fang EF, Shen HM (2024) Spautin-1 promotes PINK1-PRKN-dependent mitophagy and improves associative learning capability in an Alzheimer disease animal model. Autophagy 20:2655–2676
88. Huang S, Li A, Ling Y, Yang X, Wang J, Yuan J, Qin D, Yao X (2025) Pharmacological activation of mitophagy confers neuroprotective benefits for amyotrophic lateral sclerosis. Aging Dis. https://doi.org/10.14336/AD.2025.1224
89. Palikaras K, Lionaki E, Tavernarakis N (2015) Coordination of mitophagy and mitochondrial biogenesis during ageing in *C. elegans*. Nature 521:525–528
90. Xiao X, Zheng M, Zhao Y, Huang K, Shi Y, Ren Y, Chen L, Sun J, Ye R, Zheng Z, Ruan Q (2026) Quercetin ameliorates diminished ovarian reserve via modulating intestinal BNIP3/*dct-1*-mediated mitophagy and subsequently activating the ovarian TGF-beta signaling pathway. J Ethnopharmacol 355:120736
91. Meléndez A, Levine B (2009) Autophagy in *C. elegans*. WormBook. https://doi.org/10.1895/wormbook.1.147.1
92. Yang P, Zhang H (2014) You are what you eat: multifaceted functions of autophagy during *C. elegans* development. Cell Res 24:80–91
93. Kovacs AL, Zhang H (2010) Role of autophagy in *Caenorhabditis elegans*. FEBS Lett 584:1335–1341
94. Wei CC, Chang CH, Liao VH (2017) Anti-parkinsonian effects of beta-amyrin are regulated via LGG-1 involved autophagy pathway in *Caenorhabditis elegans*. Phytomedicine 36:118–125
95. Wang Z, Zheng P, Chen X, Xie Y, Weston-Green K, Solowij N, Chew YL, Huang XF (2022) Cannabidiol induces autophagy and improves neuronal health associated with SIRT1 mediated longevity. Geroscience 44:1505–1524
96. Martínez G, Duran-Aniotz C, Cabral-Miranda F, Vivar JP, Hetz C (2017) Endoplasmic reticulum proteostasis impairment in aging. Aging Cell 16:615–623
97. Xu Y, Park Y (2018) Application of *Caenorhabditis elegans* for research on endoplasmic reticulum stress. Prev Nutr Food Sci 23:275–281
98. Taylor RC, Hetz C (2020) Mastering organismal aging through the endoplasmic reticulum proteostasis network. Aging Cell 19:e13265
99. Xu J, Sabatino B, Yan J, Ermakova G, Doering KRS, Taubert S (2024) The unfolded protein response of the endoplasmic reticulum protects *Caenorhabditis elegans* against DNA damage caused by stalled replication forks. G3 14:jkae017
100. Kim S, Shin HJ, Do SH, Na HS (2021) Role of unfolded protein response and endoplasmic reticulum-associated degradation by repeated exposure to inhalation anesthetics in *Caenorhabditis elegans*. Int J Med Sci 18:2890–2896
101. Zhou H, Shen Y, Dong C, Feng W, Tian Y, Xiao Y (2025) Asperuloside promotes innate immunity via IRE-1/XBP-1 mediated unfolded protein response. Bioorg Chem 157:108318
102. Tong Y, Mukhamejanova Z, Zheng Y, Wen T, Xu F, Pang J (2021) Marine-derived xyloketal compound ameliorates MPP(+)-induced neuronal injury through regulating of the IRE1/XBP1 signaling pathway. ACS Chem Neurosci 12:3101–3111
103. Bett JS (2016) Proteostasis regulation by the ubiquitin system. Essays Biochem 60:143–151
104. Dikic I (2017) Proteasomal and autophagic degradation systems. Annu Rev Biochem 86:193–224
105. Zhang WH, Koyuncu S, Vilchez D (2022) Insights into the links between proteostasis and aging from *C. elegans*. Front Aging 3:854157
106. Yuan W, Wang S, Chen Y, Liang Y (2026) Exploring the ubiquitin-proteasome system in protein homeostasis and healthy aging using *Caenorhabditis elegans* as a model: a review. Int J Biol Macromol 3:151811
107. Sedlacek J, Smahelova Z, Adamek M, Subova D, Svobodova L, Kadlecova A, Majer P, Machara A, Grantz Saskova K (2025) Small-molecule activators of NRF1 transcriptional activity prevent protein aggregation. Biomed Pharmacother 183:117864

108. Ariyath A, Hone E, Fernando WMADB, Pedrini S, Martins R, Bharadwaj P (2026) USP14 inhibitor IU1 alleviates amyloid-β mediated toxicity in Alzheimer's disease cell and worm models. J Alzheimer's Dis 110:1328–1354
109. Martineau CN, Kirienko NV, Pujol N (2021) Innate immunity in *C. elegans*. Curr Top Dev Biol 144:309–351
110. Tse-Kang S, Wani KA, Pukkila-Worley R (2025) Patterns of pathogenesis in innate immunity: insights from *C. elegans*. Nat Rev Immunol 25:637–648
111. Kurz CL, Tan MW (2004) Regulation of aging and innate immunity in *C. elegans*. Aging Cell 3:185–193
112. Harding BW, Ewbank JJ (2021) An integrated view of innate immune mechanisms in *C. elegans*. Biochem Soc Trans 49:2307–2317
113. Kim DH, Ausubel FM (2005) Evolutionary perspectives on innate immunity from the study of *Caenorhabditis elegans*. Curr Opin Immunol 17:4–10
114. Zhu X, Liu F, Wu Q, Li S, Ruan G, Yang J, Yu C, Jiang N, Xiao Y, Liu Y (2022) Brevilin A enhances innate immunity and the resistance of oxidative stress in *Caenorhabditis elegans* via p38 MAPK pathway. Int Immunopharmacol 113:109385
115. Xiao Y, Zhou H, Cui Y, Zhu X, Li S, Yu C, Jiang N, Liu L, Liu F (2024) Schisandrin A enhances pathogens resistance by targeting a conserved p38 MAPK pathway. Int Immunopharmacol 128:111472
116. Shu C, Sun L, Zhang W (2016) Thymol has antifungal activity against *Candida albicans* during infection and maintains the innate immune response required for function of the p38 MAPK signaling pathway in *Caenorhabditis elegans*. Immunol Res 64:1013–1024
117. Dinh J, Angeloni JT, Pederson DB, Wang X, Cao M, Dong Y (2014) Cranberry extract standardized for proanthocyanidins promotes the immune response of *Caenorhabditis elegans* to Vibrio cholerae through the p38 MAPK pathway and HSF-1. PLoS One 9:e103290
118. Liu J, Hafting J, Critchley AT, Banskota AH, Prithiviraj B (2013) Components of the cultivated red seaweed *Chondrus crispus* enhance the immune response of *Caenorhabditis elegans* to Pseudomonas aeruginosa through the *pmk-1*, *daf-2/daf-16*, and *skn-1* pathways. Appl Environ Microbiol 79:7343–7350
119. Shu C-J, Wang W, Cao S-Q, Hou B-W, Gu S-L, Fu L, Zhang F-L, Wang D-Y (2026) 6-PPD quinone induces lifespan reduction by causing immunosuppression via DAF-16/PMK-1 signaling in *Caenorhabditis elegans*. Environ Chem Ecotoxicol 8:1176–1184
120. Fischer A, Niklowitz P, Menke T, Döring F (2014) Promotion of growth by Coenzyme Q10 is linked to gene expression in *C. elegans*. Biochem Biophys Res Commun 452:920–927
121. Liao XH, Majithia A, Huang X, Kimmel AR (2008) Growth control via TOR kinase signaling, an intracellular sensor of amino acid and energy availability, with crosstalk potential to proline metabolism. Amino Acids 35:761–770
122. Ogawa T, Isik M, Wu Z, Kurmi K, Meng J, Cho S, Lee G, Fernandez-Cardenas LP, Mizunuma M, Blenis J, Haigis MC, Blackwell TK (2024) Nutrient control of growth and metabolism through mTORC1 regulation of mRNA splicing. Mol Cell 84:4558–4575
123. Schmeisser K, Kaptan D, Raghuraman BK, Shevchenko A, Rodenfels J, Penkov S, Kurzchalia TV (2024) Mobilization of cholesterol induces the transition from quiescence to growth in *Caenorhabditis elegans* through steroid hormone and mTOR signaling. Commun Biol 7:121
124. Dempsey CM, Mackenzie SM, Gargus A, Blanco G, Sze JY (2005) Serotonin (5HT), fluoxetine, imipramine and dopamine target distinct 5HT receptor signaling to modulate *Caenorhabditis elegans* egg-laying behavior. Genetics 169:1425–1436
125. Weinshenker D, Garriga G, Thomas JH (1995) Genetic and pharmacological analysis of neurotransmitters controlling egg laying in *C. elegans*. J Neurosci 15:6975–6985
126. Horvitz HR, Chalfie M, Trent C, Sulston JE, Evans PD (1982) Serotonin and octopamine in the nematode *Caenorhabditis elegans*. Science 216:1012–1014
127. Hua X, Feng X, Hua Y-S, Wang D-Y (2023) Paeoniflorin attenuates polystyrene nanoparticle-induced reduction in reproductive capacity and increase in germline apoptosis through

suppressing DNA damage checkpoints in *Caenorhabditis elegans*. Sci Total Environ 871: 162189
128. Chaudhari SN, Mukherjee M, Vagasi AS, Bi G, Rahman MM, Nguyen CQ, Paul L, Selhub J, Kipreos ET (2016) Bacterial folates provide an exogenous signal for *C. elegans* germline stem cell proliferation. Dev Cell 38:33–46
129. Aprison EZ, Dzitoyeva S, Ruvinsky I (2023) Serotonergic signaling plays a deeply conserved role in improving oocyte quality. Dev Biol 499:24–30
130. Zhou Y, Rothe M, Schunck WH, Ruess L, Menzel R (2023) Serotonin-induced stereospecific formation and bioactivity of the eicosanoid 17,18-epoxyeicosatetraenoic acid in the regulation of pharyngeal pumping of *C. elegans*. Biochim Biophys Acta Mol Cell Biol Lipids 1868: 159304
131. Hobson RJ, Hapiak VM, Xiao H, Buehrer KL, Komuniecki PR, Komuniecki RW (2006) SER-7, a *Caenorhabditis elegans* 5-HT7-like receptor, is essential for the 5-HT stimulation of pharyngeal pumping and egg laying. Genetics 172:159–169
132. Liu H, Qin LW, Li R, Zhang C, Al-Sheikh U, Wu ZX (2019) Reciprocal modulation of 5-HT and octopamine regulates pumping via feedforward and feedback circuits in *C. elegans*. Proc Natl Acad Sci USA 116:7107–7112
133. Osuna-Luque J, Rodríguez-Ramos Á, Gámez-Del-Estal MDM, Ruiz-Rubio M (2018) Behavioral mechanisms that depend on dopamine and serotonin in *Caenorhabditis elegans* interact with the antipsychotics risperidone and aripiprazole. J Exp Neurosci 12:1179069518798628
134. Raizen DM, Lee RY, Avery L (1995) Interacting genes required for pharyngeal excitation by motor neuron MC in *Caenorhabditis elegans*. Genetics 141:1365–1382
135. Izquierdo PG, Calahorro F, Thisainathan T, Atkins JH, Haszczyn J, Lewis CJ, Tattersall JEH, Green AC, Holden-Dye L, O'Connor V (2022) Cholinergic signaling at the body wall neuromuscular junction distally inhibits feeding behavior in *Caenorhabditis elegans*. J Biol Chem 298:101466
136. Avery L, Horvitz HR (1990) Effects of starvation and neuroactive drugs on feeding in *Caenorhabditis elegans*. J Exp Zool 253:263–270
137. Worthy SE, Haynes L, Chambers M, Bethune D, Kan E, Chung K, Ota R, Taylor CJ, Glater EE (2018) Identification of attractive odorants released by preferred bacterial food found in the natural habitats of *C. elegans*. PLoS One 13:e0201158
138. Urushihata T, Takuwa H, Nishino A, Kikuchi A, Ichijo A, Obara C, Wakabayashi T, Sakata K, Matsuura T (2016) Food search strategy changes in *Caenorhabditis elegans* under chronic starvation conditions. J Exp Zool A Ecol Genet Physiol 325:409–414
139. Hale LA, Lee ES, Pantazis AK, Chronis N, Chalasani SH (2017) Altered sensory code drives juvenile-to-adult behavioral maturation in *Caenorhabditis elegans*. eNeuro 3: ENEURO.0175-16.2016
140. Pereira S, van der Kooy D (2012) Two forms of learning following training to a single odorant in *Caenorhabditis elegans* AWC neurons. J Neurosci 32:9035–9044
141. Cheng H, Liu Y, Xue Y, Shao J, Tan Z, Liu S, Duan S, Kang L (2021) Molecular strategies for intensity-dependent olfactory processing in *Caenorhabditis elegans*. Front Mol Neurosci 14: 748214
142. Matsuura T, Izumi J, Hioki M, Nagaya H, Kobayashi Y (2013) Sensory interaction between attractant diacetyl and repellent 2-nonanone in the nematode *Caenorhabditis elegans*. J Exp Zool A Ecol Genet Physiol 319:285–295
143. Kimura KD, Fujita K, Katsura I (2010) Enhancement of odor avoidance regulated by dopamine signaling in *Caenorhabditis elegans*. J Neurosci 30:16365–16375
144. Ezak MJ, Ferkey DM (2010) The *C. elegans* D2-like dopamine receptor DOP-3 decreases behavioral sensitivity to the olfactory stimulus 1-octanol. PLoS One 5:e9487
145. You YJ, Kim J, Raizen DM, Avery L (2008) Insulin, cGMP, and TGF-beta signals regulate food intake and quiescence in *C. elegans*: a model for satiety. Cell Metab 7:249–257
146. Makino M, Ulzii E, Shirasaki R, Kim J, You YJ (2021) Regulation of satiety quiescence by neuropeptide signaling in *Caenorhabditis elegans*. Front Neurosci 15:678590

147. Ma YC, Yang ZS, Ma LQ, Shu R, Zou CG, Zhang KQ (2020) YAP in epithelium senses gut barrier loss to deploy defenses against pathogens. PLoS Pathog 16:e1008766
148. Wang Y-X, Wang D-Y (2024) Transgenerational intestinal toxicity of 6-PPD quinone in causing ROS production, enhancement in intestinal permeability and suppression in innate immunity in *C. elegans*. Environ Pollut 363:125208
149. Zhu J, Dou J, Wu C, Fan G, Li T, Shen D (2023) Intestinal barrier protective study of jujube peel polyphenols/zein complexes by a combined caco-2 cell and *Caenorhabditis elegans* model: a perspective of proteomics. J Agric Food Chem 71:10097–10106
150. Song M-X, Ruan Q-L, Wang D-Y (2024) Paeoniflorin alleviates toxicity and accumulation of 6-PPD quinone by activating ACS-22 in *Caenorhabditis elegans*. Ecotoxicol Environ Saf 286: 117226
151. Zhao Y-L, Yu X-M, Jia R-H, Yang R-L, Rui Q, Wang D-Y (2015) Lactic acid bacteria protects *Caenorhabditis elegans* from toxicity of graphene oxide by maintaining normal intestinal permeability under different genetic backgrounds. Sci Rep 5:17233
152. Bae WY, Nguyen UTT, Le TAN, Tran SH, Lee S, Hong SC, Cha KH, Choi ID, Kang K (2025) *Lacticaseibacillus casei* HY2782 improves the intestinal barrier and tract environment and ultimately prolongs the lifespan of *Caenorhabditis elegans*. Food Funct 16:5871–5884
153. Yu YT, Lin HR, Chen XY, Sun ML, Wei CM, Xue MF, Gao YH, Tang WB, Zheng CY, Li S, Wang HB (2024) Dendrobium officinale phenolic extract maintains proteostasis by regulating autophagy in a *Caenorhabditis elegans* model of Alzheimer's disease. Fitoterapia 175:105924
154. Wen YP, Fu HJ, Chen Q, Lan C, Qin DL, Wu JM, Wu AG, Zhou XG (2024) Exploring the therapeutic potential of Nelumbo nucifera leaf extract against amyloid-beta-induced toxicity in the *Caenorhabditis elegans* model of Alzheimer's disease. Front Pharmacol 15:1408031
155. Sadananda G, Velmurugan JD, Subramaniam JR (2021) DMSO delays Alzheimer disease causing Abeta-induced paralysis in *C. elegans* through modulation of glutamate/acetylcholine neurotransmission. Ann Neurosci 28:55–64
156. Chen X, Zhou XY, Lan C, Fu HJ, Li ZC, Chen MY, Wen YP, Yu L, Qin DL, Wu AG, Wu JM, Zhou XG (2025) Araloside A induces Raf/MEK/ERK-dependent autophagy to mitigate Alzheimer's and Parkinson's pathology in cellular and *C. elegans* models. Mol Neurobiol 62:16061–16080
157. Chalorak P, Sornkaew N, Manohong P, Niamnont N, Malaiwong N, Limboonreung T, Sobhon P, Aschner M, Meemon K (2021) Diterpene glycosides from Holothuria scabra exert the alpha-synuclein degradation and neuroprotection against alpha-synuclein-mediated neurodegeneration in *C. elegans* model. J Ethnopharmacol 279:114347
158. Wu AG, Yong YY, He CL, Li YP, Zhou XY, Yu L, Chen Q, Lan C, Liu J, Yu CL, Qin DL, Wu JM, Zhou XG (2024) Novel 18-norspirostane steroidal saponins: extending lifespan and mitigating neurodegeneration through promotion of mitophagy and mitochondrial biogenesis in *Caenorhabditis elegans*. Mech Ageing Dev 218:111901
159. Srivastava T, Tyagi D, Fatima S, Sathyan MTV, Raj R, Sharma A, Chaturvedi M, Sinha M, Shishodia SK, Kumar D, Sharma SK, Shankar J, Satish A, Priya S (2024) A natural small molecule-mediated inhibition of alpha-synuclein aggregation leads to neuroprotection in *Caenorhabditis elegans*. J Neurochem 168:1640–1654
160. Perni M, Flagmeier P, Limbocker R, Cascella R, Aprile FA, Galvagnion C, Heller GT, Meisl G, Chen SW, Kumita JR, Challa PK, Kirkegaard JB, Cohen SIA, Mannini B, Barbut D, Nollen EAA, Cecchi C, Cremades N, Knowles TPJ, Chiti F, Zasloff M, Vendruscolo M, Dobson CM (2018) Multistep inhibition of alpha-synuclein aggregation and toxicity in vitro and in vivo by trodusquemine. ACS Chem Biol 13:2308–2319
161. Fang M, Liu Y, Gao X, Yu J, Tu X, Mo X, Zhu H, Zou Y, Huang C, Fan S (2025) Perillaldehyde alleviates polyQ-induced neurodegeneration through the induction of autophagy and mitochondrial UPR in *Caenorhabditis elegans*. Biofactors 51:e2089
162. Xu H, Jia C, Cheng C, Wu H, Cai H, Le W (2022) Activation of autophagy attenuates motor deficits and extends lifespan in a *C. elegans* model of ALS. Free Radic Biol Med 181:52–61

163. Li A, Huang S, Cao S, Lin J, Zhao L, Yu F, Huang M, Yang L, Xin J, Wen J, Yan L, Zhang K, Jiang M, Le W, Li P, Liu YU, Qin D, Lu J, Lu G, Shen H, Yao X, Fang EG, Su H (2025) Isoginkgetin antagonizes ALS pathologies in its animal and patient iPSC models via PINK1-Parkin-dependent mitophagy. EMBO Mol Med 17:3139–3173
164. Ikenaka K, Tsukada Y, Giles AC, Arai T, Nakadera Y, Nakano S, Kawai K, Mochizuki H, Katsuno M, Sobue G, Mori I (2019) A behavior-based drug screening system using a *Caenorhabditis elegans* model of motor neuron disease. Sci Rep 9:10104
165. Patten SA, Aggad D, Martinez J, Tremblay E, Petrillo J, Armstrong GA, La Fontaine A, Maios C, Liao M, Ciura S, Wen XY, Rafuse V, Ichida J, Zinman L, Julien JP, Kabashi E, Robitaille R, Korngut L, Parker JA, Drapeau P (2017) Neuroleptics as therapeutic compounds stabilizing neuromuscular transmission in amyotrophic lateral sclerosis. JCI Insight 2:e97152
166. Fardghassemi Y, Maios C, Parker JA (2021) Small molecule rescue of ATXN3 toxicity in *C. elegans* via TFEB/HLH-30. Neurotherapeutics 18:1151–1165
167. Jalles A, Vieira C, Pereira-Sousa J, Vilasboas-Campos D, Mota AF, Vasconcelos S, Ferreira-Lomba B, Costa MD, Da Silva JD, Maciel P, Teixeira-Castro A (2022) Aripiprazole offsets mutant ATXN3-induced motor dysfunction by targeting dopamine D(2) and serotonin 1A and 2A receptors in *C. elegans*. Biomedicine 10:370
168. Câmara DF, Machado ML, Arantes LP, Silva TC, Silveira TL, Leal JG, Dornelles L, Stefanello ST, Soares FAA (2019) MPMT-OX up-regulates GABAergic transmission and protects against seizure-like behavior in *Caenorhabditis elegans*. Neurotoxicology 74:272–281
169. Jones A, Barker-Haliski M, Ilie AS, Herd MB, Baxendale S, Holdsworth CJ, Ashton JP, Placzek M, Jayasekera BAP, Cowie CJA, Lambert JJ, Trevelyan AJ, White HS, Marson AG, Cunliffe VT, Sills GJ, Morgan A (2020) A multiorganism pipeline for antiseizure drug discovery: identification of chlorothymol as a novel γ-aminobutyric acidergic anticonvulsant. Epilepsia 61:2106–2118
170. Gonzalez-Lerma PX, Lloyd C, Park SJ, Dawson-Scully K (2026) Anticonvulsant effects of novel and repurposed drugs on docetaxel-induced neuropathy in *C. elegans*. PLoS One 21: e0342236
171. Risley MG, Kelly SP, Minnerly J, Jia K, Dawson-Scully K (2018) *egl-4* modulates electroconvulsive seizure duration in *C. elegans*. Invertebr Neurosci 18:8
172. Cheong MC, Artyukhin AB, You Y, Avery L (2015) An opioid-like system regulating feeding behavior in *C. elegans*. elife 4:e06683
173. Engleman EA, Steagall KB 2nd, Bredhold KE, Breach M, Kline HL, Bell RL, Katner SN, Neal-Beliveau BS (2018) *Caenorhabditis elegans* show preference for stimulants and potential as a model organism for medications screening. Front Physiol 9:1200
174. Ide S, Ikeda K (2024) *Caenorhabditis elegans* for opioid addiction research. Curr Opin Neurobiol 88:102914
175. Scott LL, Sahn JJ, Ferragud A, Yen RC, Satarasinghe PN, Wood MD, Hodges TR, Shi T, Prakash BA, Friese KM, Shen A, Sabino V, Pierce JT, Martin SF (2018) Small molecule modulators of σ2R/Tmem97 reduce alcohol withdrawal-induced behaviors. Neuropsychopharmacology 43:1867–1875
176. Naß J, Abdelfatah S, Efferth T (2021) Ursolic acid enhances stress resistance, reduces ROS accumulation and prolongs life span in *C. elegans* serotonin-deficient mutants. Food Funct 12: 2242–2256
177. Naß J, Efferth T (2021) Withanone ameliorates stress symptoms in *Caenorhabditis elegans* by acting through serotonin receptors. Pharmacopsychiatry 54:215–223
178. de la Mora MP, Gallegos-Cari A, Arizmendi-García Y, Marcellino D, Fuxe K (2020) Role of dopamine receptor mechanisms in the amygdaloid modulation of fear and anxiety: structural and functional analysis. Prog Neurobiol 90:198–216
179. Naß J, Efferth T (2021) Ursolic acid ameliorates stress and reactive oxygen species in *C. elegans* knockout mutants by the dopamine Dop1 and Dop3 receptors. Phytomedicine 81:153439

180. Giunti S, Blanco MG, De Rosa MJ, Rayes D (2024) The ketone body β-hydroxybutyrate ameliorates neurodevelopmental deficits in the GABAergic system of *daf-18*/PTEN *Caenorhabditis elegans* mutants. elife 13:RP94520
181. Izquierdo PG, Calahorro F, Ruiz-Rubio M (2013) Neuroligin modulates the locomotory dopaminergic and serotonergic neuronal pathways of *C. elegans*. Neurogenetics 14:233–242
182. Smisko WR, Haury WR, Perni M, Kalik R, Palo J, Bastien BL, Bilchak JN, Jongens TA, Kayser MS, Hart MP (2026) Allele-specific rescue of neurexin behavioral phenotypes by monoamine-targeting compounds. bioRxiv. https://doi.org/10.64898/2026.02.06.703315
183. Lee WJ, Petticord MC, Woolley JT, Robinson ZM, Schultz DC, Raizen DM (2026) Tricyclic antidepressants suppress sleep in *Caenorhabditis elegans*. Sleep 5:zsag034
184. Rossi L, Amoako K, Busack I, Golinelli L, Courtney A, Besseling J, Schafer W, Beets I, Bringmann H (2025) The neuropeptide FLP-11 induces and self-inhibits sleep through the receptor DMSR-1 in *Caenorhabiditis elegans*. Curr Biol 25:2183–2194
185. Kosel M, Wild W, Bell A, Rothe M, Lindschau C, Steinberg CE, Schunck WH, Menzel R (2011) Eicosanoid formation by a cytochrome P450 isoform expressed in the pharynx of *Caenorhabditis elegans*. Biochem J 435:689–700
186. Rivas D, Baltasar-Fernandez I, AlOkda A, Saedi AA, Karasik D, Van Raamsdonk JM, Duque G (2026) Picolinic acid, a tryptophan metabolite, exhibits anabolic effects in muscle cells and improves lifespan and movement in *C. elegans*. J Gerontol A Biol Sci Med Sci 81:glaf239
187. García-Casas P, Alvarez-Illera P, Gómez-Orte E, Cabello J, Fonteriz RI, Montero M, Alvarez J (2021) The mitochondrial Na^{+}/Ca^{2+} exchanger inhibitor CGP37157 preserves muscle structure and function to increase lifespan and healthspan in *Caenorhabditis elegans*. Front Pharmacol 12:695687
188. Carre-Pierrat M, Mariol MC, Chambonnier L, Laugraud A, Heskia F, Giacomotto J, Ségalat L (2006) Blocking of striated muscle degeneration by serotonin in *C. elegans*. J Muscle Res Cell Motil 27:253–258
189. Rohde CB, Gilleland C, Samara C, Norton S, Haggarty S, Yanik MF (2009) Microfluidic in vivo screen identifies compounds enhancing neuronal regeneration. Annu Int Conf IEEE Eng Med Biol Soc 2009:5950–5952
190. Samara C, Rohde CB, Gilleland CL, Norton S, Haggarty SJ, Yanik MF (2010) Large-scale in vivo femtosecond laser neurosurgery screen reveals small-molecule enhancer of regeneration. Proc Natl Acad Sci USA 107:18342–18347
191. Xu J, Meng X, Yang Q, Zhang J, Hu W, Fu H, Chen JW, Ma W, Chisholm AD, Sun Q, Xu S (2021) Redox-sensitive CDC-42 clustering promotes wound closure in *C. elegans*. Cell Rep 37:110040
192. Thomas-Virnig CL, Sims PA, Simske JS, Hardin J (2004) The inositol 1,4,5-trisphosphate receptor regulates epidermal cell migration in *Caenorhabditis elegans*. Curr Biol 14:1882–1887
193. Gunaydin-Akyildiz A, Yanikoglu RS, Gulec M, Alim-Toraman GO, Kuran ED, Atasoy S, Olgun A, Topcu G (2023) Emodin and aloe-emodin, two potential molecules in regulating cell migration of skin cells through the MAP kinase pathway and affecting *Caenorhabditis elegans* thermotolerance. BMC Mol Cell Biol 24:23
194. Pooranachithra M, Bhaskar JP, Murali D, Das SS, JebaMercy G, Krishnan V, Balamurugan K (2019) Unravelling the wound healing ability and mode of action of pyridine carboxamide oxime using *Caenorhabditis elegans* as potential prescreen wound model. Life Sci 235: 116859
195. Qu H, Shan K, Tang C, Cui G, Fu G, Qi Y, Cui J, Li J, Wang R, Feng N, Chen YQ (2021) A novel small-molecule fatty acid synthase inhibitor with antitumor activity by cell cycle arrest and cell division inhibition. Eur J Med Chem 219:113407
196. Ghali RE, Izadi M, Alrayyes Z, Ali TA, Uddin S, Pourkarimi E (2026) Sanguinarine activates ATM/ATR-mediated CHK-1 signaling to drive p53-dependent apoptosis in the *C. elegans* germline. Sci Rep 16(1):18731

197. Makhoul P, Galas S, Paniagua-Gayraud S, Deleuze-Masquefa C, Hajj HE, Bonnet P, Richaud M (2024) Uncovering the molecular pathways implicated in the anti-cancer activity of the imidazoquinoxaline derivative EAPB02303 using a *Caenorhabditis elegans* model. Int J Mol Sci 25:7785
198. Sánchez-Blanco A, Rodríguez-Matellán AG, Reis-Sobreiro M, Sáenz-Narciso B, Cabello J, Mohler WA, Mollinedo F (2014) *Caenorhabditis elegans* as a platform to study the mechanism of action of synthetic antitumor lipids. Cell Cycle 13:3375–3389
199. Shen P, Yue Y, Kim KH, Park Y (2017) Piceatannol reduces fat accumulation in *Caenorhabditis elegans*. J Med Food 20:887–894
200. Mihaylova LV, Savova MS, Todorova MN, Tonova V, Binev BK, Georgiev MI (2026) Cycloastragenol improves fatty acid metabolism through NHR-49/FAT-7 suppression and potent AAK-2 activation in *Caenorhabditis elegans* obesity model. Int J Mol Sci 27:772
201. Navarro-Herrera D, Aranaz P, Eder-Azanza L, Zabala M, Hurtado C, Romo-Hualde A, Martínez JA, González-Navarro CJ, Vizmanos JL (2018) Dihomo-gamma-linolenic acid induces fat loss in *C. elegans* in an omega-3-independent manner by promoting peroxisomal fatty acid beta-oxidation. Food Funct 9:1621–1637
202. Sun Q, Yue Y, Shen P, Yang JJ, Park Y (2016) Cranberry product decreases fat accumulation in *Caenorhabditis elegans*. J Med Food 19:427–433
203. Chen Y, Qin Q, Luo J, Dong Y, Lin C, Chen H, Cao Y, Chen Y, Su Z (2022) Litchi flower essential oil balanced lipid metabolism through the regulation of DAF-2/IIS, MDT-15/SBP-1, and MDT-15/NHR-49 pathway. Front Nutr 9:934518
204. Lee J, Yue Y, Park Y, Lee SH (2017) 3,3′-Diindolylmethane suppresses adipogenesis using AMPKalpha-dependent mechanism in 3T3-L1 adipocytes and *Caenorhabditis elegans*. J Med Food 20:646–652
205. Kim AT, Park Y (2025) Esculetin inhibits fat accumulation through insulin/insulin-like growth factor- and AMP-activated protein kinase-dependent pathways in *Caenorhabditis elegans*. Nutrients 17:1565
206. Zhao Y, Wu C, Bai J, Li J, Cheng K, Zhou X, Dong Y, Xiao X (2020) Fermented barley extracts with lactobacillus plantarum dy-1 decreased fat accumulation of *Caenorhabditis elegans* in a *daf-2*-dependent mechanism. J Food Biochem 44:e13459
207. Schifano E, Zinno P, Natella F, Pompa L, Sonaglia E, Spaziani S, Sharbaf M, Esposito F, Cirillo T, Squillante J, Maglione G, Mancini P, Angeloni A, Santarelli ML, Devirgiliis C, Uccelletti D (2026) Coffee silverskin supplementation alleviates high-glucose-diet-induced obesity by modulating lipogenic gene expression in *Caenorhabditis elegans* model. Molecules 31:887
208. Mouchiroud L, Houtkooper RH, Moullan N, Katsyuba E, Ryu D, Cantó C, Mottis A, Jo Y, Viswanathan M, Schoonjans K, Guarente L, Auwerx J (2013) The NAD^+/sirtuin pathway modulates longevity through activation of mitochondrial UPR and FOXO signaling. Cell 154:430–441

Interactions Between Compounds 12

Abstract

Not only the monotherapy, the possible combinational effects between compounds or drugs are also needed to be carefully considered. We here mainly introduced and discussed synergistical effects, antagonistic effects, and additive effects of compounds or drugs in *Caenorhabditis elegans*.

Keywords

Interaction · Compounds · *Caenorhabditis elegans*

12.1 Introduction

In Chap. 3, we have introduced the classification of examined compounds and drugs in *Caenorhabditis elegans*. Moreover, in Chap. 11, we have introduced and discussed pharmacological mechanisms of compounds and drugs at different aspects. Nevertheless, during the clinical use, the pharmacological interactions may exist among different drugs. The drug–drug interactions refer to the possibility that one drug can modify the action, efficacy, or safety profile of another. Such a possibility may result in the enhanced pharmacological effects, the reduced pharmacological efficacy, or the emergence of adverse effects that are not present when the drug is administrated alone. Therefore, besides the pharmacological effects of single compound or drug, the possible combinational effects among compounds or drugs also need to be carefully investigated and considered. In this chapter, mainly surrounding synergistical effects, antagonistic effects, and additive effects, we introduced and discussed the pharmacological interactions among the compounds in *C. elegans*.

D. Wang, *Pharmacology in Caenorhabditis elegans*,
https://doi.org/10.1007/978-981-92-5166-7_12

12.2 Synergistical Effects Between Compounds

The pharmacological synergistical effect suggests that the combinational effect between compounds is greater than the sum of the individual effects. This has been frequently sought in the design for the antimicrobial therapy. That is, based on assessment and screen, the desired combinations are hoped to be identified to achieve a goal, which cannot be obtained with monotherapy.

The systematic identification of synergistic effects between compounds can be performed based on both experimental identification and structure-activity prediction. For the experimental identification, it can contain the following several steps:

1. Characterization of single compound: The concentration-activity curves need to be generated for each compound, and at least the EC_{50} or IC_{50} values need to be examined.
2. Testing effects of combinational treatments: At least effects of three different concentrations should be tested, and the tested concentrations for examined compounds are suggested to be below the EC_{50} values.
3. Phenotypic readouts: Based on phenotypic analysis using examined endpoints for certain research aim, the combinational groups with synergistic effects between compounds can be identified and screened.
4. Data analysis: Software (e.g., CompuSyn) applying synergy models (Loewe Additivity, Bliss Independence, or Chou-Talalay) can be used to calculate Combination Index values.

Besides the experimental identification, the further prediction can be also performed. Based on obtained experimental data, the feedback system control (FSC) scheme was employed to predict possible effects of new drug combinations [1]. Some new potent combinations of available anthelmintics on parasitic nematodes were predicted and raised [1].

12.2.1 Identification of Anticipated Combinations Between Compounds for Certain Pharmacological Aims

In *C. elegans*, the synergistic effects between some compounds have been examined, which are reflected by several aspects. The first aspect is to identify anticipated pharmacological combinations for certain research aims. Using higher single activity (HSA) model, among the ten pairs tested, two potentially synergistic drug pairs (rifampicin combined with either rapamycin or psora-4) were identified to improve the healthspan, which involved the TGF-β signaling and SREBP lipid signaling [2]. The combination between apple peel and blueberry extracts also showed synergistical effect in extending the lifespan by modulating DAF-16 in insulin signaling pathway [3]. Besides the effect on longevity, the combination between apple peel and blueberry extracts further exhibited the synergistical effect against the oxidative stress by influencing SKN-1 signaling [4]. In addition, the combinations of extract composed of docosahexaenoic acid, *Ginkgo biloba*, D-pinitol, and ursolic acid could show the synergistic neuroprotective effects against AD hallmarks [5].

12.2.2 Resistance Management

Using the *C. elegans* assay system, the identified combination therapy is helpful for overcoming the drug resistance of bacterial and fungal pathogens. For example, the combination of ellagic acid and tetracycline could suppress the biofilm formation and increase the *C. elegans* survival after infection by *Propionibacterium acnes* [6]. The combination of caffeic acid phenethyl ester and fluconazole showed the synergistic effect against the resistant *Candida albicans* [7]. The tyrocidines synergistically enhanced the antifungal activity of amphotericin B and caspofungin and increased *C. elegans* survival after fungal infection [8].

12.2.3 Insight in Drug Mechanisms

The *C. elegans* assay platform can also enable the elucidation of drug mechanisms based on the combining test of compounds. For example, Amph could potentiate the effects of β-phenylethylamine in activating dopamine transporter and amine-gated chloride channel LGC-55 to induce corresponding behaviors, which offers a novel mechanism of the action of Amph [9].

12.2.4 Cost-Effective Screening

As a small classic model animal, the *C. elegans* system can offer a cost-effective and high-throughput alternative to mammalian models for investigating and screening the effects of drug-drug interactions. Based on the methicillin-resistant *Staphylococcus aureus-Caenorhabditis elegans* high-throughput screen, four novel antibacterial compounds were screened [10]. Among these compounds, compounds 1, 2, and 4 synergistically enhanced activity of ciprofloxacin, and compound 3 synergistically enhanced activity of erythromycin, gentamicin, streptomycin, and vancomycin against the *Staphylococcus aureus* (MRSA) [10]. In addition, based on *Caenorhabditis elegans/Galleria mellonella* sequential-screening, five novel anti-*Staphylococcus aureus* (MRSA) compounds were identified [11]. Moreover, among these compounds, 4,5,6,7-tetrabromobenzotriazole enhanced the activity of doxycycline and oxacillin, and 4,4′,4′-(4-propyl-[1H]-pyrazole-1,3,5-triyl)trisphenol, (1S,2S)-2-[2-[[3-(1H-benzimidazol-2-yl)propyl]methylamino]ethyl]-6-fluoro-1,2,3,4-tetrahydro-1-(1-methylethyl)-2-naphthalenyl cyclopropanecarboxylate dihydrochloride, 3-[2-[2-chloro-4-[3-(2,6-dichlorophenyl)-5-(1-methylethyl)-4-isoxazolyl]methoxy]phenyl]ethenyl] benzoic acid, and N-(cyclopropylmethoxy)-3,4,5-trifluoro-2-[(4-iodo-2-methylphenyl)amino] benzamide could synergize with doxycycline, polymyxin-B, gentamicin, and erythromycin, respectively, to be against the *Staphylococcus aureus* MRSA-MW2 [11].

12.3 Antagonistic Effects Between Compounds

The pharmacological antagonistic effect suggests that one drug can inhibit or reverse the effect of another. Paying attention to this aspect can be helpful for understanding and avoiding the possible clinical treatment failure. For the experimental identification of this kind of drug–drug interaction, the procedures are the same as those for assessing synergistic effects between compounds.

The values of this aspect of research for clinical treatment and management at least include: (1) elucidation of mechanisms of drug resistance, (2) avoidance of therapeutic interference, and (3) validation of drug targets.

12.3.1 Establishment of an Antagonism Relationship

Firstly, the interaction analysis can help us confirm the pharmacological antagonistic effects between compounds. The volatile anesthetics of halothane (2.1%) and enflurane (4%) were antagonistic to neurotransmitter GABA [12]. Nevertheless, the ED50s of these volatile anesthetics could not be changed by GABA (150 mM), which suggested that GABA did not mediate the halothane or enflurane sensitivity in nematodes [12]. In addition, the effect of caffeine intake in inhibiting mitochondrial activity and lipogenesis could be suppressed by phosphoethanolamine supplementation, which implied that the observed inhibition in mitochondrial activity and lipogenesis is partially due to the low phosphoethanolamine level [13].

12.3.2 Examination of Receptor Antagonism

The second aim for investigating pharmacological antagonistic effects is to examine the receptor antagonism. This suggests that one drug binds to a receptor and blocks the action of another drug by acting as an agonist at the same receptor. Naloxone and CTOP are two opioid receptor antagonists, and their combinations with certain compounds were tested [14]. It was observed that the effects of morphine, endomorphin 1, and endomorphin 2 in affecting the thermal avoidance behavior could be reversed by Naloxone and CTOP, suggesting that these opioid antagonists can reverse the overdose of some compounds, such as the morphine [14]. In *C. elegans*, glutamate-gated chloride channels (GluCls) act as targets of ivermectin (IVM) and moxidectin (MOX), and these two compounds are the corresponding positive allosteric modulators [15]. Moreover, the effect of MOX in inhibiting GLC-2 receptors was 10-fold more than that of IVM [15]. Additionally, in the Met291Gln mutant, IVM could be converted into an agonist; however, MOX remained its role of antagonist [15].

12.3.3 Induction of Chemical Inactivation

The chemical inactivation suggests that one compound may physically or chemically bind to another and then reduce its activity and even bioavailability. For example, the calcium chelation abolished urolithin A-induced mitophagy, which in turn blocked the beneficial effect of urolithin A in modulating muscle function and lifespan [16]. Besides this, certain metal ions chelating may also affect pharmaceutical ingredients and prevent their absorption in the gut lumen.

12.4 Additive Effects Between Compounds

The pharmacological additive effect suggests that the observed combined effect equals the sum of the individual effects. For example, after assessment of activities of combinations of citral with anidulafungin, amphotericin B, and fluconazole, the combination of citral with anidulafungin showed the additive interaction against 11 of the 19 *Candida auris* isolates [17]. In addition, among the examined 16 compounds from the literature, some of them could be combined to exhibit the additive effect in extending the lifespan [18].

12.5 Perspectives

In this chapter, we introduced and discussed synergistical effects, antagonistic effects, and additive effects of compounds or drugs (Fig. 12.1). One of the goals of understanding these interactions is to manage them—sometimes deliberately combining drugs for synergistic benefit (e.g., in cancer therapy or treat of pathogen resistance). Another goal is to avoid the combinations that lead to toxicity or treatment failure. For the introduced three kinds of drug–drug interactions, we did not cover all of the aspects due to the lack of support by evidence. For example, for the antagonistic effects between compounds, sometimes it also involves the pathway interference, which means that two compounds act on converging signaling pathways with opposing net effects. More importantly, if one compound can alter the absorption, distribution, metabolism, or excretion of another, this compound may even increase the toxicity of another. Furthermore, the antagonism framework can also be extended to the chemical–genetic interactions. This means that a compound may antagonize the effect of a genetic mutation, which can enable the systematic mapping of gene–drug interaction networks.

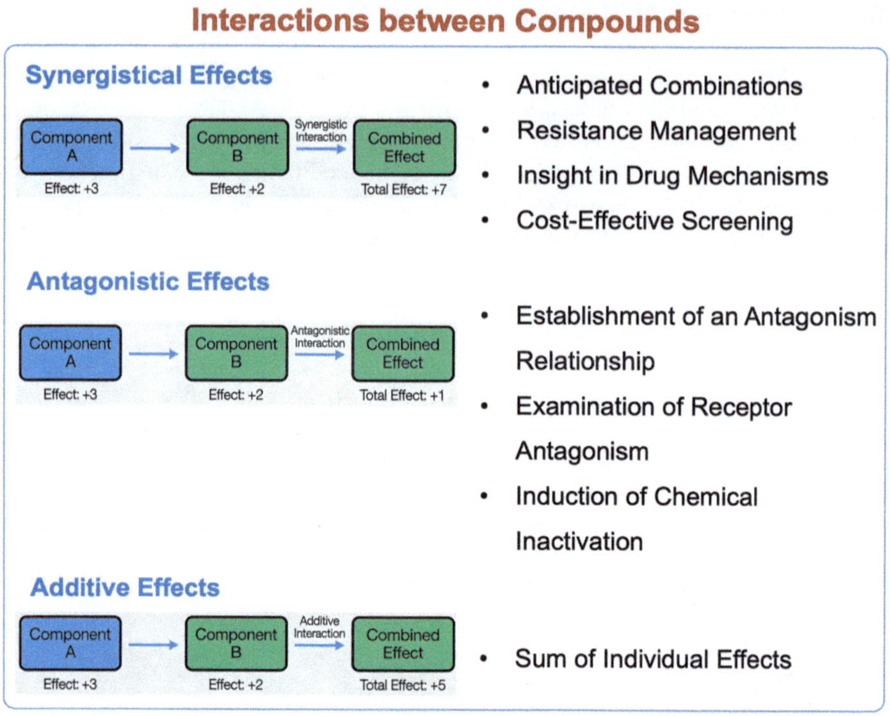

Fig. 12.1 The synergistical effects, antagonistic effects, and additive effects of compounds or drugs in *Caenorhabditis elegans*

References

1. Ding X, Njus Z, Kong T, Su W, Ho C, Pandey S (2017) Effective drug combination for *Caenorhabditis elegans* nematodes discovered by output-driven feedback system control technique. Sci Adv 3:eaao1254
2. Admasu TD, Batchu KC, Barardo D, Ng LF, Lam VYM, Xiao L, Cazenave-Gassiot A, Wenk MR, Tolwinski NS, Gruber J (2018) Drug synergy slows aging and improves healthspan through IGF and SREBP lipid signaling. Dev Cell 47:67–79
3. Song B, Wang H, Xia W, Zheng B, Li T, Liu RH (2020) Combination of apple peel and blueberry extracts synergistically induced lifespan extension via DAF-16 in *Caenorhabditis elegans*. Food Funct 11:6170
4. Song B, Zheng B, Li T, Liu RH (2020) SKN-1 is involved in combination of apple peels and blueberry extracts synergistically protecting against oxidative stress in *Caenorhabditis elegans*. Food Funct 11:5409
5. Griñán-Ferré C, Bellver-Sanchis A, Olivares-Martín M, Bañuelos-Hortigüela O, Pallàs M (2021) Synergistic neuroprotective effects of a natural product mixture against AD hallmarks and cognitive decline in *Caenorhabditis elegans* and an SAMP8 mice model. Nutrients 13:2411
6. Sivasankar C, Maruthupandiyan S, Balamurugan K, Bhaskar JP, Krishnan V, Pandian SK (2016) A combination of ellagic acid and tetracycline inhibits biofilm formation and the associated virulence of *Propionibacterium acnes in vitro* and *in vivo*. Biofouling 32:397–410

7. Sun L, Liao K, Hang C (2018) Caffeic acid phenethyl ester synergistically enhances the antifungal activity of fluconazole against resistant *Candida albicans*. Phytomedicine 40:55–58
8. Troskie AM, Rautenbach M, Delattin N, Vosloo JA, Dath M, Cammu BPA, Thevissen K (2014) Synergistic activity of the tyrocidines, antimicrobial cyclodecapeptides from *Bacillus aneurinolyticus*, with amphotericin B and caspofungin against *Candida albicans* biofilms. Antimicrob Agents Chemother 58:3697–3707
9. Safratowich BD, Hossain M, Bianchi L, Carvelli L (2014) Amphetamine potentiates the effects of β-phenylethylamine through activation of an amine-gated chloride channel. J Neurosci 34: 4686–4691
10. Tharmalingam N, Rajmuthiah R, Kim W, Fuchs BB, Jeyamani E, Kelso MJ, Mylonakis E (2018) Antibacterial properties of four novel hit compounds from a methicillin-resistant *Staphylococcus aureus-Caenorhabditis elegans* high-throughput screen. Microb Drug Resist 24: 666–674
11. Khader R, Tharmalingam N, Mishra B, Felix L, Ausubel FM, Kelso MJ, Mylonakis E (2020) Characterization of five novel anti-MRSA compounds identified using a whole-animal *Caenorhabditis elegans/Galleria mellonella* sequential-screening approach. Antibiotics (Basel) 9:449
12. Boswell MV, Morgan PG, Sedensky MM (1990) Interaction of GABA and volatile anesthetics in the nematode *Caenorhabditis elegans*. FASEB J 4:2506–2510
13. Min H, Youn E, Kim J, Son SY, Lee CH, Shim Y (2020) Effects of phosphoethanolamine supplementation on mitochondrial activity and lipogenesis in a caffeine ingestion *Caenorhabditis elegans* model. Nutrients 12:3348
14. Nieto-Fernandez F, Andrieux S, Idrees S, Bagnall C, Pryor SC, Sood R (2009) The effect of opioids and their antagonists on the nocifensive esponse of *Caenorhabditis elegans* to noxious thermal stimuli. Invertebr Neurosci 9:195–200
15. Kaji MD, Noonan JD, Geary TG, Beech RN (2022) Structural mechanism underlying the differential effects of ivermectin and moxidectin on the *C. elegans* glutamate-gated chloride channel GLC-2. Biomed Pharmacother 145:112380
16. Roussos A, Kitopoulou K, Borbolis F, Ploumi C, Gianniou DD, Li Z, He H, Tsakiri E, Borland H, Kostakis IK, Samiotaki M, Trougakos IP, Bohr VA, Palikaras K (2025) Urolithin A modulates inter-organellar communication via calcium-dependent mitophagy to promote healthy ageing. Autophagy 21:3097–3122
17. de-la-Fuente I, Guridi A, Jauregizar N, Eraso E, Quindós G, Sevillano E (2023) In vitro and in vivo activity of citral in combination with amphotericin B, anidulafungin and fluconazole against *Candida auris* isolates. J Fungi (Basel) 9:648
18. Phelps GB, Morin J, Pinto C, Schoenfeldt L, Guilmot S, Ocampo A, Perez K (2025) Comprehensive evaluation of lifespan-extending molecules in *C. elegans*. Aging Cell 24:e14424

13 Identification of Active Compounds in Plant Extracts and Active Components in Traditional Chinese Medicine Formula

Abstract

Plant extracts (including Traditional Chinese Medicines) and Traditional Chinese Medicine Formula represent complex drugs or biological samples used for clinical treatment. We first introduced and discussed the identification of pharmacological active compounds in plant extracts. Moreover, the identification of pharmacological active components in Traditional Chinese Medicine formula was further introduced and discussed.

Keywords

Identification · Active compounds · Active components · Plant extracts · Traditional Chinese Medicine formula · *Caenorhabditis elegans*

13.1 Introduction

For the pharmacological mechanisms of compounds and drugs, we have introduced and discussed underlying molecular mechanisms for observed pharmacological effects of compounds and drugs in Chap. 11. Moreover, in Chap. 12, we introduced and discussed a certain aspect of pharmacological mechanisms of compounds and drugs, namely drug–drug interactions. Actually, there are still some other aspects of pharmacological mechanisms for drugs. For extracts of plant extracts (including Traditional Chinese Medicines) and Traditional Chinese Medicine Formula, the underlying mechanisms for their pharmacological effects are largely due to the effects of their active compounds and the combination of active compounds. For this reason, we here further introduced and discussed the identification of pharmacological active compounds in plant extracts and the identification of pharmacological active components in Traditional Chinese Medicine formula using *Caenorhabditis elegans* as the animal model.

D. Wang, *Pharmacology in Caenorhabditis elegans*,
https://doi.org/10.1007/978-981-92-5166-7_13

13.2 Identification of Pharmacological Active Compounds in Plant Extracts

In *C. elegans*, the aim for identification of pharmacological active compounds in plant extracts is to discover bioactive compounds derived from plants, which can account for the observed beneficial effects of plant extracts. The detailed identification process can contain the following several steps:

1. Plant material preparation: This step contains collection, drying, fractionation of plant tissues, and extraction (e.g., ethanol or water).
2. Fingerprinting of active compounds in plants: This can be identified by liquid chromatography-mass spectrometry (LC-MS) or gas chromatography-mass spectrometry (GC-MS). For example, the fingerprints of active compounds in *Rumex crispus* L. extracts, and *Asparagus cochinchinensis* (Lour.) Merr. have been identified [1, 2].
3. Bioassay of activity of compounds: With the concern of different research aims, the compounds with the anticipated pharmacological effects need to be further identified.
4. Assessment of combinational effects of compounds: Sometimes, it is also needed to evaluate the combinational effects of compounds with anticipated pharmacological effects. That is, the possible synergistic or additive effects among the identified active compounds are also needed to be further determined.

In *C. elegans*, exposure to graphene oxide, a carbon-based nanomaterial, caused multiple aspects of toxicity [3–8]. Treatment with *Glycyrrhizae radix* extract could extend lifespan and inhibit oxidative damage in *C. elegans* [9]. Moreover, toxicity of graphene oxide in reducing lifespan, decreasing brood size, and inhibiting locomotion behavior could also be suppressed by *Glycyrrhizae radix* extract (Fig. 13.1) [10]. With the *Glycyrrhizae radix* extract as the example, we introduced the identification of its active compounds and their function in affecting graphene oxide toxicity. Based on high-performance liquid chromatography (HPLC) analysis, the main active components in *Glycyrrhizae radix* extract were glycyrrhizic acid, liquiritin, and isoliquiritin (Fig. 13.2) [10]. The 250 mg/mL *Glycyrrhizae radix* extract contained 9.7 mg/mL glycyrrhizic acid, 1.5 mg/mL liquiritin, and 0.5 mg/mL isoliquiritin on a mass/volume basis (Fig. 13.2) [10].

Among these three active components identified in the GR *Glycyrrhizae radix* extract, treatment with glycyrrhizic acid (9.7 mg/mL), one of triterpenoid saponins, could inhibit the toxicity of graphene oxide on lifespan, reproduction, and locomotion behavior, and its pharmacological effect was comparable to that of 250 mg/mL *Glycyrrhizae radix* extract (Figs. 13.2 and 13.3) [10]. Different from this, the toxicity of graphene oxide on lifespan, reproduction, and locomotion behavior was not obviously altered by treatment with liquiritin (1.5 mg/mL) and isoliquiritin (0.5 mg/mL) (Fig. 13.4) [10]. The pharmacological effect of glycyrrhizic acid in suppressing graphene oxide toxicity on lifespan, reproduction, and locomotion behavior was concentration dependent [10]. For the cellular basis of

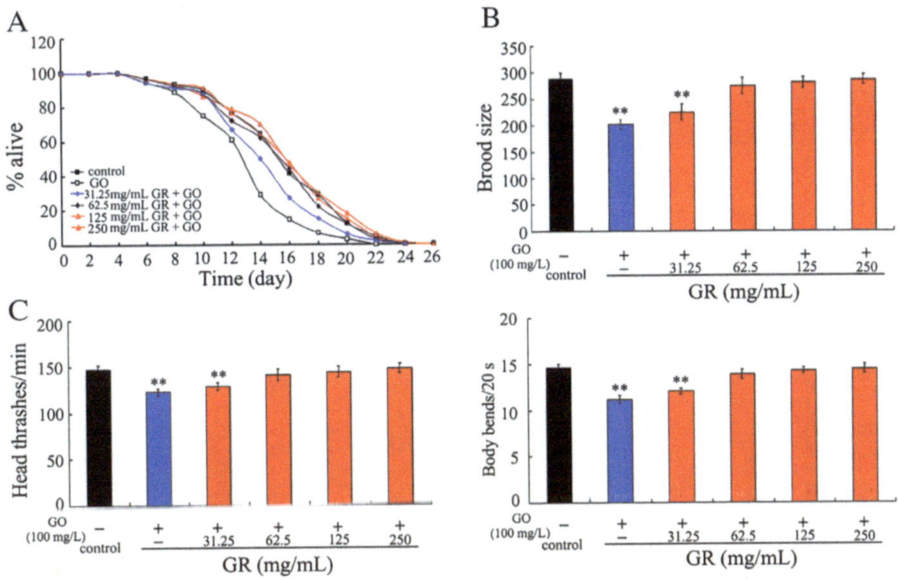

Fig. 13.1 Effects of pretreatment with GR on GO toxicity in exposed wild-type nematodes [10]. (**a**) Effects of pretreatment with GR on lifespan in GO (100 mg/L)-exposed nematodes. (**b**) Effects of pretreatment with GR on brood size in GO-exposed nematodes. (**c**) Effects of pretreatment with GR on head thrash and body bend in GO-exposed nematodes. GR *Glycyrrhizae radix*, GO graphene oxide. Nematodes were pretreated with GR from L1-larvae to L4-larvae in the presence of food, and then exposed to GO from L4-laevae for 24-h. Bars represent means ± S.E.M. $^{**}P < 0.01$ vs. control

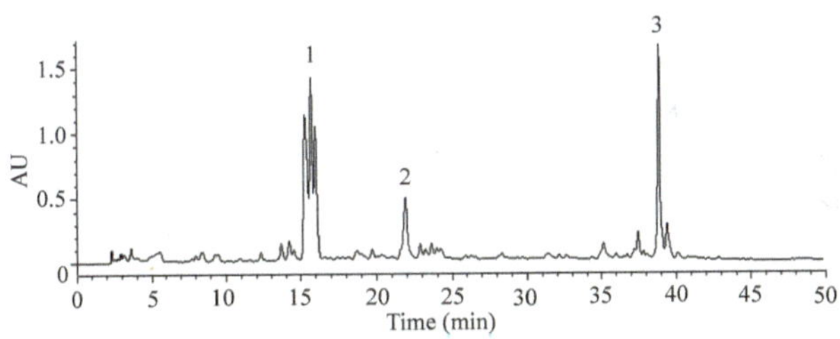

Fig. 13.2 Chemical components in GR analyzed by HPLC [10]. *1* liquiritin, *2* isoliquiritin, *3* glycyrrhizic acid. GR *Glycyrrhizae radix*

pharmacological effect of glycyrrhizic acid, it was observed that treatment with the glycyrrhizic acid, but not the liquiritin and the isoliquiritin, could reduce oxidative damage (Fig. 13.5) [10]. Treatment with glycyrrhizic acid further rescued the graphene oxide induced dysregulation in expression of genes encoding Mn-SODs and mitochondrial complex components (Fig. 13.5) [10]. Moreover, accumulation of graphene oxide in the body of nematodes could be blocked by both *Glycyrrhizae radix* extract and glycyrrhizic acid (Fig. 13.6) [10]. These observations suggested

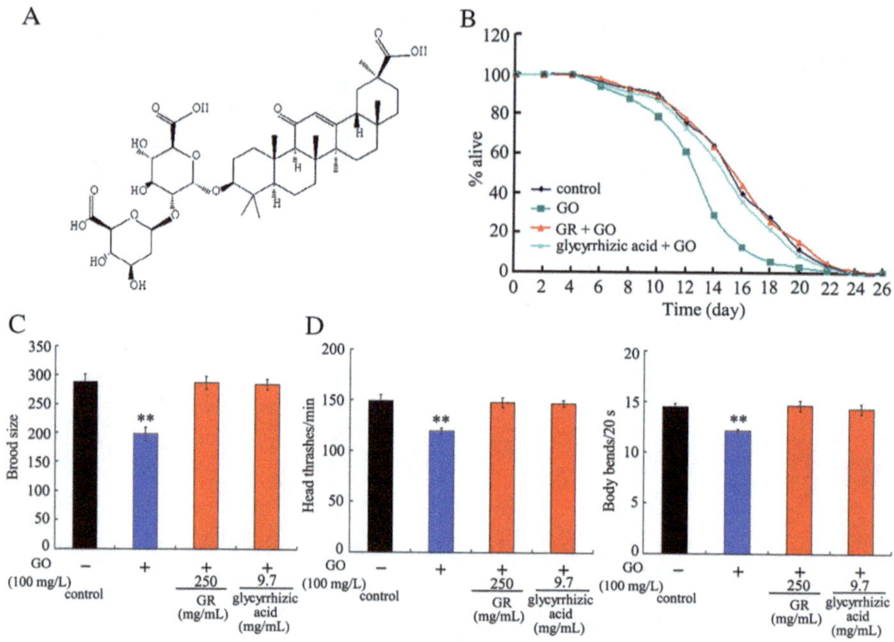

Fig. 13.3 Effects of pretreatment with glycyrrhizic acid on GO toxicity in exposed wild-type nematodes [10]. (**a**) Chemical structure of glycyrrhizic acid. (**b**) Effects of pretreatment with glycyrrhizic acid (9.7 mg/mL) on lifespan in GO (100 mg/L)-exposed nematodes. (**c**) Effects of pretreatment with glycyrrhizic acid on brood size in GO-exposed nematodes. (**d**) Effects of pretreatment with glycyrrhizic acid on head thrash and body bend in GO-exposed nematodes. GR *Glycyrrhizae radix*, GO graphene oxide. Nematodes were pretreated with glycyrrhizic acid from L1-larvae to L4-larvae in the presence of food, and then exposed to GO from L4-laevae for 24-h. Bars represent means ± S.E.M. $^{**}P < 0.01$ vs. control

glycyrrhizic acid as the key active compound to mediate the pharmacological effect of *Glycyrrhizae radix* extract against graphene oxide toxicity. For the underlying molecular mechanism for observed pharmacological effect of glycyrrhizic acid, it was raised that glycyrrhizic acid prevented graphene oxide toxicity and translocation by modulating certain microRNAs (such as *mir-360*) to control activity and function of their downstream targeted genes [10].

13.3 Identification of Pharmacological Active Components in Traditional Chinese Medicine Formula

Traditional Chinese Medicine formula refers to the prescribed combination of medicinal substances (herbs, minerals, or animal parts), which is formulated according to the theoretical principles of Traditional Chinese Medicine for the aim of treating clinical diseases. Using *C. elegans* as an animal model, pharmacological

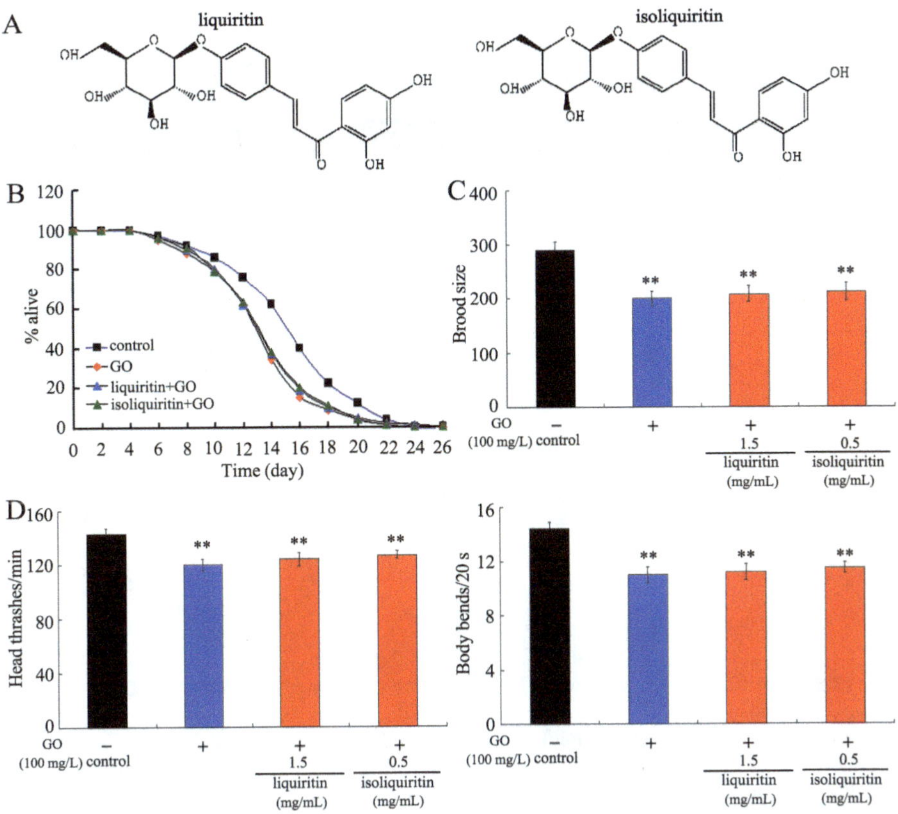

Fig. 13.4 Effects of pretreatment with liquiritin or isoliquiritin on toxicity of GO in exposed wild-type nematodes [10]. (**a**) Chemical structures of liquiritin and isoliquiritin. (**b**) Effects of pretreatment with liquiritin (1.5 mg/mL) or isoliquiritin (0.5 mg/mL) on lifespan in GO (100 mg/L)-exposed nematodes. (**c**) Effects of pretreatment with liquiritin or isoliquiritin on brood size in GO-exposed nematodes. (**d**) Effects of pretreatment with liquiritin or isoliquiritin on head thrash and body bend in GO-exposed nematodes. Nematodes were pretreated with liquiritin or isoliquiritin from L1-larvae to L4-larvae in the presence of food, and then nematodes were exposed to GO from L4-laevae for 24-h. GO, graphene oxide. Bars represent means ± S.E.M. $^{**}P < 0.01$ vs control

effects of different Traditional Chinese Medicine formula have been assessed and confirmed [11–23]. The detailed identification process for active components in Traditional Chinese Medicine formula is largely similar to that for identification of active compounds in plants. The detailed identification process includes the following several steps:

1. Preparation of Traditional Chinese Medicine formula: This step contains collection, drying, fractionation of plant tissues, extraction (e.g., ethanol or water), and pharmaceutical processing of formula.

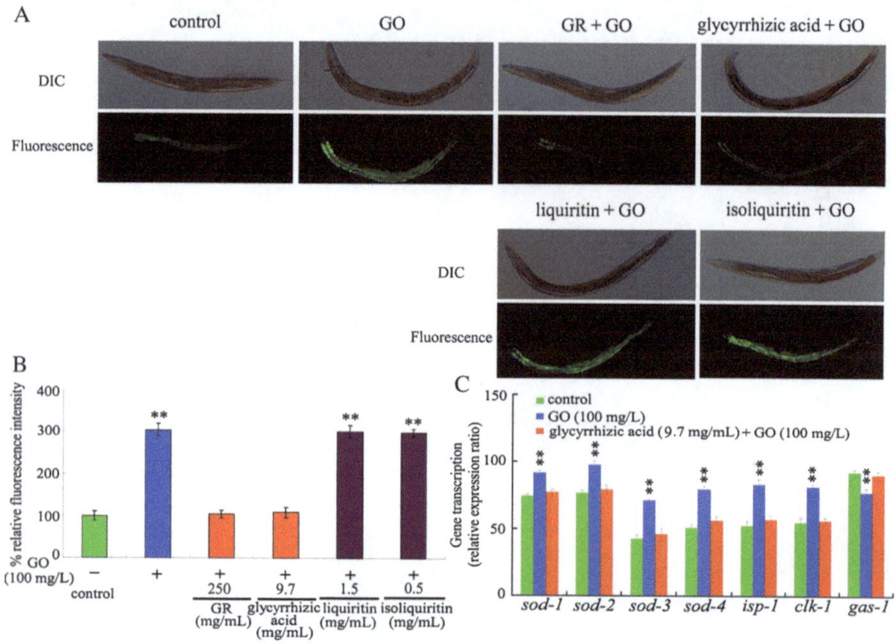

Fig. 13.5 Effects of pretreatment with GR or its active components on intestinal ROS production induced by GO in exposed wild-type nematodes [10]. (**a**) The pictures showing intestinal ROS production in nematodes. GO, 100 mg/L; GR, 250 mg/mL; glycyrrhizic acid, 9.7 mg/mL; liquiritin, 1.5 mg/mL; isoliquiritin, 0.5 mg/mL. (**b**) Comparison of relative fluorescence intensity for signals labeling ROS production in intestine of nematodes. (**c**) Comparison of expression levels of genes required for the control of oxidative stress. The results were expressed as the relative expression ratio between targeted gene and *tba-1* gene encoding a tubulin protein. GR *Glycyrrhizae radix*, GO graphene oxide. Nematodes were pretreated with GR, glycyrrhizic acid, liquiritin, or isoliquiritin from L1-larvae to L4-larvae in the presence of food, and then exposed to GO from L4-laevae for 24-h. Bars represent means ± S.E.M. $^{**}P < 0.01$ vs control

2. Fingerprinting of active compounds in Traditional Chinese Medicine formula: This can also be identified by HPLC, LC-MS, or GC-MS. At this step, all the main active compounds for each Traditional Chinese Medicine included in the examined formula should be covered.
3. Pharmacological activity bioassay: Using *C. elegans* as the animal model, pharmacological effects of active compounds or Traditional Chinese Medicines included in the examined formula can be evaluated using corresponding endpoints with the concern of different research aims.
4. Assessment of combinational effects of active compounds: The combinational effects (especially the possible synergistic effects) among the identified active compounds in the examined formula are needed to be determined.
5. Assessment of combinational effects of Traditional Chinese Medicines: Based on assessment of combinational effects of active compounds, the corresponding candidate Traditional Chinese Medicines in the formula can be confirmed.

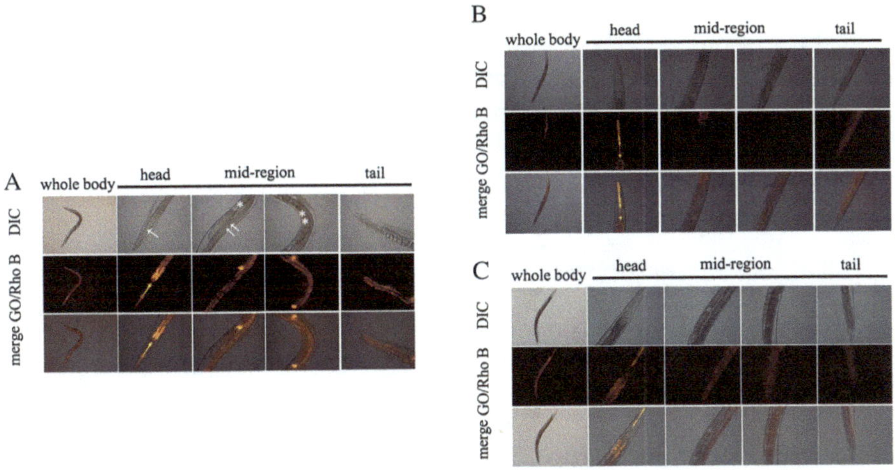

Fig. 13.6 Effects of pretreatment with GR (250 mg/mL) or glycyrrhizic acid (9.7 mg/mL) on translocation of GO/Rho B (100 mg/L) in wild-type nematodes [10]. (**a**) Nematodes were exposed to GO/Rho B from L4-laevae for 24-h. (**b**) Nematodes were pretreated with GR from L1-larvae to L4-larvae in the presence of food, and then exposed to GO/Rho B from L4-laevae for 24-h. (**c**) Nematodes were pretreated with glycyrrhizic acid from L1-larvae to L4-larvae in the presence of food, and then exposed to GO/Rho B from L4-laevae for 24-h. Arrowheads indicate the pharynx (single-arrowhead) and intestine (double-arrowhead), and asterisks indicate the spermatheca (*) and gonad (**). GR *Glycyrrhizae radix*, GO graphene oxide

Moreover, the combinational effects among Traditional Chinese Medicines in the formula are needed to be further examined.

In the clinical, treatment with Xuebijing injection, a Traditional Chinese Medicine formula, could improve the outcomes of sepsis patients and reduce their case fatality rate [24–28]. Using *C. elegans* as the animal model, it was reported that the bacterial pathogen caused reduction in lifespan could be inhibited by different doses of Xuebijing treatment, and the pharmacological effect of Xuebijing treatment was dose-dependent (Fig. 13.7) [29]. Accompanied with this, accumulation of *P. aeruginosa* PA14 in the body of *C. elegans* was obviously reduced by different doses of Xuebijing treatment (Fig. 13.8) (L. Zhang and D.-Y. Wang, personal communication). In addition, *P. aeruginosa* PA14 induced decrease in expressions of *lys-1* and *lys-8* could be significantly inhibited by Xuebijing during the first three-day treatment period, suggesting the potential effect of Xuebijing in suppressing bacterial pathogen-induced immunosuppression (Fig. 13.9) (L. Zhang and D.-Y. Wang, personal communication). Xuebijing treatment could not only confer the resistance to bacterial pathogen infection, but also confer the resistance to fungal pathogen infection, such as the *Candida albicans* infection. Xuebijing treatment could also extend the lifespan in *C. albicans*-infected nematodes (Fig. 13.10) (L. Zhang and D.-Y. Wang, personal communication). Besides the effects on the hosts, Xuebijing treatment could further inhibit biofilm formation of *P. aeruginosa*

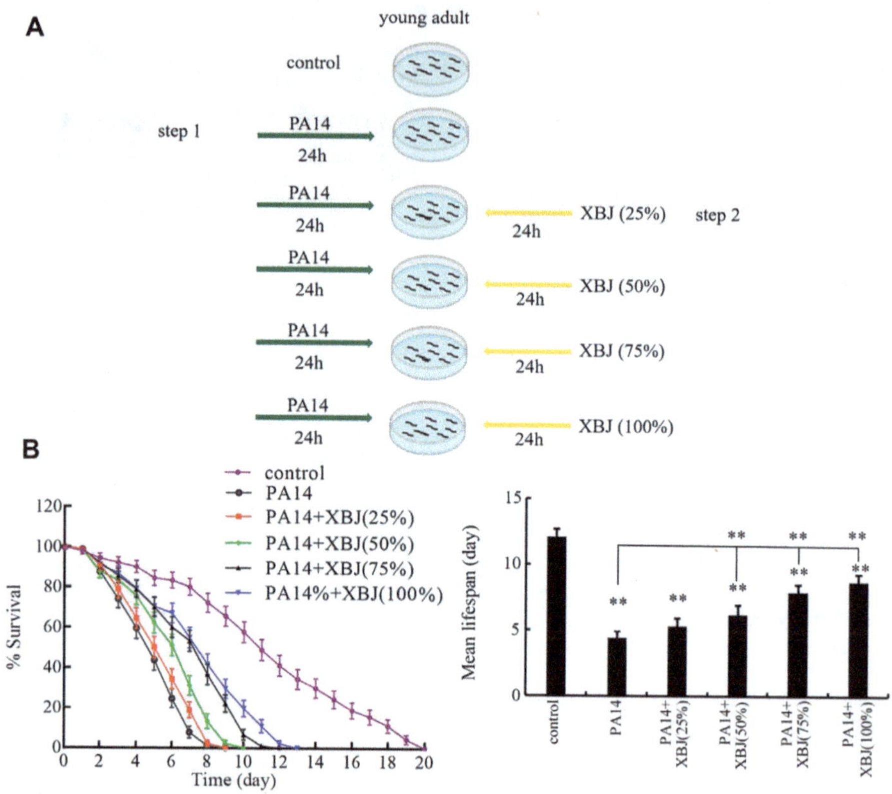

Fig. 13.7 Effect of Xuebijing post-treatment against *P. aeruginosa* infection [29]. (**a**) Diagram showing the method for pharmacological post-treatment of Xuebijing after *P. aeruginosa* infection. (**b**) Effect of Xuebijing post-treatment on lifespan of nematodes after *P. aeruginosa* infection. XBJ Xuebijing. $^{**}P < 0.01$ vs control (if not specially indicated)

PA14 by decreasing expressions of related virulence genes (*lasA*, *lasB*, *rhlA*, *rhlC*, *phzA*, *phzM*, *phzH*, and *phzS*), as well as the quorum sensing-related genes (*lasI*, *lasR*, *rhlI*, *rhlR*, *pqsA*, and *pqsR*) (Fig. 13.11) [29].

Xuebijing is prepared based on a combination of *Carthamus tinctorius* L. (Carthami Flos, Honghua), *Paeonia lactiflora* Pall. (Paeoniae Radix Rubra, Chishao), *Ligusticum chuanxiong* Hort. (Chuanxiong Rhizoma, Chuanxiong), *Salvia miltiorrhiza* Bge. (Salviae Miltiorrhizae Radix Et Rhizoma, Danshen), and *Angelica sinensis* (Oliv.) Diels (Angelicae Sinensis Radix, Danggui) [29]. Based on the HPLC analysis, fingerprint of Xuebijing indicated that protocatechuic aldehyde acts as marker compound of Danshen, hydroxysafflor yellow A acts as marker compound of Honghua, paeoniflorin acts as marker compound of Chishao, and ferulic acid and senkyunolide act as marker compounds of Chuanxiong and Danggui (Fig. 13.12) [29]. Among the active compounds in Xuebijing, treatment with rosmarinic acid, tanshinone II A, senkyunolide I, salvianolic acid A, salvianolic

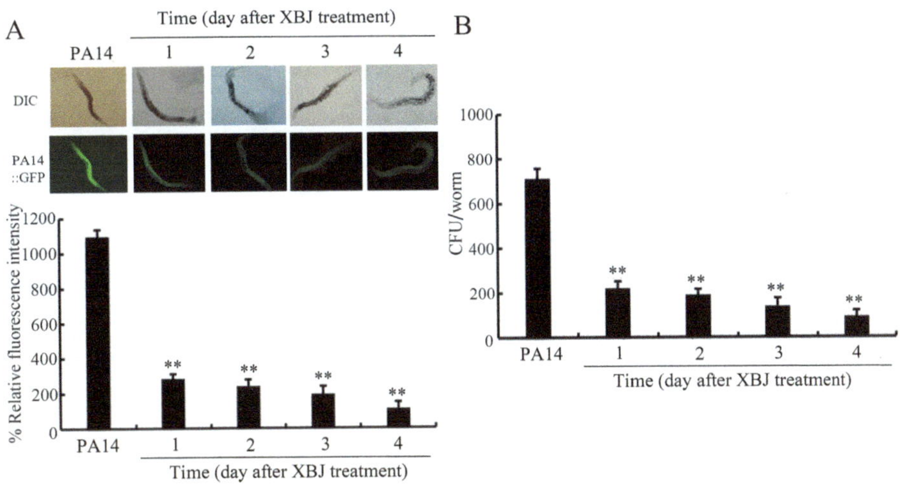

Fig. 13.8 Effect of 100% XBJ treatment on PA14::GFP accumulation in intestinal lumen (**a**) and PA14 CFU (**b**) in nematodes at immunosuppression stage (day-5) after *P. aeruginosa* infection. XBJ Xuebijing. $^{**}P < 0.01$ vs PA14

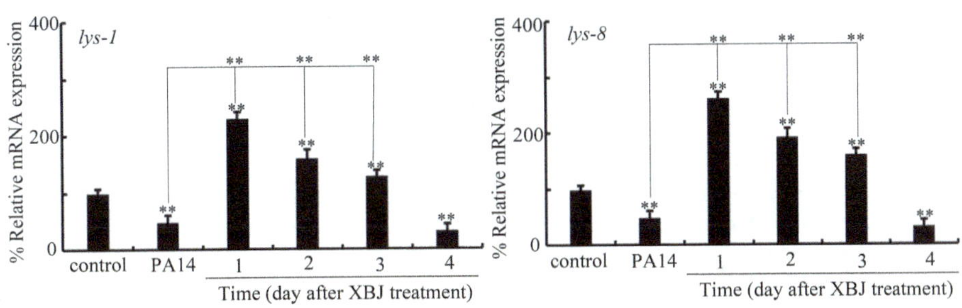

Fig. 13.9 Effect of 100% XBJ treatment on expression of antimicrobial genes (*lys-1* and *lys-8*) in nematodes at immunosuppression stage (day-5) after *P. aeruginosa* infection. Control, without PA infection. XBJ Xuebijing. $^{**}P < 0.01$ vs control (if not specially indicated)

acid B, hydroxysafflor yellow A, albiflorin, and salvianic acid A solium could not affect the lifespan in *P. aeruginosa* PA14-infected nematodes (Fig. 13.13) (L. Zhang and D.-Y. Wang, personal communication). In contrast, the lifespan in *P. aeruginosa* PA14-infected nematodes could be significantly extended by procatechuic acid, protocatechuic aldehyde, paeoniflorin, benzoylpaeoniflorin, ferulic acid, and oxypaeoniflorin (Fig. 13.13) (L. Zhang and D.-Y. Wang, personal communication). In addition, the *P. aeruginosa* PA14 accumulation in intestinal lumen could be reduced by treatment with procatechuic acid, protocatechuic aldehyde, paeoniflorin, benzoylpaeoniflorin, ferulic acid, and oxypaeoniflorin (Fig. 13.14) (L. Zhang and D.-Y. Wang, personal communication). Moreover, the biofilm formation of *P. aeruginosa* PA14 could be further inhibited by treatment with procatechuic acid, protocatechuic aldehyde, paeoniflorin, benzoylpaeoniflorin,

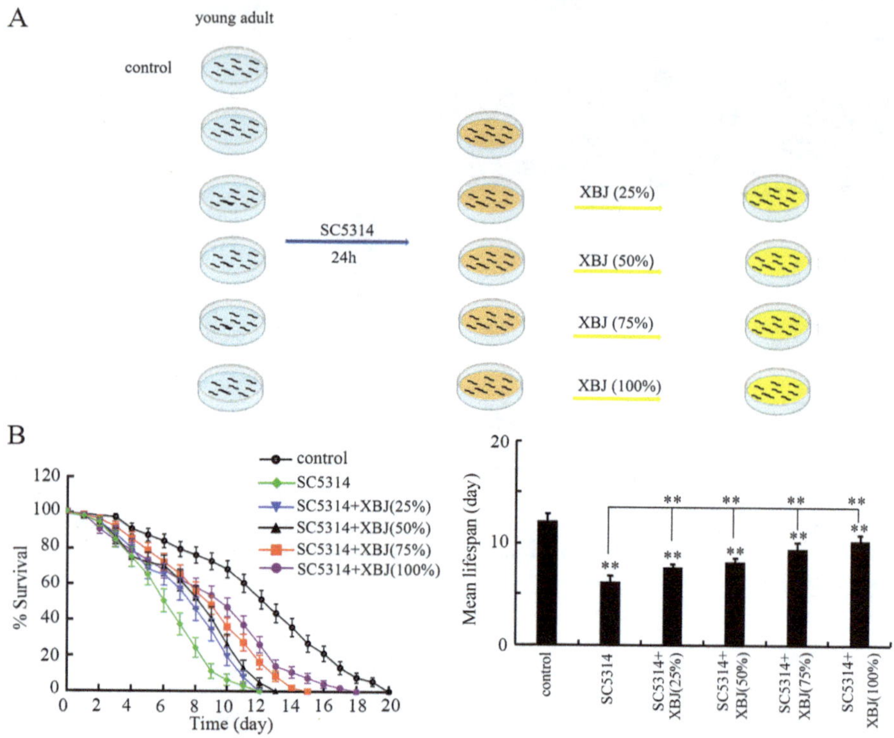

Fig. 13.10 Effect of Xuebijing post-treatment against fungal infection. (**a**) A diagram shows the method for pharmacological post-treatment of Xuebijing after fungal infection. (**b**) Effect of Xuebijing post-treatment on lifespan of nematodes after *C. albicans* infection. XBJ Xuebijing. $^{**}P < 0.01$ vs control (if not specially indicated)

ferulic acid, and oxypaeoniflorin (Fig. 13.15) (L. Zhang and D.-Y. Wang, personal communication). These observations suggested the procatechuic acid, protocatechuic aldehyde, paeoniflorin, benzoylpaeoniflorin, ferulic acid, and oxypaeoniflorin as the crucial active compounds in Xuebijing to account for the clinical pharmacological effects of this Traditional Chinese Medicine formula.

13.4 Perspectives

Using *C. elegans* as the animal model, many plant extracts have been shown to exhibit certain aspects of pharmacological effects [30–33]. In this chapter, we first used *Glycyrrhizae radix* extract as the example of plant extracts to introduce and discuss the identification of pharmacological active compounds in plant extracts. In the *Glycyrrhizae radix* extract, the glycyrrhizic acid was identified as the key active compound to mediate its pharmacological effect against the toxicity of graphene oxide. However, if the identified active compounds are two or more compounds in

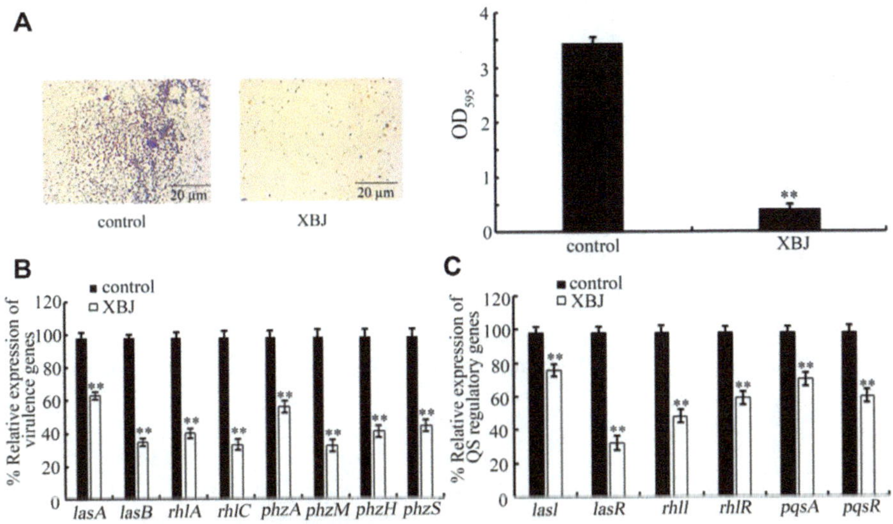

Fig. 13.11 Effect of Xuebijing post-treatment on biofilm formation (**a**), expression of virulence genes (**b**), and expression of QS-related genes (**c**) of PA14 [29]. The 100% Xuebijing was used for pharmacological treatment. XBJ Xuebijing. $^{**}P < 0.01$ vs control

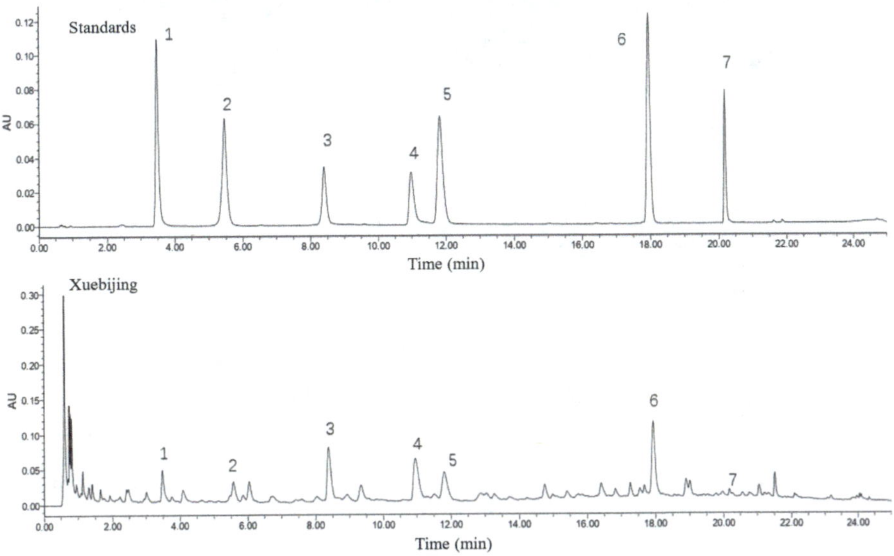

Fig. 13.12 Biochemical fingerprint of Xuebijing [29]. The fingerprint was analyzed by reverse phase high-performance liquid chromatography (RP-HPLC). The separation was performed on a Alltech C18 column with a mobile phase composed of 1% methanoic acid and acetonitrile. The column temperature was set at 30 °C, and the flow rate was 1.0 mL/min. The detective wavelength was set 280 nm. The injection volume was 10 μL. *1* Protocatechuic aldehyde, *2* Chlorogenic acid, *3* Hydroxysafflor yellow A, *4* Paeoniflorin, *5* Ferulic acid, *6* Senkyunolide, and *7* Salvianolic acid B

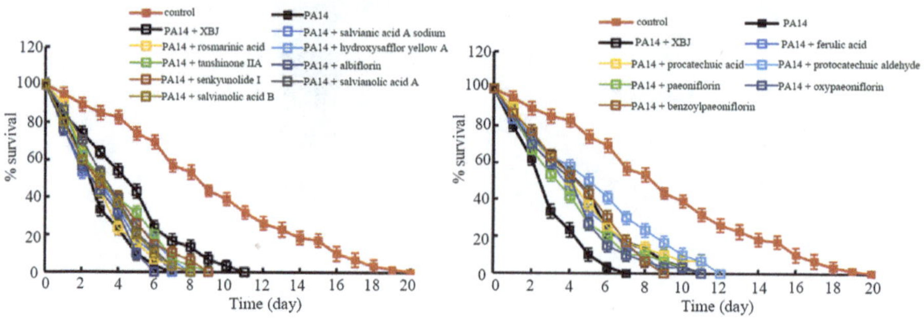

Fig. 13.13 Identification of active components in Xuebijing in extending lifespan in *P. aeruginosa* PA14-infected nematodes. XBJ Xuebijing. Treatment concentration of active compounds was 1 mg/L

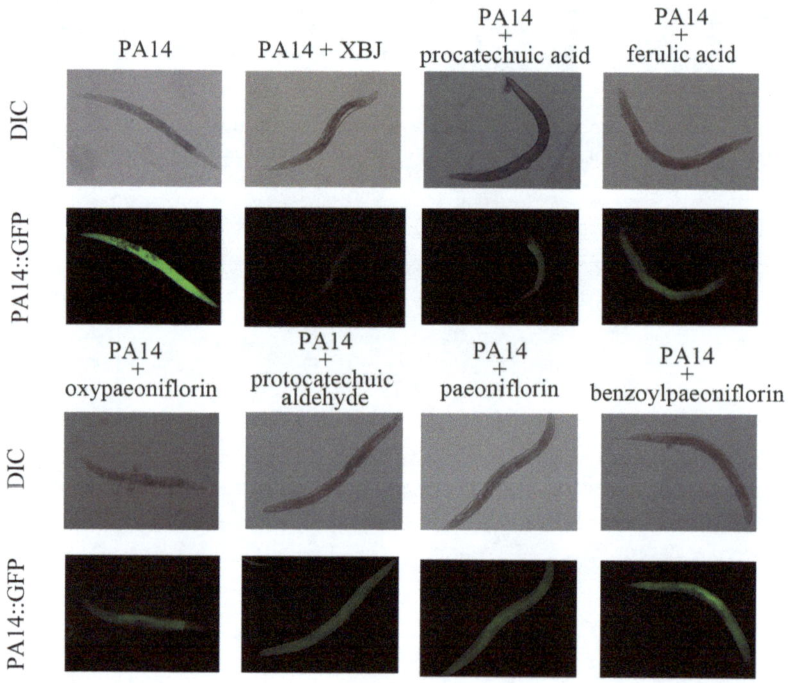

Fig. 13.14 Identification of active components in Xuebijing in reducing *P. aeruginosa* PA14::GFP accumulation in intestinal lumen. XBJ Xuebijing. Treatment concentration of active compounds was 1 mg/L

plants extracts, the further combinational effects of identified active compounds need to be further determined.

Moreover, we used Xuebijing as the example of Traditional Chinese Medicine formula to introduce and discuss the identification of pharmacological active components in Traditional Chinese Medicine formula. The procatechuic acid,

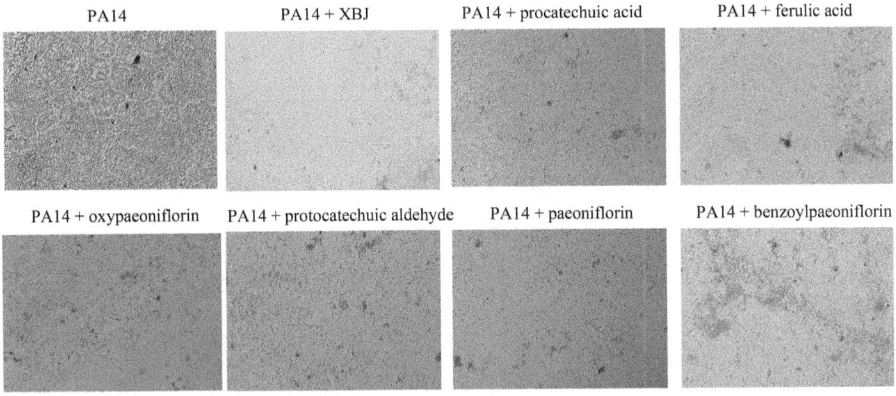

Fig. 13.15 Identification of active components in Xuebijing in inhibiting biofilm formation of *P. aeruginosa* PA14. XBJ Xuebijing. Treatment concentration of active compounds was 1 mg/L

protocatechuic aldehyde, paeoniflorin, benzoylpaeoniflorin, ferulic acid, and oxypaeoniflorin have been identified as important active compounds in Xuebijing to mediate its pharmacological effects. Nevertheless, both the combinational effects among these identified active compounds and the combinational effects among Traditional Chinese Medicines in Xuebijing are also needed to be further examined.

References

1. Idris OA, Wintola OA, Afolayan AJ (2022) Anthelmintic potency of *Rumex crispus* L. extracts against *Caenorhabditis elegans* and non-targeted identification of the bioactive compounds. Saudi J Biol Sci 29:541–549
2. Wan R, Cai Z, Liu Z, You J, Wang L, Tang X, Chen Y, Wang Q, Ou L, Zhang C (2025) Identification of active anti-aging ingredients from *Asparagus cochinchinensis* based on the spectrum-effect relationship. BMC Complement Med Ther 25:428
3. Wu Q-L, Yin L, Li X, Tang M, Zhang T, Wang D-Y (2013) Contributions of altered permeability of intestinal barrier and defecation behavior to toxicity formation from graphene oxide in nematode *Caenorhabditis elegans*. Nanoscale 5:9934–9943
4. Wu Q-L, Zhao Y-L, Li Y-P, Wang D-Y (2014) Molecular signals regulating translocation and toxicity of graphene oxide in nematode *Caenorhabditis elegans*. Nanoscale 6:11204–11212
5. Wu Q-L, Zhao Y-L, Fang J-P, Wang D-Y (2014) Immune response is required for the control of *in vivo* translocation and chronic toxicity of graphene oxide. Nanoscale 6:5894–5906
6. Zhao Y-L, Yu X-M, Jia R-H, Yang R-L, Rui Q, Wang D-Y (2015) Lactic acid bacteria protects *Caenorhabditis elegans* from toxicity of graphene oxide by maintaining normal intestinal permeability under different genetic backgrounds. Sci Rep 5:17233
7. Zhao Y-L, Wu Q-L, Wang D-Y (2016) An epigenetic signal encoded protection mechanism is activated by graphene oxide to inhibit its induced reproductive toxicity in *Caenorhabditis elegans*. Biomaterials 79:15–24
8. Zhao Y-L, Zhi L-T, Wu Q-L, Yu Y-L, Sun Q-Q, Wang D-Y (2016) p38 MAPK-SKN-1/Nrf signaling cascade is required for intestinal barrier against graphene oxide toxicity in *Caenorhabditis elegans*. Nanotoxicology 10:1469–1479

9. Ruan Q-L, Qiao Y, Zhao Y-L, Xu Y, Wang M, Duan J-A, Wang D-Y (2016) Beneficial effects of *Glycyrrhizae radix* extract in preventing oxidative damage and extending the lifespan of *Caenorhabditis elegans*. J Ethnopharmacol 177:101–110
10. Zhao Y-L, Jia R-H, Qiao Y, Wang D-Y (2016) Glycyrrhizic acid, active component from *Glycyrrhizae radix*, prevents toxicity of graphene oxide by influencing functions of microRNAs in nematode *Caenorhabditis elegans*. Nanomedicine: Nanotechnol. Biol Med 12:735–744
11. Gao G, Zhang X, Cui Z, Fan M, Yan Y, Huang Y, Shi Y, Ma H, Wang Z, Su Y, Zhang Z, Xie Z (2025) Shenghui decoction inhibits neuronal cell apoptosis to improve Alzheimer's disease through the PDE4B/cAMP/CREB signaling pathway. Phytomedicine 137:156366
12. Zhang W, Zhi D, Ren H, Wang D, Wang X, Zhang Z, Fei D, Zhu H, Li H (2016) Shengmai formula ameliorates pathological characteristics in AD *C. elegans*. Cell Mol Neurobiol 36: 1291–1302
13. Meng F, Li J, Rao Y, Wang W, Fu Y (2017) A Chinese herbal formula, Gengnianchun, ameliorates beta-amyloid peptide toxicity in a *Caenorhabditis elegans* model of Alzheimer's disease. Evid Based Complement Alternat Med 2017:7480980
14. Zhou L, Yang C, Liu Z, Chen L, Wang P, Zhou Y, Yuan M, Zhou LT, Wang X, Zhu LQ (2024) Neuroprotective effect of the traditional decoction Tian-Si-Yin against Alzheimer's disease via suppression of neuroinflammation. J Ethnopharmacol 321:117569
15. Xiao L, Li H, Tian J, Jin N, Zhang J, Yang F, Zhou L, Wang Q, Huang Z (2020) The traditional formula Kai-Xin-San alleviates polyglutamine-mediated neurotoxicity by modulating proteostasis network in *Caenorhabditis elegans*. Rejuvenation Res 23:207–216
16. Zhao R, Zhao T, Shi T, Feng C, Li H, Xu Y, Zhang Q, Yin X, Yao J, Zhang G, Xia X (2025) Elucidating the anti-aging bioactive compounds and mechanisms of Jingfang Granules formula based on network pharmacology and *Caenorhabditis elegans*. Biogerontology 26:127
17. Chen W, Liu S, Xu G, Liu X, Shi Y, Wang G, Ning Y, Lu Z, Cao Y, Zhao Y (2025) Zuogui and Yougui pills extend lifespan and improve ageing biomarkers via kaempferol-mediated mitophagy in *Caenorhabditis elegans*. Biogerontology 26:172
18. Hai Y, Ren K, Hou WQ, Cao HS, Zhang YR, Li ZM, Wang SQ, Yang W, Liu DL (2024) Hypoglycemic TCM formulas (Huangqi-Gegen drug pair) have the potential as an Alzheimer's disease. Phytomedicine 130:155723
19. Lin H, Wu S, Weng Z, Wang H, Shi R, Tian M, Wang Y, He H, Wang Y, Liu X, Jian Z, Wei F, Wang P, Zhang L, Liu Y, Guo Q, Chen C, Yang W (2021) Tianma formula alleviates dementia via ACER2-mediated sphingolipid signaling pathway involving Abeta. Evid Based Complement Alternat Med 2021:6029237
20. Zeng L, Yang Z, Yun T, Fan S, Pei Z, Chen Z, Sun C, Xu F (2019) Antiaging effect of a Jianpiyangwei formula in *Caenorhabditis elegans*. BMC Complement Altern Med 19:313
21. Wang X, Zhang L, Gao X, Zhang L, Yu J, Liu Y, Fang M, Yan Y, Chen L, Du J, Guan H, Huang C, Fan S (2025) Liu Jun Zi decoction extends lifespan and healthspan through p16/p21 signaling in *Caenorhabditis elegans*. Biogerontology 26:201
22. Zhao T, Hou Y, Feng C, Zhao R, Li H, Zhao P, Wang M, Ren J, Meng Y, Xu K, Liu J, Yin X, Zhang G, Yao J, Xia X (2025) Jingfang granule extends lifespan and healthspan in *Caenorhabditis elegans*: insights from RNA-seq analysis of genetic mechanisms. Biogerontology 26:168
23. Chen L, Wu K, Shou C, Zhang L, Tu M, Wang B, Cai Y, Zheng Z, Sun J, Ruan Q, Guo J (2026) Bushen Huoxue decoction alleviates bisphenol a-induced infertility through the PMK-1 mitogen-activated protein kinases signaling pathway and downstream mitochondrial unfolded protein response in *Caenorhabditis elegans*. Front Pharmacol 16:1713681
24. Lou X, Chen H, Shi N, Yu R, Li S, Yang Y, Liu S, Xie J, Qiu H (2025) Treatment effects of Xuebijing injection in patients with sepsis by clinical phenotype: a post hoc analysis of the EXIT-SEP trial. EClinicalMedicine 86:103341
25. Liu S, Yao C, Xie J, Liu H, Wang H, Lin Z, Qin B, Wang D, Lu W, Ma X, Liu Y, Liu L, Zhang C, Xu L, Zheng R, Zhou F, Liu Z, Zhang G, Zhou L, Liu J, Fei A, Zhang G, Zhu Y, Qian K, Wang R, Liang Y, Duan M, Wu D, Sun R, Wang Y, Zhang X, Cao Q, Yang M, Jin M,

Song Y, Huang L, Zhou F, Chen D, Liang Q, Qian C, Tang Z, Zhang Z, Feng Q, Peng Z, Sun R, Song Z, Sun Y, Chai Y, Zhou L, Cheng C, Li L, Yan X, Zhang J, Huang Y, Guo F, Li C, Yang Y, Shang H, Qiu H, EXIT-SEP Investigators (2023) Effect of an herbal-based injection on 28-day mortality in patients with sepsis: the EXIT-SEP randomized clinical trial. JAMA Intern Med 183:647–655

26. Li C, Wang P, Li M, Zheng R, Chen S, Liu S, Feng Z, Yao Y, Shang H (2021) The current evidence for the treatment of sepsis with Xuebijing injection: bioactive constituents, findings of clinical studies and potential mechanisms. J Ethnopharmacol 265:113301
27. Liu S, Yao C, Zhang J, Yang Y, Qiu H, Investigators EXIT-SEP (2019) Efficacy of Xuebijing Injection for Sepsis (EXIT-SEP): protocol for a randomised controlled trial. BMJ Open 9: e028664
28. Huang W, Wang M (2025) Advancing traditional Chinese medicine for sepsis-challenges beyond clinical evidence. Crit Care 29:222
29. Zhang L, Wang Y-X, Cao C, Zhu Y-K, Huang W, Yang Y, Qiu H-B, Liu S-Q, Wang D-Y (2022) Beneficial effect of Xuebijing against *Pseudomonas aeruginosa* infection in *Caenorhabditis elegans*. Front Pharmacol 13:949608
30. Chen X, Bahramimehr F, Shahhamzehei N, Fu H, Lin S, Wang H, Li C, Efferth T, Hong C (2024) Anti-aging effects of medicinal plants and their rapid screening using the nematode *Caenorhabditis elegans*. Phytomedicine 129:155665
31. Dyshlyuk LS, Dmitrieva AI, Drozdova MY, Milentyeva IS, Prosekov AY (2022) Relevance of bioassay of biologically active substances (BAS) with geroprotective properties in the model of the nematode *Caenorhabditis elegans* in *in vivo* experiments. Curr Aging Sci 15:121–134
32. Chen JC, Wang R, Wei CC (2024) Anti-aging effects of dietary phytochemicals: from *Caenorhabditis elegans*, *Drosophila melanogaster*, rodents to clinical studies. Crit Rev Food Sci Nutr 64:5958–5983
33. Chattopadhyay D, Thirumurugan K (2018) Longevity promoting efficacies of different plant extracts in lower model organisms. Mech Ageing Dev 171:47–57

High-Throughput Drug Screening

14

Abstract

An important tendency for the pharmacological study in *Caenorhabditis elegans* is to perform the high-throughput drug screening. Surrounding this aspect, we first introduced the pharmacological screening in small compound library. Moreover, the pharmacological screening in microfluidic system, the chemogenomic-based pharmacological screening, and the pharmacological screening among structurally related compounds were introduced and discussed.

Keywords

High throughput · Drug screening · *Caenorhabditis elegans*

14.1 Introduction

Caenorhabditis elegans is not only suitable for assessment of pharmacological effects of single compounds and drugs or their combinations, but also can be employed for the aim of high-throughput drug screening. The following several properties of *C. elegans* make it suitable to perform the high-throughput drug screening:

1. Large-scale screening and cost-efficiency: The *C. elegans* high-throughput drug screening can be performed with the multi-well plate formats (96 or 384 wells). The *C. elegans* are approximately 1 mm in length, which allows thousands of individuals to be cultured in a well plate. In addition, the liquid culture assays for *C. elegans* can allow for complete automation by integrating the use of robotic systems to handle dispensing of worms, compounds, and even readouts in a relatively short timeframe.

D. Wang, *Pharmacology in Caenorhabditis elegans*,
https://doi.org/10.1007/978-981-92-5166-7_14

2. Assessment at the whole-organism level: *C. elegans* can fill the gap between cells and mammals for the pharmacological study. That is, *C. elegans* can allow the screening for compounds that require metabolic activation by the gut or bioavailability of the compound that can penetrate biological barriers and reach target tissues. In addition, instead of simply measuring cell death or a single biochemical reaction, high-throughput drug screening in *C. elegans* can quantify a series of complex phenotypes, such as behaviors, physiological states, and longevity.
3. Genetic conservation: *C. elegans* shares a high degree of genetic conservation with humans. Meanwhile, genetically, it is far easier to manipulate. It was predicted that approximately 60–80% of human genes have a functional ortholog in *C. elegans*.
4. Established physiological and pathological models: For this aspect, we have introduced in Chap. 4.
5. Body transparency and in vivo fluorescent imaging: One of the most significant technical advantages for high-throughput drug screening is that *C. elegans* body is transparent throughout its life. This can allow the automated visualization of subcellular structures inside this living animal without dissection. Especially, the generated fluorescent reporters are helpful for the quantification of drug effects, such as the effects on mitochondrial structure and function, autophagosome formation, and neurodegeneration.
6. Phenotypic screening with the unbiased discovery: Rather than targeting a specific protein, *C. elegans* is more suitable for the phenotypic screening. In addition, due to the test on a whole animal, novel mechanisms of action may be identified for the examined pharmacological effects of compounds, which cannot be discovered in a cell-based biochemical screen.
7. Simplicity for certain aspects of complex biology: Certain biological processes that are difficult to be examined in mammals might be easy to be quantified in *C. elegans*. For example, *C. elegans* has a lifespan of only 2–3 weeks, which allows the rapidly screening of thousands of compounds for their ability to extend lifespan or improve health span.

14.2 Pharmacological Screening in Small Compound Library

The small compound library refers to the curated collection of chemical molecules, typically ranging from hundreds to thousands of compounds. The small compound libraries can be classified into three groups:

1. Containing structurally diverse compounds: This kind of libraries can maximize the probability of finding novel hits.
2. Natural product-derived: This kind of libraries usually focus on specific classes (e.g., kinase inhibitors or FDA-approved drugs for repurposing).
3. Containing compounds from plants, fungi, or marine organisms.

In *C. elegans*, high-throughput drug screening is a systematic experimental approach to evaluate the biological activity, efficacy, and toxicity (for the aim of safety assessment) of a collection of chemical compounds in the small compound library, to identify the "hits"—compounds that produce a desired phenotypic change, and to provide preliminary data on bioavailability and safety.

During the high-throughput drug screening in the small compound library, after the library preparation, the compounds are added in wells of multi-well plates at designed concentrations and with the replicates and the controls. Again, the synchronized *C. elegans* populations (L1-larvae, L4-larvae, or young adults selected based on certain research aims) are dispensed into the wells of multi-well plates. Compounds and worms can be added either manually or robotically. For the readout and detection, besides the analysis of specific phenotypes, the automated imaging systems, plate readers for luminescence or fluorescence, and behavioral tracking software can be employed to quantify the phenotypic response of worms. Moreover, after the hit identification, the further validation is needed to confirm the dose-dependent response for the candidate compounds and to examine the mechanism of action.

14.2.1 Phenotypic Screening

For the high-throughput drug screening, this is the most common approach. That is, the candidate compounds can be screened based on their ability to alter a visible phenotype in *C. elegans* [1]. At least the following several kinds of phenotypic screening have been examined:

1. Screening of cell death inhibitors: Using *tra-2(ar221);xol-1(y9) mig-24p::Venus* as the screening strain, among the 23,797 compounds, 11 compounds were found to block the onset of linker cell death [2]. For example, treatment with tyrphostin A9 caused the linker cell persistence [2]. The further experiments indicated that 5 of the 11 candidate compounds were toxic and induced the lethality [2].
2. Screening of anti-aging compounds: Median lifespan, maximum lifespan, and survival curve can be used as the related readouts. Based on the survival curve analysis, among 88,000 chemicals, mianserin, a drug to treat the antidepressant, was identified to extend the lifespan when given only during the *C. elegans* adulthood [3]. Treatment with mianserin inhibited serotonin receptor and octopamine receptor, and its pharmacological effect in extending lifespan was reduced or abolished by mutations of genes governing serotonin synthesis and reuptake at synapses, which suggested its role as the serotonin receptor antagonist [3]. Moreover, the pharmacological effect of mianserin was found to link to dietary restriction [3].
3. Screening of drugs improving health span: Some physiological endpoints during the aging can be used to reflect the health span. In order to find the drugs improving health span, a high-content screen was performed to identify natural

compounds that can modulate the mitochondrial stress [4]. Transgenic markers of GST-4::GFP and HSP-6::GFP were used as the markers to reflect the mitochondrial stress response, and isobavachalcone, manzamine A, kahalalide F, and lutein were identified to have the consistent intervention effects [4]. After treatment with kahalalide F and lutein, not only the health span was improved, but also resistance to the heat stress and the oxidative stress was induced [4].

4. Screening of drugs having neuroprotective effects: Considering that Parkinson's disease (PD)-like symptoms can be caused by neurotoxin MPTP and its active metabolite MPP^+, *C. elegans* MPP^+ model of Parkinson's disease was generated [5]. In this generated model, degeneration in dopaminergic neurons, reduction in mobility, and increase in lethality can be observed [6]. Based on automated recording and quantification, in this disease model, at least lisuride (dopamine receptor agonist), apomorphine (dopamine receptor agonist), and rottlerin (protein kinase C inhibitor) were identified to ameliorate the Parkinson's disease-like defects [5]. Besides this, considering that brain accumulation of RNA-binding protein TDP-43 is linked to occurrence of amyotrophic lateral sclerosis (ALS) and frontotemporal lobar degeneration (FTLD), transgene of TDP-43 was used to generate ALS disease model [6]. In GABAergic motor neurons of transgenic strain expressing human TDP-43 (A315T), UNC-25::GFP expression was reduced [6]. Combinational use of transgenic PC12 cells and *C. elegans*, a high content screen was performed, and LDN-0130436 was identified to be able to rescue the reduction in UNC-25::GFP expression in GABAergic motor neurons and the defect in locomotion in transgenic strain expressing human TDP-43 (A315T) [6].
5. Screening of neuroactive drugs and muscle function modulators: For the aim of neuroactive drugs screening, high-throughput drug screening was performed in the FDA-approved compound library using generated *C. elegans* models of human Mendelian diseases by CRISPR genome-editing [7]. In this FDA-approved compound library, two compounds (liranaftate and atorvastatin) were identified to rescue the defects in behaviors in the model of UNC80 deficiency, a severe neurodevelopmental disorder [7]. For the aim of muscle function modulators screening, high-throughput drug screening was performed among 1040 chemical compounds in nematodes with mutation of *smn-1* encoding homologue of survival motor neuron (SMN) protein and slow swimming phenotype [8]. The *smn-1* mutants can be used to reflect the spinal muscular atrophy, an autosomal recessive genetic disorder [8]. After the screening, 4-aminopyridine, gaboxadol hydrochloride, and N-acetylneuraminic acid were identified to rescue behavioral defects in *smn-1* mutants [8].
6. Screening of cognition improving drugs: In *C. elegans*, RNA-binding protein Musashi (MSI) functions as translational regulator of forgetting [9]. Based on high-throughput natural products screening, some inhibitors of MSI were identified [9]. Among them, ellagic acid improved both short- and long-term associative memory, and its role as the molecular target of MSI was confirmed based on genetic analysis [9].

7. Screening of drugs enhancing regeneration: Following microsurgeries at single-axon precision, *C. elegans* were treated with 100 small molecule compounds [10]. Based on this screening, some compounds were observed to affect the neuronal regeneration [10]. Among them, staurosporine (a PKC kinase inhibitor) could significantly enhance neuronal regeneration in a neuronal type-specific manner [10].
8. Screening of ion channel modulators: Based on high-throughput drug screening in 14,100 small molecules library, 308 compounds were found to induce different phenotypic changes in *C. elegans* [11]. Among them, nemadipine-A with the similar structure of 1,4-dihydropyridines (DHPs) could induce the deficits in egg-laying and related morphology [11]. Genetic analysis identified the EGL-19, a L-type calcium channel α1-subunit, as the candidate target of nemadipine-A, suggesting its role as the novel calcium channel antagonist [11]. Moreover, based on high-throughput drug screening in the long-QT syndrome model expressing ether-a-go-go-related gene (hERG) carrying trafficking-defective mutation A561V and having deficits in egg-laying and locomotion, Prostratin and ingenol-3,20-dibenzoate were identified to activate PKCε signaling and promote hERGA561V trafficking to plasma membrane by phosphorylating S606 at the pore of potassium channel [12]. Thus, these two identified ion channel modulators were suggested to be useful for treating the channelopathies [12].
9. Screening of anti-infective compounds and immunomodulators: To identify compounds against bacterial pathogen infection, high-throughput drug screening was performed among 37,200 compounds and natural product extracts to screen those enhancing *C. elegans* survival after infection with *Enterococcus faecalis* [13]. Based on automated recording and image analysis, 28 compounds and extracts were identified to have anti-infective effect [13]. In addition, 6 structural classes of them cured the infected nematodes, but did not influence bacterial pathogen growth [13]. High-throughput drug screening of compounds against fungal pathogen infection was also performed. In a library containing 1280 diversity-oriented synthesis derived small molecules and 1948 bioactive compounds, 19 compounds were identified to enhance the survival of nematodes after *Candida albicans* infection [14]. Among them, 7 compounds are clinically used antifungal compounds, and 12 compounds are new identified antifungal agents, including 3 immunosuppressive drugs [14].
   Moreover, the immunomodulators were identified based on high-throughput drug screening. Among 1120 compounds, 45 compounds were observed to increase PMK-1::GFP expression [15]. Among them, the further confirmation indicated that the colistin could protect against bacterial pathogen infections by *Yersinia pestis* and *Pseudomonas aeruginosa*, but could not reduce bacterial pathogen burden [15]. The innate immune response activated by the colistin was not only mediated by p38/PMK-1 signaling, but also by DAF-16 signaling and SKN-1 signaling [15].
10. Screening of potential anti-obesity drugs: In the Prestwick Chemical Library of 1200 FDA-approved drugs, 9 compounds were identified to reduce fat content

within 2 weeks based on semi-automated whole-organism approach [16]. Midodrine, vinpocetine, fenoprofen, and lamivudine were further confirmed to exhibit significant promise for reduction in fat levels [16]. Moreover, in *nhr-49* mutants with "overweight" phenotype, their fat levels could be further reduced by these identified compounds [16].

11. Screening of drugs promoting collagen secretion: In a small-scale of drug screening, among 614 compounds, Danshensu, Lawsone, and Sanguinarine were identified to elevate expression of COL-12 in transgenic strain carrying P*col-19- col-12::GFP* [17], implying their possible role in modulating the collagen secretion.

14.2.2 Target-Based Screening

Target-based screening refers to the drug screening utilizing genetically engineered strains to identify and to evaluate compounds for their ability to modulate specific molecular targets or pathways. Unlike the aims of phenotypic screening, which asks "what does this compound do?", target-based screening will ask "does this compound affect the pathway of interest?" As described above, the high-throughput drug screening based on the effects on GST-4::GFP and HSP-6::GFP can also be considered as the target-based screening [4]. Besides these, the SKN-1 inhibitors were also identified based on high-throughput drug screening [18]. Nevertheless, as introduced and discussed in Chap. 9, the G-protein coupled receptors-based, ion channels-based, nuclear hormone receptors-based, and enzymes-based are suggested to be further systematically performed. In *C. elegans*, based on high-throughput drug screening, some compounds were identified as the ion channel modulators [11, 12].

14.3 Pharmacological Screening in Microfluidic System

Pharmacological screening in microfluidic systems can allow the high-throughput and the automated drug screening with precise environmental control and analytical capabilities. Microfluidic chips can physically restrain the *C. elegans* using certain techniques (such as physical constriction, cooling, or anesthetic gas diffusion) without killing them, and are suitable for high-resolution microscopy-based observation of subcellular structures. In this assay system, rapid and programmable switching between different drug concentrations, buffers, and stimuli can be performed, and hundreds of individual worms can be analyzed simultaneously.

Using the microfluidic system, the following several kinds of drug screening have been examined:

1. Screening of anesthetic drugs: In the microfluidic system, based on high-throughput and real-time behavior analysis on the motility, the dose–response relation of an anesthetic drugs was identified [19].

2. Screening of drugs against the neurodegenerative disease: With the aid of microfluidic system, in the SC_APP model, norbenzomorphans were identified to inhibit the neurodegeneration with the σ2R/Tmem97 binding potential [20].
3. Screening of anti-infective drugs: By assessing survival of worms in the microfluidic system, anti-infective activity of aloe-emodin, rhein, and emodin was assessed in *Caenorhabditis elegans–Staphylococcus aureus* infection model [21].

14.4 Chemogenomic-Based Pharmacological Screening

Chemogenomic-based pharmacological screening refers to the systematic approach to identify molecular targets, mechanisms of action, and genetic determinants of drugs by integrating chemical biology with functional genomics. At this aspect, the forward genetics and the reverse genetics are suggested to be integrated with chemical biology to perform the high-throughput drug screening.

Based on chemogenomic-based pharmacological screening, the following several kinds of drug screening have been examined:

1. Drug-target identification based on RNAi screening: To identify genetic suppressor of antipsychotic drug (APD) targets, genome-wide RNA interference (RNAi) screen for Suppressors of Clozapine-induced Larval Arrest (*scla* genes) was performed [22]. Based on the screen, 40 suppressors including α-like nicotinic acetylcholine receptor (nAChR) homolog ACR-7 were identified [22]. The further validation indicated that, on the one hand, clozapine-induced larval arrest could be suppressed by mutation of *acr-7*, and rescued by expression of full-length *acr-7* [22]. On the other hand, the clozapine-induced larval arrest was phenocopied by nAChR agonists, and partially blocked by nAChR antagonists [22]. The further obtained evidence suggested that clozapine can lead to tetanus of pharyngeal muscle with the consequent larval arrest by activating ACR-7 [22].
2. Drug target identification based on mutant screen: Using the available alleles, PINK-1, SDHA-1, SDHB-1, SDHD-1, and MEV-1 were identified as the possible targets of CID 2747322 with the anthelmintic activity [23].

14.5 Pharmacological Screening Among Structurally Related Compounds

Pharmacological screening among structurally related compounds refers to the systematic approach to evaluate chemical analogs (derivatives, congeners, or stereoisomers) sharing a common core scaffold, to optimize the drug potency, to dissect the mechanisms of action, and to assess the possible off-target effects. The systematic structure–activity relationship analysis in *C. elegans* can bridge the gap between in vitro biochemical assays and mammalian preclinical evaluation by enabling the refinement of lead compounds via the correlation of specific chemical substituents, stereochemical configurations, or scaffold modifications with defined biological outcomes at the whole-animal level. In *C. elegans*, the antioxidative activity of 30 xanthone derivatives was screened [24]. Although all the examined xanthone derivatives showed the antioxidative activity, 1,3,5,8-tetrahydroxy-9H-xanthen-9-one was identified as the most active antioxidant to eliminate excessive intracellular reactive oxygen species and to activate antioxidant enzymes (superoxide dismutase, catalase, and glutathione peroxidase) [24]. DAF-16, AGE-1, AKT-1, and SIR-2.1 were identified as the possible targets of 1,3,5,8-tetrahydroxy-9H-xanthen-9-one [24]. Besides this, the drug–protein interactions were further predicted, and 1147 drugs were ranked by their likelihood of targeting human aging-related proteins [25]. For targets of each candidate drug, their association with longevity was established based on pathways, functions, and protein interactions, and a ranked list of drugs was further calculated [25]. Based on these aspects of analysis, tanespimycin was identified as the top-ranked anti-aging compound by acting as an inhibitor of HSP-90 [25]. In *C. elegans*, the further experimental validation of pro-longevity effect of tanespimycin via HSP-90 was performed [25].

14.6 Perspectives

In this chapter, surrounding the high-throughput drug screening in *C. elegans*, we mainly introduced and discussed the pharmacological screening in small compound library, the pharmacological screening in microfluidic system, the chemogenomic-based pharmacological screening, and the pharmacological screening among structurally related compounds (Fig. 14.1). For the pharmacological screening in small compound library, besides the introduced phenotypic screening and the target-based screening, the high-throughput drug screening with the concerns on bioavailability, chemical metabolism, and compound solubility and stability are also suggested to be examined. For the chemogenomic-based pharmacological screening, the comparative chemogenomics strategy can also be employed to perform the related high-throughput drug screening. For example, based on this strategy, potential drug targets in the metazoan pathogen were predicted [26]. For the pharmacological screening among structurally related compounds, the network pharmacology approach can also be considered. Based on analysis of network pharmacology, rapamycin, allantoin, trichostatin A, and LY-294002 were identified as new candidate caloric restriction mimetics [27].

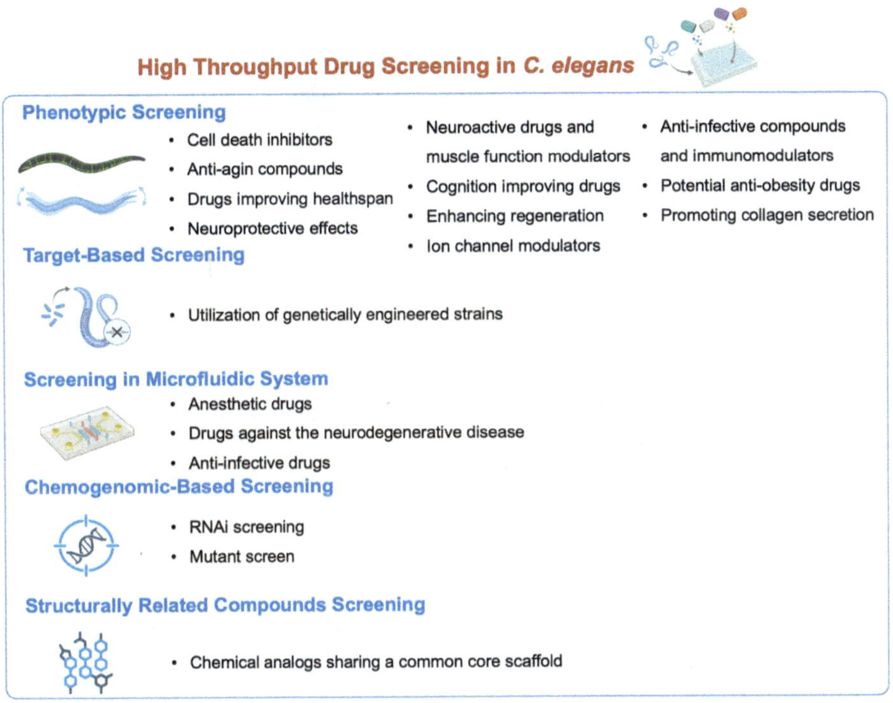

Fig. 14.1 High-throughput drug screening performed in *Caenorhabditis elegans*

References

1. Dyle MC, Dierdorff JM, Adams CM (2014) Systems-based discovery of tomatidine as a natural small molecule inhibitor of skeletal muscle atrophy. J Biol Chem 289:14913–14924
2. Schwendeman AR, Shaham S (2016) A high-throughput small molecule screen for *C. elegans* linker cell death inhibitors. PLoS One 11:e0164595
3. Petrascheck M, Ye X, Buck LB (2009) A high-throughput screen for chemicals that increase the lifespan of *Caenorhabditis elegans*. Ann NY Acad Sci 1170:698–701
4. Maglioni S, Arsalan N, Hamacher A, Afshar S, Schiavi A, Beller M, Ventura N (2021) High-content C. *elegans* screen identifies natural compounds impacting mitochondria-lipid homeostasis and promoting healthspan. Cells 11:100
5. Braungart E, Gerlach M, Riederer P, Baumeister R, Hoener MC (2004) *Caenorhabditis elegans* MPP$^+$ model of Parkinson's disease for high-throughput drug screenings. Neurodegener Dis 1: 175–118
6. Boyd JD, Lee P, Feiler MS, Zauur N, Liu M, Concannon J, Ebata A, Wolozin B, Glicksma MA (2014) A high content screen identifies novel compounds that inhibit stress-induced TDP-43 cellular aggregation and associated cytotoxicity. J Biomol Screen 19:44–56
7. O'Brien TJ, Barlow IL, Feriani L, Brown AEX (2025) High-throughput tracking enables systematic phenotyping and drug repurposing in *C. elegans* disease models. elife 12:RP92491
8. Sleigh JN, Buckingham SD, Esmaeili B, Viswanathan M, Cuppen E, Westlund BM, Sattelle DB (2011) A novel *Caenorhabditis elegans* allele, *smn-1(cb131)*, mimicking a mild form of

spinal muscular atrophy, provides a convenient drug screening platform highlighting new and pre-approved compounds. Hum Mol Genet 20:245–260
9. Balsiger T, Hagmann R, Huynh K, Solis P, Hamburger M, Papassotiropoulos A, Teufel R, Garo E, Stetak A (2025) Natural product screening reveals nanomolar Musashi2 inhibitor reducing forgetting in *Caenorhabditis elegans*. Biomed Pharmacother 193:118811
10. Samara C, Rohde CB, Gilleland CL, Norton S, Haggarty SJ, Yanik MF (2010) Large-scale in vivo femtosecond laser neurosurgery screen reveals small-molecule enhancer of regeneration. Proc Natl Acad Sci USA 107:18342–18347
11. Kwok TCY, Ricker N, Fraser R, Chan AW, Burns A, Stanley EF, McCourt P, Cutler SR, Roy PJ (2006) A small-molecule screen in *C. elegans* yields a new calcium channel antagonist. Nature 441:91–95
12. Jiang Q, Li K, Lu W, Li S, Chen X, Liu X, Yuan J, Ding Q, Lan F, Cai S (2018) Identification of small-molecule ion channel modulators in *C. elegans* channelopathy models. Nat Commun 9:1038
13. Moy TI, Conery AL, Larkins-Ford J, Wu G, Mazitschek R, Casadei G, Lewis K, Carpenter AE, Ausubel FM (2009) High throughput screen for novel antimicrobials using a whole animal infection model. ACS Chem Biol 4:527–533
14. Okoli I, Coleman JJ, Tempakakis E, An WF, Holson E, Wagner F, Conery AL, Larkins-Ford J, Wu G, Stern A, Ausubel FM, Mylonakis E (2009) Identification of antifungal compounds active against *Candida albicans* using an improved high-throughput *Caenorhabditis elegans* assay. PLoS One 4:e7025
15. Cai Y, Cao X, Aballay A (2014) Whole-animal chemical screen identifies colistin as a new immunomodulator that targets conserved pathways. MBio 5:e01235–e01214
16. Haerkens F, Kikken C, Kirkels L, van Amstel M, Wouters W, van Doornmalen E, Francke C, Hughes S (2022) A new use for old drugs: identifying compounds with an anti-obesity effect using a high through-put semi-automated *Caenorhabditis elegans* screening platform. Heliyon 8:e10108
17. Fang J, Wu X, Meng X, Xun D, Xu S, Wang Y (2022) Discovery of natural small molecules promoting collagen secretion by high-throughput screening in *Caenorhabditis elegans*. Molecules 27:8361
18. Leung CK, Wang Y, Malany S, Deonarine A, Nguyen K, Vasile S, Choe KP (2013) An ultra high-throughput, whole-animal screen for small molecule modulators of a specific genetic pathway in *Caenorhabditis elegans*. PLoS One 8:e62166
19. Migliozzi D, Cornaglia M, Mouchiroud L, Uhlmann V, Unser MA, Auwerx J, Gijs MAM (2019) Multimodal imaging and high-throughput image-processing for drug screening on living organisms on-chip. J Biomed Opt 24:021205
20. Monda S, Hegarty E, Sahn JJ, Scott LL, Gökçe SK, Martin C, Ghorashian N, Satarasinghe PN, Iyer S, Sae-Lee W, Hodges TR, Pierce JT, Martin SF, Ben-Yakar A (2018) High-content microfluidic screening platform used to identify σ2R/Tmem97 binding ligands that reduce age-dependent neurodegeneration in *C. elegans* SC_APP model. ACS Chem Neurosci 9:1014–1026
21. Yang J, Chen Z, Ching P, Shi Q, Li X (2013) An integrated microfluidic platform for evaluating *in vivo* antimicrobial activity of natural compounds using a whole-animal infection model. Lab Chip 13:3373–3382
22. Saur T, DeMarco SE, Ortiz A, Sliwoski GR, Hao L, Wang X, Cohen BM, Buttner EA (2013) A genome-wide RNAi screen in *Caenorhabditis elegans* identifies the nicotinic acetylcholine receptor subunit ACR-7 as an antipsychotic drug target. PLoS Genet 9:e1003313
23. Mathew MD, Mathew ND, Miller A, Simpson M, Au V, Garland S, Gestin M, Edgley ML, Flibotte S, Balgi A, Chiang J, Giaever G, Dean P, Tung A, Roberge M, Roskelley C, Forge T, Nislow C, Moerman D (2016) Using *C. elegans* forward and reverse genetics to identify new compounds with anthelmintic activity. PLoS Negl Trop Dis 10:e0005058

24. Tang Y, Liu J, Wang H, Li Y, Liu Z, Chen H (2019) 1,3,5,8-Tetrahydroxy-9H-xanthen-9-one exerts its antiageing effect through the regulation of stress-response genes and the MAPK signaling pathway. Arch Pharm Chem Life Sci 352:1900100
25. Fuentealba M, Donertaş HM, Williams R, Labbadia J, Thornton JM, Partridge L (2019) Using the drug-protein interactome to identify anti-ageing compounds for humans. PLoS Comput Biol 15:e1006639
26. Caffrey CR, Rohwer A, Oellien F, Marhofer RJ, Braschi S, Oliveira G, McKerrow JH, Selzer PM (2009) A comparative chemogenomics strategy to predict potential drug targets in the metazoan pathogen, *Schistosoma mansoni*. PLoS One 4:e4413
27. Calvert S, Tacutu R, Sharifi S, Teixeira R, Ghosh P, de Magalhaes JP (2016) A network pharmacology approach reveals new candidate caloric restriction mimetics in *C. elegans*. Aging Cell 15:256–266

FSC
www.fsc.org
MIX
Papier | Fördert gute Waldnutzung
FSC® C083411

Kolibri 360 GmbH
Ferdinand-Jühlke-Straße 7
99095 Erfurt, Deutschland
produktsicherheit@kolibri360.de